PCR Detection of Microbial Pathogens

METHODS IN MOLECULAR BIOLOGY™

John M. Walker, SERIES EDITOR

METHODS IN MOLECULAR BIOLOGY™

PCR Detection of Microbial Pathogens

Edited by

Konrad Sachse, Dr. RER. NAT.

Federal Institute for Health Protection
of Consumers and Veterinary Medicine (BgVV),
Jena, Germany

and

Joachim Frey, PhD

University of Bern,
Bern, Switzerland

Humana Press ✳ Totowa, New Jersey

Cover illustration: Background photo shows a confocal laser scanning microscopic image of cells infected with chlamydiae (kindly provided by Angela Berndt). Foreground gel shows the products of chlamydia-specific PCR separated on agarose (taken from Fig. 3 in Chapter 8; photo provided by K. Sachse). Illustrations of "Taq" polymerase synthesizing new strands of DNA and double helices in foreground schematically depict molecular processes underlying PCR. Cover design concept by Mandy Elschner.

Cover design by Patricia F. Cleary.

Production Editor: Mark J. Breaugh.

For additional copies, pricing for bulk purchases, and/or information about other Humana titles, contact Humana at the above address or at any of the following numbers: Tel.: 973-256-1699; Fax: 973-256-8341; E-mail: humana@humanapr.com; Website: http://humanapress.com

Printed in the United States of America. 10 9 8 7 6 5 4 3 2 1

Library of Congress Cataloging in Publication Data

PCR detection of microbial pathogens / edited by Konrad Sachse and Joachim Frey.
      p. ; cm. -- (Methods in molecular biology ; v. 216)
   Includes bibliographical references and index.
   ISBN 1-58829-049-2 (alk. paper)
      1. Diagnostic microbiology. 2. Polymerase chain reaction--Diagnostic use. 3. Molecular diagnosis. I. Sachse, Konrad. II. Frey, Joachim. III. Methods in molecular biology (Totowa, N.J.) ; v. 216.
      [DNLM: 1. Microbiological Techniques. 2. Polymerase Chain Reaction--methods. QW 25 P3476 2003]
   QR67 .P37 2003
   616'.01--dc21
                                                                    2002024051

Preface

The polymerase chain reaction (PCR) has become an increasingly important tool in microbial diagnosis in recent years, mainly because of its rapidity, high sensitivity, and specificity. In a number of instances, particularly with noncultivatable and slow-growing microorganisms, reliable identification and differentiation became possible only after the introduction of PCR-based methods.

Because the number of published procedures is large and steadily rising, there is a need for critical evaluation, comparison of performance, and eventually also standardization of methods to enable laboratory diagnosticians to select the optimal methodology.

PCR Detection of Microbial Pathogens is designed to provide an overview of the possibilities and problems connected with the use of PCR-based identification and detection of important bacterial and other microbial pathogens, among them several zoonotic agents. It is intended to provide microbiologists and biochemists the opportunity to extend their knowledge of state-of-the-art detection procedures, as well as pre-PCR sample processing and various general aspects of PCR. The chapters of this volume have been written by investigators well conversant with molecular detection and identification of microorganisms. They have contributed their extensive practical experience to the protocols, not only concerning PCR methodology itself, but also the highly important area of pre-amplification sample processing.

The various chapters of *PCR Detection of Microbial Pathogens* contain a wealth of useful practical hints and recommendations that may be helpful for optimization or the development of new PCR detection systems. Readers are therefore also invited to consult those chapters not dealing overtly with their "favorite pathogen," since they might otherwise miss interesting details of preparation procedures, as well as cross-references on safety issues and molecular epidemiology.

The editors have intentionally limited the contents of the volume to "classical" PCR methodology, thus omitting real-time PCR and its steadily growing area of application in order to leave ample space for pre-amplification procedures and explanation of epidemiological issues.

Veterinary diagnosticians and food analysts considering the introduction of PCR methods in their laboratory may regard *PCR Detection of Micro-*

bial Pathogens as an aid to orientation and decision-making and, we greatly hope, find standard protocols that are adaptable to their specific needs. Finally, this compiled knowledge on microbial diagnostic PCR can help vets, public health, and veterinary officers to recognize the potential of DNA-based diagnostic methods and their practical possibilities.

Konrad Sachse, DR. RER. NAT.
Joachim Frey, PhD

Contents

Contributors

NATHALIE ARRICAU-BOUVERY • *Pathologie Infectieuse et Immunologie, INRA Tours-Nouzilly, Nouzilly, France*

JOHN BASHIRUDDIN • *Veterinary Laboratories Agency - Weybridge, New Haw, Addlestone, Surrey, UK*

MUSTAPHA BERRI • *Pathologie Infectieuse et Immunologie, INRA Tours-Nouzilly, Nouzilly, France*

MAGNE BISGAARD • *Department of Veterinary Microbiology, The Royal Veterinary and Agricultural University, Frederiksberg, Denmark*

BETSY J. BRICKER • *National Animal Disease Center, Agricultural Research Service, United States Department of Agriculture, Ames, IA*

HENRIK CHRISTENSEN • *Department of Veterinary Microbiology, The Royal Veterinary and Agricultural University, Frederiksberg, Denmark*

NIGEL COOK • *Central Science Laboratory, York, UK*

MARIA DAHLENBORG • *Applied Microbiology, Center for Chemistry and Chemical Engineering, Lund University, Lund, Sweden*

MARK D. ENGLEN • *Antimicrobial Resistance Research Unit, Richard B. Russell Agricultural Research Center, Agricultural Research Service, USDA, Athens, GA*

DARLA R. EWALT • *National Veterinary Services Laboratories, Animal and Plant Health Inspection Service, Veterinary Services, United States Department of Agriculture, Ames, IA*

PAULA J. FEDORKA-CRAY • *Antimicrobial Resistance Research Unit, Richard B. Russell Agricultural Research Center, Agricultural Research Service, USDA, Athens, GA*

JOACHIM FREY • *Institute for Veterinary Bacteriology, University of Berne, Berne, Switzerland*

PETER GALLIEN • *National Veterinary Reference Laboratory for E. Coli, Federal Institute for Health Protection of Consumers and Veterinary Medicine (BgVV), Dessau, Germany*

REINER HELMUTH • *National Salmonella Reference Laboratory, Federal Institute for Health Protection of Consumers and Veterinary Medicine (BgVV), Berlin, Germany*

JEFFREY HOORFAR • *Danish Veterinary Institute, Copenhagen, Denmark*

HELMUT HOTZEL • *Division 4: Bacterial Animal Diseases and Control of Zoonoses, Federal Institute for Health Protection of Consumers and Veterinary Medicine (BgVV), Jena, Germany*

M. SIOBHAN HUGHES • *Veterinary Sciences Division, Department of Agriculture and Rural Development, Belfast, UK*

MAZHAR I. KHAN • *Department of Pathobiology and Veterinary Science, College of Agriculture and Natural Resources, University of Connecticut, Storrs, CT*

RICKARD KNUTSSON • *Applied Microbiology, Center for Chemistry and Chemical Engineering, Lund University, Lund, Sweden*

MARYLÈNE KOBISCH • *French Food Safety Agency (AFSSA), Unité Mycoplasmologie - Bacteriologie, Ploufragan, France*

GIUSEPPE LA ROSA • *Laboratory of Parasitology, Istituto Superiore di Sanità, Rome, Italy*

SCOTT R. LADELY • *Antimicrobial Resistance Research Unit, Richard B. Russell Agricultural Research Center, Agricultural Research Service, USDA, Athens, GA*

JESPER LARSEN • *Department of Veterinary Microbiology, The Royal Veterinary and Agricultural University, Frederiksberg, Denmark*

RANCE B. LEFEBVRE • *Pathology Microbiology and Immunology, Department of Veterinary Medicine, University of California at Davis, Davis, CA*

JOHN W. LESTER • *Pathology Microbiology and Immunology, Department of Veterinary Medicine, University of California at Davis, Davis, CA*

CHARLOTTA LÖFSTRÖM • *Department of Applied Microbiology, Center for Chemistry and Chemical Engineering, Lund University, Lund, Sweden*

PETER STEPHENSEN LÜBECK • *Danish Veterinary Institute, Copenhagen, Denmark*

BURKHARD MALORNY • *National Salmonella Reference Laboratory, Federal Institute for Health Protection of Consumers and Veterinary Medicine (BgVV), Berlin, Germany*

JENS G. MATTSSON • *Department of Parasitology (SWEPAR), National Veterinary Institute, Uppsala, Sweden*

SYDNEY D. NEILL • *Veterinary Sciences Division, Department of Agriculture and Rural Development, Belfast, UK*

LOUISE O'CONNOR • *The National Diagnostics Centre, BioResearch Ireland, The National University of Ireland, Galway, Ireland*

JOHN ELMERDAHL OLSEN • *Department of Veterinary Microbiology, The Royal Veterinary and Agricultural University, Frederiksberg, Denmark*

MICHEL R. POPOFF • *Unité des Toxines Bactériennes, CNR Anaérobies, Institut Pasteur, Paris, France*

EDOARDO POZIO • *Laboratory of Parasitology, Istituto Superiore di Sanità, Rome, Italy*

PETER RÅDSTRÖM • *Department of Applied Microbiology, Center for Chemistry and Chemical Engineering, Lund University, Lund, Sweden*

ANNIE RODOLAKIS • *Pathologie Infectieuse et Immunologie, INRA Tours-Nouzilly, Nouzilly, France*

KONRAD SACHSE • *Division 4: Bacterial Animal Diseases and Control of Zoonoses, Federal Institute for Health Protection of Consumers and Veterinary Medicine (BgVV), Jena, Germany*

ROBIN A. SKUCE • *Veterinary Sciences Division, Department of Agriculture and Rural Development, Belfast, UK*

MALCOLM J. TAYLOR • *Veterinary Sciences Division, Department of Agriculture and Rural Development, Belfast, UK*

JONATHAN M. WASTLING • *Division of Infection and Immunity, Institute of Biomedical and Life Sciences, University of Glasgow, Glasgow, UK*

PETRA WOLFFS • *Department of Applied Microbiology, Center for Chemistry and Chemical Engineering, Lund University, Lund, Sweden*

I

REVIEWS

1

Specificity and Performance
of Diagnostic PCR Assays

Konrad Sachse

1. Introduction

The undisputed success of detection assays based on the polymerase chain reaction (PCR) has been largely due to its rapidity in comparison to many conventional diagnostic methods. For instance, detection and identification of mycobacteria, chlamydiae, mycoplasmas, brucellae, and other slow-growing bacteria can be accelerated from several days to a single working day when clinical samples are directly examined. Other microbial agents that are difficult to propagate outside their natural host often remain undetected by techniques relying on cultural enrichment, thus rendering PCR the only viable alternative to demonstrate their presence. Additionally, there is the enormous potential of DNA amplification assays with regard to sensitivity and specificity.

Nowadays, when a new PCR assay is introduced into a laboratory, the diagnostician expects it to facilitate the examination of clinical samples without pre-enrichment and to allow specific differentiation between closely related species or subtypes at the same time. While there can be no doubt that the potential to fulfill these demanding criteria is actually inherent in PCR-based methods, and the present volume contains convincing evidence of this in **Chapters 5 – 22,** there is often a need for critical evaluation of a given methodology, not only in the case of obvious failure or underperformance, but also when certain parameters have to be optimized to further improve performance or reduce costs.

The present chapter is designed to discuss the importance of key factors in PCR detection assays and provide an insight into basic mechanisms underlying the amplification of DNA templates from microbial sources. Besides the quality of the nucleic acid template (*see* **Chapter 2**) there are several other crucial parameters deciding over the performance of a detection method, e.g., the target region, primer sequences, and efficiency of amplification.

From: *Methods in Molecular Biology, vol. 216: PCR Detection of Microbial Pathogens: Methods and Protocols*
Edited by: K. Sachse and J. Frey © Humana Press Inc., Totowa, NJ

2. Selection of Target Sequences

2.1. General Criteria

The choice of the genomic region to be amplified will determine the specificity of detection from the outset. Obviously, a genomic DNA segment characteristic for the respective microorganism or a group of species has to be selected, and knowledge of the nucleotide sequence is practically indispensable to assess its suitability. With the steadily increasing amount of publicly accessible DNA sequence data it is no longer a problem to check a given sequence for its degree of homology to other organisms.

The sensitivity of the detection assay is connected with the nature of the target region via the efficiency of primer binding (*see* **Subheading 3.2.2.**), which determines the efficiency of amplification. The finding that different primer pairs for the same gene can exhibit up to 1000-fold differences in sensitivity *(1)* illustrates the extent of this relationship. Likewise, primer pairs flanking different genomic regions can be expected to perform differently in amplification reactions.

As the length of the PCR product has an inverse correlation to the efficiency of amplification *(2–4)*, relatively short targets do not only facilitate high sensitivity of detection, but are also preferable for quantitative PCR assays. Furthermore, genomic regions of shorter size can be expected to remain intact at conditions of moderate DNA degradation, thus making detection more robust and less dependent on the use of fresh sample material. Hence, there seems to be consensus among most workers that the optimum size of PCR fragments for detection purposes is between 100 and 300 bp.

While the first detection methods of the 1980s and early 1990s had to rely on randomly chosen target sequences the vast majority of currently used targets is well characterized. An overview on the various categories of target sequences used in PCR detection assays for bacteria is given in **Table 1**. These data will be discussed in the following paragraphs.

2.2. Ribosomal RNA Genes

The ribosomal (r) RNA gene region has emerged as the most prominent target in microbial detection. Among the assays reviewed in the present chapter, about 50% are based on sequences of rRNA genes, i.e., rDNA. Their popularity is certainly due to the fact that the region represents a versatile mix of highly conserved and moderately to highly variable segments. Moreover, rRNA gene sequences are now known for virtually all microorganisms of veterinary and human health interest.

The structure of the rRNA operon in bacteria, as schematically depicted in **Fig. 1** comprises three gene sequences and two spacer regions.

Table 1
Target Sequences Used in PCR Detection Assays of Microorganisms

Target gene region	Organism	Authors (Reference)
16S rRNA	*Campylobacter* spp.	van Camp et al. (7)
		Giesendorf et al. (8)
		Metherell et al. (9)[H]
		Cardarelli-Leite et al. (10)[R]
	Chlamydia pneumoniae,	Messmer et al. (13)[M]
	C. psittaci, C. trachomatis	Madico et al. (14)[M]
	Clostridium perfringens	Wang et al. (20)
	Leptospira spp.	Heinemann et al. (11)[R]
	Mycobacterium spp.	Kox et al. (15)[H]
		Oggioni et al. (16)[N]
	Mycoplasma capripneumoniae (F38)	Bascunana et al. (17)[R]
	Mycoplasma conjunctivae	Giacometti et al. (18)
	Mycoplasma mycoides subsp. *mycoides* SC	Persson et al. (18)[R]
	Group B streptococci	Lammler et al. (12)[R]
	Yersinia enterocolitica	Lantz et al. (21)
	many bacterial species	Greisen et al. (6)
18S rRNA	*Cryptococcus neoformans*	Prariyachatigul et al. (119)
16–23S intergenic spacer	*Campylobacter jejuni/coli*	O'Sullivan et al. (30)[H]
	Chlamydiaceae spp.	Everett and Andersen (27)[R]
	Clostridium difficile	Cartwright et al. (31)
	Cryptococcus neoformans	Mitchell et al. (32)
	Cryptococcus neoformans	Rappelli et al. (33)
	Listeria spp.	Drebót et al. (34)
	Listeria monocytogenes	O'Connor et al. (35)[H]
	Mycobacterium spp.	Park et al. (36)
	Mycoplasma spp.	Uemori et al. (37)
	Pasteurella multocida serotype B:1	Brickell et al. (38)

(continued)

Table 1
Target Sequences Used in PCR Detection Assays of Microorganisms (*continued*)

Target gene region	Organism	Authors (Reference)
16—23S *intergenic spacer*	*Pseudomonas* spp.	Gill et al. (*39*)
	Staphylococcus spp., *Streptococcus* spp.	Forsman et al. (*41*)
	Streptococcus milleri	Whiley et al. (*40*)
	many bacterial species	Gürtler & Stanisich (*28*)
		Scheinert et al. (*29*)
23S rRNA	*Campylobacter* spp.	Fermer & Olsson (*25*)[R]
	Campylobacter spp.	Eyers et al. (*26*)
rRNA genomic region[a]	*Trichinella* spp.	Zarlenga et al. (*117*)[M]
omlA (outer membrane lipoprotein)	*Actinobacillus pleuropneumoniae*	Gram et al. (*50*)
tbpA + tbpB	*Actinobacillus pleuropneumoniae* serotypes	de la Puente-Redondo et al. (*51*)[R]
*apxIV*A (toxin)	*Actinobacillus pleuropneumoniae*	Schaller et al. (*49*)
bcsp31	*Brucella* spp.	Baily et al. (*56*)
hippuricase gene	*Campylobacter jejuni*	Englen et al. (*57*)
GTPase	*Campylobacter* spp.	van Doorn et al. (*58*)[H]
flaA (flagellin)	*Campylobacter jejuni/coli*	Oyofo et al. (*52*)
*ompA/omp*1(major outer membrane protein)	*Chlamydia psittaci, C. trachomatis, C. pneumoniae, C. pecorum*	Kaltenböck et al. (*53*)[N]
plC (phospholipase C, α-toxin)	*Clostridium perfringens*	Fach et al. (*42*)[N]
α-, β-, ε-toxins	*Clostridium perfringens*	Buogo et al. (*43*)
α-, β-, ε-, ι-toxins and enterotoxin	*Clostridium perfringens*	Meer et al. (*44*)[M]
URA5	*Cryptococcus neoformans*	Tanaka et al. (*59*)[N]
stx (shiga-like toxins)	*Escherichia coli* (EHEC)	Karch et al. (*45*)
*uidA + eaeA + stx*1 + *stx*2 + *ehxA*	*Escherichia coli* (EHEC)	Feng et al. (*46*)[M]

hly (hemolysin)	*Listeria monocytogenes*	Furrer et al. (*60*)
*inl*A + *inl*B (internalins)	*Listeria monocytogenes*	Ericsson et al. (*69*)[R]
iap gene	*Listeria monocytogenes*	Manzano et al. (*73*)[R]
hsp65 (heat shock protein)	*Mycobacterium avium* complex	Hance et al. (*66*)[H]
hsp65	*Mycobacterium* spp.	Steingrube et al. (*67*)
		Taylor et al. (*68*)
*opp*D+F (oligopeptide permease)	*Mycoplasma bovis*	Pinnow et al. (*63*)
		Hotzel et al. (*64*)
*uvr*C (uv repair gene)	*Mycoplasma bovis*	Subramaniam et al. (*65*)
p36 (cytosolic protein) + p46 (membrane protein	*Mycoplasma hyopneumoniae*, *M. hyorhinis*	Caron et al. (*54*)
dtx (dermonecrotic toxin)	*Pasteurella multocida*	Hotzel et al. (*47*)
psl (P6-like protein)	*Pasteurella multocida*	Kasten et al. (*55*)[H]
*lkt*A (leukotoxin)	*Pasteurella haemolytica*, *P. trehalosi*	Fisher et al. (*48*)
rfb genes (abequose and paratose synthase)	*Salmonella* serogroups	Luk et al. (*61*)
		Hoorfar et al. (*62*)
*inv*A (invasion-associated protein)	*Salmonella* serovars	Rahn et al. (*70*)
B1 gene	*Toxoplasma gondii*	Wastling et al. (*72*)
*yad*A (virulence gene)	*Yersinia enterocolitica*	Lantz et al.(*21*)
IS 1111	*Coxiella burnetii*	Willems et al. (*74*)
		Schrader et al. (*75*)
IS 1533	*Leptospira interrogans*	Redstone et al. (*79*)
IS 6110 + direct repeat	*Mycobacterium bovis*	Roring et al. (*76*)[H]
IS 6110	*Mycobacterium tuberculosis* complex	Thierry et al. (*77*)[H]
		de Lassence et al. (*78*)
IS 6110 + direct repeat	*Mycobacterium tuberculosis* complex	Mangiapan et al. (*79*)
repetitive genomic sequences*	*Trichinella* spp.	Appleyard et al. (*81*)

* = Target region not specified, [M], Multiplex PCR.
Recommended subsequent steps for verification and/or characterization: [H], Hybridization using specific probe, [N], Nested amplification, [R], Restriction enzyme analysis.

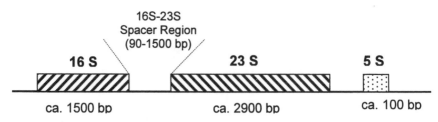

Fig. 1. Structural organization of ribosomal RNA genes in bacteria .

Due to its manageable size of approx 1500 bp, the 16S rRNA gene has become the best characterized part of the operon with more than 33,000 sequences from bacterial sources alone available on the GenBank® database *(5)*. Many studies of genetic relatedness leading to the construction of phylogenetic trees are based on sequence analysis of the 16S region. An extensive diagnostic system based on 16S rRNA gene amplification was proposed by Greisen et al. *(6)* for differentiation of many pathogenic bacteria including *Campylobacter* spp., *Clostridium* spp., *Lactococcus lactis*, *Listeria monocytogenes*, *Staphylococcus aureus* and *Streptococcus agalactiae*. PCR detection systems based on 16S rDNA target sequences were also used for identification of *Campylobacter (7–10)*, *Leptospira (11)* or streptococci *(12)*, as well as species differentiation within chlamydiae *(13,14)*, mycobacteria *(15,16)*, mycoplasmas *(17–19)* or identification of *Clostridium perfringens (20)* and *Yersinia enterocolitica (21)*.

However, it must be noted that detection and differentiation based on 16S rDNA can be hampered by significant intraspecies sequence heterology, as reported for *Riemerella anatipestifer (22)*, or by high homology between related species, e.g., in the case of *Mycoplasma bovis/Mycoplasma agalactiae (23)* and *Bacillus anthracis/Bacillus cereus/Bacillus thuringiensis (24)*.

The gene of the RNA of the large ribosomal unit, the 23S rDNA, has been used less frequently for diagnostic purposes so far, perhaps because of its greater size. The number of complete bacterial 23S rRNA gene sequences available from databases is still very small compared to 16S data. However, considering the extent of sequence variation known at present, there is probably also a great potential for species differentiation in this genomic region. Examples of 23S rRNA sequences serving as target region for species identification include assays for campylobacter *(25,26)* and chlamydiae *(27)*.

Located between the two major ribosomal rRNA genes, the 16S–23S intergenic spacer region (also called internal transcribed spacer) can be an attractive alternative target. Besides sequence variation, it is the size variation that renders this segment suitable for identification and differentiation. Spacer length was found to vary between 60 bp in *Thermoproteus tenax* and 1529 bp in *Bartonella elizabethae (28)*. In a systematic study, Scheinert et al. *(29)* were able to distinguish 55 bacterial species, among them 18 representatives of *Clostridium* and 15 of *Mycoplasma*, on the basis of PCR-amplified 16S–23S spacer segment lengths. Other authors developed assays for *Campylobacter* spp. *(30)*, chlamydiae *(27)*, *Clostridium difficile (31)*, *Cryptococcus neoformans (32,33)*, *Listeria* spp. *(34,35)*, mycobacteria *(36)*, mycoplasmas *(37)*, *Pasteurella multocida (38)*, *Pseudomonas* spp. *(39)*, streptococci, and staphylococci *(40,41)*.

Although the above-mentioned examples clearly illustrate the broad applicability of rDNA-based PCR assays, the feasibility of any new assay has to be examined first by sequence alignments, as lack of sufficient sequence variation in the operon region may not allow the development of genus- or species-specific assays with particular groups of microorganisms. Moreover, the diagnostic potential of this target region is usually insufficient for intraspecies differentiation. If the isolates are to be differentiated for medical purposes, e.g., according to serotype or virulence factors, other target sequences are usually preferable.

2.3. Protein Genes

Many PCR assays targeting protein genes were developed in an effort to genetically replicate conventional typing methods based on phenotypic properties, such as serological reactivity, enzymatic or toxigenic activity. In contrast to rDNA amplification assays, they are usually specially designed for a particular microbial species or a small group of related organisms. The only notable exception would include methods based on largely universal housekeeping protein genes, e.g., elongation factor EF-Tu, DNA repair enzymes, DNA-binding proteins, etc., that are present in all organisms and whose sequences are phylogenetically interrelated in a manner comparable to rRNA genes.

The lower part of **Table 1** shows the wide variety of protein-encoding sequences used for diagnostic purposes. Toxin genes naturally lend themselves as targets because, in many instances, they were among the first genes cloned from the respective microbes, thus they are usually well characterized. It is, therefore, not surprising that PCR assays for toxigenic bacteria, such as *Clostridium perfringens (42–44)*, *Escherichia coli (45,46)*, *Pasteurella multocida (47)*, *Pasteurella/Mannheimia hemolytica (48)*, and *Actinobacillus pleuropneumoniae (49)* were based on this category of genes. Another frequently used target are the genes of surface antigens or outer membrane proteins, which were described in connection with detection methods for *Actinobacillus pleuropneumoniae (50,51)* campylobacter *(52)*, chlamydiae *(53)*, porcine mycoplasmas *(54)*, *Pasteurella multocida (55)*, and brucellae *(56)*. Furthermore, there are reports of genes coding for cellular enzymes *(57–62)*, essential transporters *(63,64)*, DNA repair enzymes *(65)*, heat shock proteins *(66–68)*, invasion factors *(69,70)* and various virulence factors *(71–73)* being used in PCR assays.

Apart from the potential to fine-tune specificity of detection as mentioned above, the most evident advantage from the utilization of protein gene-based PCR assays is the concomitant information provided on toxins, surface antigens, or other virulence markers, as these factors are supposed to be directly involved in pathogenesis. In this respect, such tests deliver more evidence on a given microorganism than just confirming its presence in a sample.

2.4. Repetitive Elements

Some microorganisms possess repetitive sequences or insertion elements. Since these segments are present in multiple copies the idea of targeting them appears straightforward. Indeed, this is a favorable prerequisite for the development of highly sensitive detection methods. In the literature, amplification assays based on repetitive elements were reported for *Coxiella burnetii* *(74,75)*, *Mycobacterium bovis* *(76)*, the *Mycobacterium tuberculosis* complex *(77–79)* *Leptospira interrogans (80)*, and trichinellae *(81)*. In combination with sequence-specific DNA capture prior to amplification, a detection limit of one mycobacterial genome was attained *(79)*.

3. Efficiency of the Amplification Reaction

3.1. Early, Middle, and Late Cycles

DNA amplification by PCR is based on a cyclical enzymatic reaction, where the products (amplicons) of the previous cycle are used as substrate for the subsequent cycle. Thus, in theory, the number of target molecules is expected to increase exponentially, i.e. double, after each cycle. As the efficiency of the reaction is not 100% in practice, the real amplification curves are known to deviate from the exponential shape *(82–85)*. The course of DNA amplicon production during 30 cycles in an ideal and a real PCR is illustrated in **Fig. 2.** The extent of deviation from the theoretical product yield is determined by the efficiency of amplification, which can be approx assessed by **Equation 1** *(86)*:

$$Y = (1 + \varepsilon)^n \qquad [\text{Eq.1}]$$

where Y is the amplification yield (expressed as quotient of the number of molecules of PCR product and the initial number of target molecules), n is the number of cycles, and ε is the mean efficiency of all cycles with $0 \leq \varepsilon \leq 1$.

The reaction efficiency may, in principle, assume a different value in each cycle. The parameters affecting ε include the concentration of DNA polymerase, dNTPs, $MgCl_2$, DNA template, primers, temperatures of denaturation,

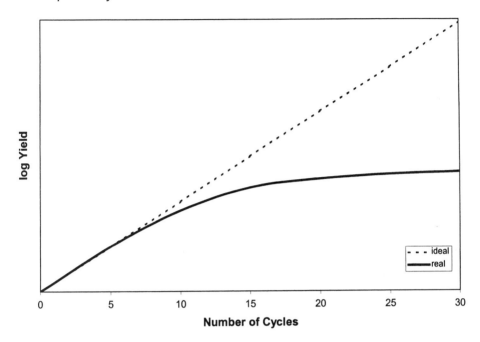

Fig. 2. Accumulation of amplification product in the course of a PCR assay. The broken line corresponds to an ideal kinetics of DNA synthesis (efficiency $\varepsilon = 1$), and the curve shows the course of a real PCR with $\varepsilon < 1$.

annealing and strand synthesis, number of cycles, ramping times, as well as the presence of inhibitors and background DNA.

In the first few cycles, when relatively few DNA template molecules are available, primers act predominantly as screening probes that hybridize independently to complementary sites *(84)*. Moreover, it is often overlooked that the first cycle generates DNA strands longer than the interprimer segment, the number of which grows arithmetically in successive cycles *(87)*. The situation changes in the middle cycles as more amplified product (of correct size) with terminal annealing sites is present and primers assume their role as amplification vectors. Regarding the yield of amplified product, a typical PCR will first be exponential, then go through a quasilinear phase, and finally reach a plateau. The plateau effect *(82,83)* is the result of a marked shift of the overall mass balance in favor of the reaction product. A complex of features seems to be responsible for the attainment of the plateau rather than a single factor or parameter, as readdition of presumably exhausted reagents (dNTPs, primers, DNA polymerase, $MgCl_2$) at late cycles did not cause the reaction to proceed with increased efficiency *(88)*.

Since there appears to be a natural limit of product concentrations in the order of $10^{-7} M$ *(89)* where the amount of accumulated amplicon can no longer be increased significantly, there is no benefit in terms of final yield from running further cycles. On the contrary, the 5'–3' exonuclease activity of the DNA polymerase may cause measurable loss of product if the reaction is extended way beyond the quasilinear phase.

3.2. Factors Influencing Kinetics and Yield of DNA Amplification

3.2.1. Primer-to-Template Ratio

In the course of the reaction, the mass balance between the reaction partners changes after each cycle. The crucial parameter in this dynamic system is the amount of amplified product. As it accumulates with each cycle the reaction efficiency decreases steadily. Using a continuous mathematical model based on the law of mass action, Schnell and Mendoza *(85)* showed that the reaction efficiency can be close to 100% only as long as there is very little product in the system. As the reaction proceeds until the concentrations of initial DNA (primers + dNTPs + template) and amplified product are the same, efficiency drops to 50% before approaching zero upon saturation of the system in the plateau phase (*see* **Fig. 3**). The decrease of the primer-to-template ratio during late cycles also promotes self-annealing of amplicons, which results in a drop of the number of free primer binding sites.

Another effect of the rise of product DNA concentration is the reduction in the efficiency of duplex denaturation. Target DNA concentrations typically are 10^3 to 10^6 copies at the beginning and can increase to 10^{12} after approximately 40 cycles. The melting temperature (T_m) of a DNA duplex, however, is known to be elevated at higher concentrations, and the effect is measurable at product concentrations corresponding to the plateau phase of PCR *(83)*.

These considerations can help to explain why PCR amplification assays often do not work when too much sample DNA is added to the reaction mixture, a problem that is usually solved by dilution of the sample. Inefficient denaturation can be avoided by strictly adhering to protocols providing for sufficiently high denaturation temperatures (94° to 95°C) and denaturation times between 30 and 60 s. Knowledge of the kinetic characteristics of the plateau phase also helps to understand that the extension of amplification protocols far beyond 40–45 cycles would make an assay more vulnerable to non-specific amplification.

3.2.2. Efficiency of Primer Annealing

Optimal primer design seeks to achieve both high specificity (i.e. exclusive amplification of the selected target) and efficiency (i.e., high yield of amplified DNA through the selection of thermodynamically efficient primer binding

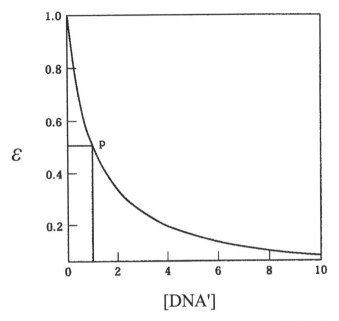

Fig. 3. Variation of the efficiency ε of a PCR as a function of [DNA'], the ratio of amplified product to initial template concentration *(85)*. Point p, where [DNA']=1, corresponds to an efficiency of ε = 0.5 (courtesy of Academic Press, Ltd.).

sites) (*see* **Chapter 4**), although it will often be necessary to accept compromised solutions for the sake of versatility and practicability.

Specificity of amplification is mainly determined by the annealing temperature (T_{ann}) and, to a lesser degree, primer length. The T_{ann} of a PCR assay is usually set within a few degrees of primer melting temperatures T_m. For a given reaction, the optimal T_{ann} can be calculated using **Equation 2** *(2, 90)*:

$$T_{ann} = 0.3\ T_{m\text{-}primers} + 0.7\ T_{m\text{-}product} - 14.9 \qquad \text{[Eq. 2]}$$

where the midpoint primer melting temperature is $T_{m\text{-}primers} = 0.5\ (T_{m\text{-}primer1} + T_{m\text{-}primer2})$, and product melting temperature is $T_{m\text{-}product} = 81.5 + 0.41(\%\ G+C) + 16.6\ log[K^+] - 675/L$ (with %G + C, molar percentage of guanosine-plus-cytosine, [K+], molar potassium ion concentration; L - length of amplicon in base pairs). The addition of co-solvents, such as dimethyl sulfoxide and formamide, allows to conduct the reaction at lower T_{ann} and was shown to improve both yield and specificity in selected cases *(90)*.

If all other conditions are optimal, a rise in T_{ann} can increase yield since primer-template mismatches are further reduced and co-synthesis of unspecific products is further suppressed. An example of the effect of T_{ann} on specificity is shown in **Fig. 4.**

In the context of identification of microbial species, genera, or serotypes, however, the expected intragenus and intraspecies variation in the target sequence region has to be taken into account. This may require the amplification assay to be run at suboptimal T_{ann} at the expense of specificity of detection. If the extent of target sequence variation is well-documented, a set of degenerate primers can solve the diagnostic problem *(91)*. Although the usefulness of such systems has been demonstrated *(53)* they need to be tested extensively before being introduced into routine diagnosis, as their performance is difficult to predict.

Optimal primer length varies between 18 and 24 nucleotides. The base composition of primer oligonucleotides should be as balanced as possible, for instance, a guanosine-plus-cytidine (G+C) content around 50% would lead to T_m values in the range of 56–62°C, thus allowing favorable annealing conditions. Within a given primer pair, the G+C content and T_m values should be no more than a few units apart to insure efficient amplification.

In the initial phase of PCR, nonspecific binding of primers and also primer–dimer formation may present problems, because the collision frequency of primer and template is still relatively low. Moreover, so-called jumping artifacts may be encountered as a result of single-stranded DNA fragments partially extended from one priming site annealing to a homologous target elsewhere in the genome, which finally leads to nonspecific product *(84)*. Secondary structures of the template (loops, hairpins) are also known to cause aberrant products as a result of DNA polymerase jumping, i.e., leaving nonlinear segments unamplified *(92,93)*.

An efficient way to deal with early-cycle mispriming effects is touch-down PCR *(94),* where the initial annealing temperature is gradually decreased with each cycle until the optimum value is reached. In the case of low target copy number samples, however, it is often helpful to run the first few cycles at lower-than-optimum T_{ann} and thus tolerate the concurrent synthesis of a certain proportion of unspecific product, and then raise T_{ann} to its optimum for the rest of cycles (touch-up PCR). A continuous increase of $\Delta T_{ann} = 1°C$ per cycle was shown to lead to higher product yield compared to a constant-T_{ann} protocol *(2)*.

As can be expected, the plateau phase exhibits more and more unfavorable conditions for efficient primer binding. Although the concentration of primers is only slightly lower than at the start, because of excess amounts present in the reaction mixture, the dramatically reduced primer-to-template ratio (from 2×10^7 to 19 after 10^6-fold amplification, *[84])]* results in a slowing down of the primer–template complex formation and, finally, saturation.

The probability of primer–dimer formation can already be reduced in the process of primer selection, e.g., by excluding primer sequences complementary to each other, particularly at 3'-ends *(95)*.

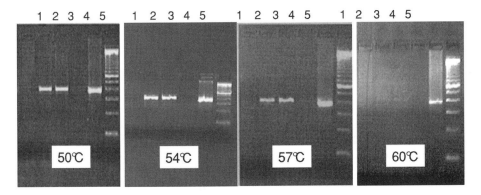

Fig. 4. Effect of the annealing temperature on the specificity of amplification. Two samples of avian feces were subjected to DNA extraction and examined by nested PCR for *Chlamydia psittaci*. The annealing temperature of the second amplification was varied from 50–60°C, all other experimental details were as in **Chapter 8**. Lane 1, sample A; lane 2, sample B; lane 3, reagent control; lane 4, DNA of strain C1 (*C. psittaci*); and lane 5, DNA size marker (100 bp ladder). Note that the correct amplification product in lane 4 is at 390 bp, whereas nonspecific bands in lanes 1 and 2 at 50°, 54° and 57°C are near 400 bp. In this case, running the PCR assay at suboptimal annealing temperature would lead to false positive results.

3.2.3. Enzyme-to-Template Ratio

The amount of DNA polymerase present in the reaction can also be a limiting factor contributing to loss of efficiency *(96)*. Schnell and Mendoza *(87)* singled out the ratio of free (unbound) to total enzyme as the most important parameter determining efficiency ε at the *i*th cycle and proposed a mathematical relationship given in **Equation 3**:

$$\varepsilon_i = 1 - \frac{[ET_i]}{[E_i] - [ET_i]}$$ [Eq. 3]

where $[E_i]$ is the DNA polymerase concentration and $[ET_i]$ is the concentration of the enzyme-template complex at the *i*th cycle.

While there is a great excess of enzyme molecules vs template DNA in the early cycles, typically in the order of 10^5, the ratio will be <1 after 10^6-fold amplification *(84)*. The consequence of product molecules outnumbering the polymerase is a reduction of yield per cycle in the linear and plateau phases. Under these conditions, it is no longer possible to have one enzyme molecule anchored to each primer–template complex in the extension step, which reduces the number of (full-length) amplicons generated per cycle and allows more nonspecific products of shorter length to be synthesized. It would be possible to partially compensate this drop in efficiency by prolonging the extension step in the later cycles or increasing the enzyme concentration of the reaction. The latter, however, has to be optimized empirically as excessive enzyme often stimulates the co-synthesis of nonspecific products.

Although thermal stability of commercially available DNA polymerases has been steadily improved over the last decade (*see* **Chapter 2**) there is a measurable decrease of its activity in the late cycles. This is another reason not to recommend high numbers of cycles in PCR detection assays, indeed, 30–35 cycles proved sufficient in most applications.

4. Methodologies to Improve the Performance of Amplification Assays

When using a PCR assay for clinical samples containing only a few cells of the pathogen, one has to be aware of the special kinetic conditions prevailing in such a reaction mixture. At the start of the reaction as few as 1 to 100 copies of the target sequence may be present, and the crucial primer annealing step is particularly difficult to control. This may require additional measures to insure that primer oligonucleotides specifically bind to as many targets as possible and, at the same time, avoid nonspecific priming of DNA synthesis.

4.1. Hot-Start PCR

The basic idea of hot-start PCR is to reduce nonspecific amplification in the initial phase by releasing active enzyme only immediately before the first primer binding step *(97)*. This approach is designed to prevent primer–dimer formation, mispriming, and spontaneous initiation of DNA strand synthesis, most of which occur already at room temperature between the operations of mixing reagents and actually starting the PCR run *(98,99)*. The most common variant of hot-start PCR involves thermostable DNA polymerases (*see* **Chapter 2**) that are supplied in an inactive form and require a 10-min heating at 94°– 96°C for activation *(100, 101)*. Many protocols of "conventional" amplification assays can be improved in terms of specificity by adaptation to the hot-start procedure and its special reagents.

Another approach involves so-called loop primers that carry additional 5' tails causing the oligonucleotides to self-anneal or oligomerize at ambient temperature *(102)*. When the reaction mixture is heated, the primers are linearized only at elevated temperatures, thus initiating a hot start. To facilitate specificity of amplification, the protocol begins with six touch-up cycles, where T_{ann} is gradually increased from 60° to 72°C. Hot-start conditions were also shown to be created through the addition of short double-stranded DNA fragments *(103)*.

4.2. Nested PCR

The use of nested PCR can often solve detection problems associated with clinical tissue samples containing low copy numbers of target against a high background of host tissue DNA and inhibitors of DNA polymerase. A typical protocol would begin with a first round of amplification (30–35 cycles) using the outer primer set, and then subject a small aliquot of first-round product to a second run with fresh reagents using the inner primer pair. This approach was shown to be more successful than diluting and reamplifying with the same primers *(104)*. The position of inner primers is often the determining factor of the assay's sensitivity and specificity.

In theory, even under optimal conditions, any PCR will reach a natural plateau (*see* **Subheading 3.1.**), so that one should not expect a gain in yield (i.e., sensitivity) from reamplifying the diluted product of a previous reaction. Nevertheless, there are numerous examples in which the necessary sensitivity of detection was attained only thanks to nested PCR. This is mainly due to the fact that many PCR detection assays are not completely optimized. Besides, clinical and field samples often contain inhibitors (*see* **Chapter 2**), which prevent the first PCR from proceeding at high efficiency. In these circumstances, a second round of amplification can make all the difference.

From the kinetic point of view, the second PCR could be seen as a continuation of the first reaction with strongly reduced amounts of product, a measure that can be expected to increase efficiency and yield as discussed in **Subheading 3.2.1.**

In diagnostic applications, nested PCR represents a powerful tool for differentiation, because of the possibility to select outer primer binding sites in a genomic segment common to a group of organisms, e.g., family, genus, or species, and place the inner primers at sites that are specific for individual species, serovars, or biovars. In this area of application, nested PCR can provide valuable data for epidemiology and taxonomy. Another advantage is that reamplification of nonspecific products from the first round is minimized because of the utilization of different primers in the second round.

As far as routine diagnosis is concerned, there are various reservations about the introduction of nested PCR assays owing to its particular vulnerability to product carryover *(71)*. The fact that there are many laboratories using nested protocols for routine purposes also shows its practicability, but it will certainly remain a domain of the more experienced and specialized laboratories.

4.3. Multiplex PCR

The possibility to use several primer pairs, each having a particular specificity, in the same reaction *(105)* adds a multidimensional perspective to the diagnostic potential of PCR. Such a procedure allows simultaneous detection of two or more different microbial agents in a single sample and the inclusion of internal controls. As multiplex PCR involves a far more complex reaction system than the normal simplex mode, its performance is more difficult to predict and can be assessed only after extensive trials.

Users must be aware that, in principle, each of these parallel amplification reactions will proceed with their own kinetics and efficiency, thus resulting in different sensitivities and specificities for each target. Even in comparatively simple competitive PCR systems for relative quantitation, where sample DNA and internal standard template have identical primer binding sites it is not certain that both components will be amplified with the same efficiency *(3, 106–108)*. Although the situation is far more complex in a multiplex assay involving different target sequences of different organisms, two important conclusions that emerged from experience in competitive PCR are of general significance: (*i*) An inverse exponential relationship exists between template size and efficiency of amplification *(3)*; and (*ii*) the formation of heteroduplexes between different amplicons becomes likely when two or more target sequences share homologous segments *(109,110)*. Also, primers can be expected to interact with each other, thus having a detrimental effect on the assay's performance. A recent study by Bercovich et al. *(111)* revealed that the relative amounts of primers had a major effect on yields in duplex amplification. Similarly, the optimal T_{ann} of a duplex assay will not necessarily be equal to the arithmetic mean value of the individual simplex assay's annealing temperatures.

A systematic study of experimental parameters in multiplex PCR led Henegariu et al. *(112)* to propose a general step-by-step protocol for optimization. These authors stressed the crucial importance of balancing relative concentrations among primer pairs, as well as between magnesium chloride and dNTPs. Compared to simplex PCR, more time is necessary to complete strand synthesis, so therefore, extension steps should generally be longer. Other important parameters include T_{ann} and concentration of the reaction buffer.

The number of multiplex PCR detection systems for microbial agents described in the literature is not very large. Some examples will be discussed in the following.

With the aim to detect bacterial agents associated with meningitis, a seminested multiplex strategy was applied for simultaneous detection of *Neisseria meningitidis, Haemophilus influenzae, Streptococcus pneumoniae, Streptococcus agalactiae*, and *Listeria monocytogenes* from cerebrospinal fluid *(113,114)*. The first round of amplification involves universal eubacterial primers binding to conserved segments of the 16S rRNA gene, one of which is replaced by a set of species-specific primers in the second (multiplex) PCR.

A procedure for screening of respiratory samples was developed by Tong et al. *(115)*, who used primers from the P1 adhesin gene of *Mycoplasma pneumoniae* and the *omp*A genes of *Chlamydia pneumoniae* and *C. psittaci* to detect these pathogens. In this case, the optimized conditions of the multiplex assay were not significantly different from the individual assays.

To differentiate three species of chlamydiae, i.e., *C. pneumoniae*, *C. psittaci,* and *C. trachomatis*, a genus-specific 16S rRNA gene segment was amplified in the first round followed by multiplex nested amplification *(13)*. The method was recommended for clinical human and avian samples, such as throat swabs, lung tissue, and feces. For the same species, another multiplex assay targeting the 16S rRNA gene and 16S–23S spacer region and based on touch-down enzyme time-release PCR was reported *(14)*. This methodology was chosen to insure high specificity. Identification could easily be done according to amplicon size, and detection limits were below 0.1 inclusion-forming units.

Feng and Monday *(46)* described a multiplex assay for detection and differentiation of enterohemorrhagic *E. coli* serotypes. Primers were placed in five different gene regions (*uid*A, *eae*A, *stx*1, *stx*2, *ehx*A) and results were consistent with the accepted classification of genotypes.

A group of avian mycoplasmas, i.e., *Mycoplasma gallisepticum, M. synoviae, M. iowae*, and *M. meleagridis*, were shown to be simultaneously detectable by multiplex amplification of nonspecified species-specific target sequences *(116)*.

The usefulness of PCR in genotyping *Clostridium perfringens* was demonstrated by Meer and Songer *(44)* who developed a multiplex assay targeting the genes coding for the major toxins, i.e., α, β, ε, ι, and enterotoxin. As genotypes determined by this method coincided with the results of phenotypic assays in 99% of all cases, the PCR approach represents a simpler and faster alternative to conventional methods.

Eight genotypes of *Trichinella*, some of them representing separate species, were differentiated in a PCR using 5 primer pairs located in ribosomal rRNA genes and spacer regions *(117)*. The size of the amplicon(s) was characteristic for each type.

Undoubtedly the demand for multiplex assays is going to increase further in the near future, not only because of the potential to reduce costs and raise throughput, but also in the light of current developments in the area of DNA array technology, which will provide new powerful detection systems.

5. Practical Implications of Routine Use of PCR in the Diagnostic Laboratory

Whenever a new PCR detection assay is introduced, verification of its findings remains an indispensable demand. The identity of a given amplification product has to be confirmed by such tests as Southern blot hybridization using a specific probe, restriction enzyme analysis, subsequent nested PCR, or DNA sequencing.

As any other new diagnostic tool, the PCR detection assay requires careful validation before it can be adopted as an official method. The ideal approach would include culture as the main reference, often called gold standard, as well as an established antigen enzyme-linked immunosorbent assay (ELISA). While this appears feasible for some prominent and well-investigated bacterial agents, such as salmonella, *E. coli*, campylobacter, or listeria, there are serious obstacles to validation in other cases. For instance, comparative studies on PCR detection of animal chlamydiae (e.g. avian and bovine *C. psittaci*, porcine *C. trachomatis*) often remain incomplete, because many strains do not grow in cell culture. Similarly, any ELISA test for these agents would face the same dilemma if subjected to validation. There is no way to validate the better-performing PCR methodology against conventional standard methods that are obviously inferior in terms of sensitivity and specificity. In such a situation, the pragmatic approach should prevail and PCR should be accepted as the most suitable method available at present, provided that there is comprehensive evidence on specificity and sensitivity of detection.

Once a PCR assay has been accepted and used in diagnostic laboratories, the question of how to deal with the data and its consequences has to be addressed.

The high sensitivity of PCR inevitably leads to a greater number of positive samples in comparison to conventional methods. As a rule, the agent will be detected over a longer period in the course of infection. The fact that the presence of a pathogen can be confirmed already in the incubation period, i.e., before the onset of the host's immune response and the appearance of clinical symptoms, represents a considerable advantage and allows necessary control measures to be taken at an earlier stage.

Similarly, asymptomatic animals harboring and shedding the pathogen are also more likely to be identified as a consequence of highly sensitive detection by PCR. This implies the possibility to follow the actual epidemiological status of a herd more closely. As high sample throughput can be achieved with 96-well microtiter plates and other formats, PCR lends itself as a powerful tool for herd diagnosis.

The fact that PCR tests may detect DNA from nonviable or dead microbial cells is occasionally interpreted as a weak point. The question of whether such a finding really represents a false positive result is difficult to answer unambiguously. Of course, one has to be careful with PCR data from animals undergoing antibiotic therapy, as they could test positive because of some remaining dead cells or DNA of the infectious agent, which are of no clinical significance *(118)*. On the other hand, micoorganisms identified by PCR in excrements of intermittent shedders, even if nonviable or nonculturable, can provide important evidence on the presence of a pathogen that would have remained undetected by other methods. Repeated PCR positivity in the face of antibiotic treatment and/or serological nonreactivity and bacteriological sterility may mean chronicity or inadequacy of treatment.

One of the main consequences for epidemiology arising from the increasing availability of PCR-based diagnostic data could be the realization that the pathogen in question was more abundant than previously assumed and that it persisted to a certain degree in apparently healthy hosts. It is certainly nothing new for experienced diagnosticians and practitioners that the mere occurrence of certain pathogens cannot always be associated with clinical disease, but this point is rarely addressed in textbooks. Therefore, increasing amounts of PCR-based epidemiological data are going to help recall (and confirm) the well-known thesis that the presence of the pathogen is a necessary yet insufficient condition to produce disease. It will certainly further mutual understanding between practitioners, veterinary officers, laboratory vets, conventional and molecular diagnosticians, if the latter manage to get this message across.

As a major limitation, it has to be emphasized that the pathogenic potential of a given microbial strain is not adequately covered by most current PCR tests, which merely show the presence of one gene or target sequence. Present and future research aimed at a new generation of assays for simultaneous identification of a whole complex of pathogenicity factors connected with the microbial agent, the animal host, and the environment will certainly provide more satisfactory answers.

References

1. He, Q., Marjamäki, M., Soini, H., Mertsola, J., and Viljanen, M. K. (1994) Primers are decisive for sensitivity of PCR. *BioTechniques* **17**, 82–87.
2. Rychlik, W., Spencer, W. J., and Rhoads, R. E. (1990) Optimization of the annealing temperature for DNA amplification in vitro. (Erratum in Nucleic Acids Res 1991; 19:698) *Nucleic Acids Res.* **18**, 6409–6412.
3. McCulloch, R. K., Choong, C. S., and Hurley, D. M. (1995) An evaluation of competitor type and size for use in the determination of mRNA by competitive PCR. *PCR Meth. Appl.* **4**, 219–226.
4. Toouli, C. D., Turner, D. ., Grist, S. A., and Morley, A. A. (1999) The effect of cycle number and target size on polymerase chain reaction amplification of polymorphic repetitive sequences. *Anal. Biochem.* **280**, 324–326.
5. GenBank®: (http://www.ncbi.nlm.nih.gov/entrez/query.fcgi).
6. Greisen, K., Loeffelholz, M., Purohit, A., and Leong, D. (1994) PCR primers and probes for the 16S rRNA gene of most species of pathogenic bacteria, including bacteria found in cerebrospinal fluid. *J. Clin. Microbiol.* **32**, 335–351.

7. van Camp, G., Fierens, H., Vandamme, P., Goossens, H., Huyghebaert, A., and de Wachter, R. (1993) Identification of enterpathogenic *Campylobacter* species by oligonucleotide probes and polymerase chain reaction based on 16S rRNA genes. *Syst. Appl. Microbiol.* **16,** 30–36.

8. Giesendorf, B .A. J., Quint, W. G. V., Henkens, M. H. C., Stegmann, H., Huf, F. A., and Niesters, H. G. M. (1992) Rapid and sensitive detection of *Campylobacter* ssp. in chicken products by using the polymerase chain reaction. *Appl. Environ. Microbiol.* **58,** 3804–3808.

9. Metherell, L. A., Logan, J. M. J., and Stanly, J. (1999) PCR-enzyme-linked immunosorbent assay for detection and identification of *Campylobacter* species: Application to isolates and stool samples. *J. Clin. Microbiol.* **37,** 433–435.

10. Cardarelli-Leite, P., Blom, K., Patton, C. M., Nicholson, M. A., Steigerwalt, A. G., Hunter, S. B., Brenner, D. J., Barrett, T. J., and Swaminathan, B. (1996) Rapid identification of *Campylobacter* species by restriction fragment length polymorphism analysis of a PCR-amplified fragment of the gene coding for 16S rRNA. *J. Clin. Microbiol.* **34,** 62–67.

11. Heinemann, M. B., Garcia, J. F. Nunes, C. M., et al. (2000) Detection and differentiation of *Leptospira* spp. serovars in bovine semen by polymerase chain reaction and restriction fragment length polymorphism. *Vet. Microbiol.* **73,** 261–267.

12. Lammler, C. Abdulmawjood, A., and Weiss, R. (1998) Properties of serological group B streptococci of dog , cat and monkey origin. *Zentralbl. Veterinärmed.* **49,** 561–566.

13. Messmer, T. O., Skelton, S. K., Moroney, J. F., Daugharty, H. and Fields, B. S. (1997) Application of a nested, multiplex PCR to psittacosis outbreaks. *J. Clin. Microbiol.* **35,** 2043–2046.

14. Madico, G., Quinn, T. C., Bomann, J., and Gaydos, C. A. (2000) Touchdown enzyme release-PCR for detection and identification of *Chlamydia trachomatis, C. pneumoniae,* and *C. psittaci* using the 16S and 16S–23S spacer rRNA genes. *J. Clin. Microbiol.* **38,** 1985–1093.

15. Kox, L. F. F., van Leeuwen, J., Knijper, S., Jansen, H. M., and Kolk, A. H. J. (1995) PCR assay based on DNA coding for rRNA for detection and identification of mycobacteria in clinical samples. *J. Clin. Microbiol.* **33,** 3225–3233.

16. Oggioni, M. R., Fattorini, L., Li, B., et al. (1995) Identification of *Mycobacterium tuberculosis* complex, *Mycobacterium avium* and *Mycobacterium intracellulare* by selective nested polymerase chain reaction. *Mol. Cell. Probes* **9,** 321–326.

17. Bascunana, C. R., Mattsson, J. G., Bölske, G., and Johansson, K.-E. (1994) Characterization of the 16S rRNA genes from *Mycoplasma* sp. strain F38 and development of an identification system based on PCR. *J. Bacteriol.* **179,** 2577–2586.

18. Giacometti, M., Nicolet, J., Johansson, K. -E., Naglic, T., Degiorgis, M. P., and Frey, J. (1999) Detection and identification of *Mycoplasma conjunctivae* in infectious keratoconjunctivitis by PCR based on the 16S rRNA gene. *Zentralbl. Veterinärmed.* B **46,** 173–180.

19. Persson, A., Pettersson, B., Bölske, G., and Johansson, K.-E. (1999) Diagnosis of contagious bovine pleuropneumonia by PCR-laser-induced fluorescence and PCR-restriction endonuclease analysis based on the 16S rRNA genes of *Mycoplasma mycoides* subsp. *mycoides* SC. *J. Clin. Microbiol.* **37,** 3815–3821.

20. Wang, R.F., Cao, W.W., Franklin, W., Campbell, W., and Cerniglia, C.E. (1994) A 16S rDNA-based PCR method for rapid and specific detection of *Clostridium perfringens* in food. *Mol. Cell. Probes* **8,** 131–138.

21. Lantz, P. G., Knutsson, R., Blixt, Y., Al Soud, W. A., Borch, E., and Radström P. (1998) Detection of pathogenic *Yersinia enterocolitica* in enrichment media and pork by a multiplex PCR: a study of sample preparation and PCR-inhibitory components. *Int. J. Food Microbiol.* **45,** 93–105.

22. Subramaniam, S., Chua, K. L., Tan, H. M., Loh, H., Kuhnert, P., and Frey, J. (1997) Phylogenetic position of *Riemerella anatipestifer* based on 16S rRNA gene sequences. *Int. J. Syst. Bacteriol.* **47,** 562–565.

23. Mattsson, J. G., Guss, B., and Johansson, K. -E. (1994) The phylogeny of *Mycoplasma bovis* as determined by sequence analysis of the 16S rRNA gene. *FEMS Microbiol. Lett.* **115,** 325–328.

24. Helgason, E., Okstad, O.A., Caugant, D.A., et al. (2000) *Bacillus anthracis, Bacillus cereus,* and *Bacillus thuringiensis* - One species on the basis of genetic evidence. *Appl. Environm. Microbiol.* **66,** 2627–2630.

25. Fermér, C. and Olsson Engvall, E. (1999) Specific PCR identification and differentiation of the thermophilic campylobacters, *Campylobacter jejuni, C. coli, C. lari,* and *C. upsaliensis. J. Clin. Microbiol.* **37,** 3370–3373.

26. Eyers, M., Chapelle, S., van Camp, G., Goossens, H., and de Wachter, R. (1993) Dis- crimination among thermophilic *Campylobacter* species by polymerase chain reaction amplification of 23S rRNA gene fragments. *J. Clin. Microbiol.* **31,** 3340–3243.

27. Everett, K. D. E., and Andersen, A. A. (1999) Identification of nine species of the *Chlamydiaceae* using RFLP-PCR. *Int. J. Syst. Bacteriol.* **49,** 217–224.

28. Gürtler, V. and Stanisich, V. A. (1996) New approaches to typing and identification of bacteria using the 16S–23S rDNA spacer region. *Microbiology* **142,** 3–16.

29. Scheinert, P., Krausse, R., Ullmann, U., Soller, R., and Krupp, G. (1996) Molecular differentiation of bacteria by PCR amplification of the 16S–23S rRNA spacer. *J. Microbiol. Meth.* **26,** 103–117.

30. O'Sullivan, N. A., Fallon, R., Carroll, C., Smith, T., and Maher, M. (2000) Detection and differentiation of *Campylobacter coli* in broiler chicken samples using a PCR/DNA probe membrane based colorimetric detection assay. *Mol. Cell. Probes* **14,** 7–16.

31. Cartwright, C. P., Stock, F., Beekmann, S. E., Williams, E. C., and Gill, V. J. (1995) PCR amplification of rRNA intergenic spacer regions as a method for epidemiologic typing of *Clostridium difficile. J. Clin. Microbiol.* **33,** 184–187.

32. Mitchell, T. G., Freedman, E. Z., White, T. J., and Taylor, J. W. (1994) Unique oligonucleotide primers in PCR for identification of *Cryptococcus neoformans. J. Clin. Microbiol.* **32,** 253–255.

33. Rappelli, P., Are, R., Casu, G., Fiori, P. L., Cappuccinelli, P., and Aceti, A. (1998) Development of a nested PCR for detection of *Cryptococcus neoformans* in cerebrospinal fluid. *J. Clin. Microbiol.* **36,** 3438–3440.

34. Drebót, M., Neal, S., Schlech, W., and Rozee, K. (1996) Differentiation of *Listeria* isolates by PCR amplicon profiling and sequence analysis of 16S–23S rRNA internal transcribed spacer loci. *J. Appl. Bact.* **80,** 174–178.

35. O'Connor, L., Joy, J., Kane, M., Smith, T. and Maher, M.(2000) Rapid polymerase chain reaction/DNA probe membrane-based assay for the detection of *Listeria* and *Listeria monocytogenes* in food. *J. Food Prot.* **63,** 337–342.

36. Park, H., Jang, H., Kim, C., Chung, B., Chang, C. L., Park, S. K., and Song, S. (2000) Detection and identification of mycobacteria by amplification of the internal transcribed spacer regions with genus- and species-specific PCR primers. *J. Clin. Microbiol.* **38,** 4080–4085.

37. Uemori, T., Asada, K., Kato, I., and Harasawa, R. (1992) Amplification of the 16S–23S spacer region in rRNA operons of mycoplasmas by the polymerase chain reaction. *System Appl. Microbiol.* **15,** 181–186.

38. Brickell, S. K., Thomas, L. M., Long, K. A., Panaccio, M., and Widders, P. R. (1998) Development of a PCR test based on a gene region associated with the pathogenicity of *Pasteurella multocida* serotype B:2, the causal agent of haemorrhagic septicaemia in Asia. *Vet. Microbiol.* **59,** 295–307.

39. Gill, S., Belles-Isles, J., Brown, G., Gagné, S., Lemieux, C., Mercier, J. -P., and Dion, P. (1994) Identification of variability of ribosomal DNA spacer from *Pseudomonas* soil isolates. *Can. J. Microbiol.* **40,** 541–547.

40. Whiley, R. A., Duke, B., Hardie, J. M., and Hall, L. M. C. (1995) Heterogeneity among 16S–23S rRNA intergenic spacers of species within the *Streptococcus milleri* group. *Microbiology* **141,** 1461–1467.

41. Forsman, P., Tilsala-Timisjarvi, A., and Alatossava, T. (1997) Identification of staphylococcal cause of bovine mastitis using 16S–23S rRNA spacer regions. *Microbiology* **143,** 3491–3500.

42. Fach, P. and Guillou, J.P. (1993) Detection by in vitro amplification of the alphatoxin (phospholipase C) gene from *Clostridium perfingens. J. Appl. Bacteriol.* **74,** 61–66.

43. Buogo, C., Capaul, S., Häni, H., Frey, J., and Nicolet, J. (1995) Diagnosis of *Clostridium perfringens* type C enteritis in pigs using a DNA amplification technique (PCR) *Zentralbl. Veterinärmed.* **42,** 51–58.

44. Meer, R. R. and Songer J. G. (1997) Multiplex polymerase chain reaction assay for genotyping *Clostridium perfringens. Am. J. Vet. Res.* **58,** 702–705.

45. Karch, H., and Meyer, T. (1989) Single primer pair for amplifying segments of distinct Shiga-like toxin genes by polymerase chain reaction. *J. Clin. Microbiol.* **27,** 2751–2757.

46. Feng, P., and Monday S. R. (2000) Multiplex PCR for detection of trait and virulence factors in enterohemorrhagic *Escherichia coli* serotypes. *Mol. Cell. Probes* **14,** 333–337.

47. Hotzel, H., Erler, W., and Schimmel, D. (1997) Detection of dermonecrotic toxin genes in *Pasteurella multocida* strains using the polymerase chain reaction (PCR). *Berl. Münch. Tierärztl. Wochenschr.* **110,** 139–142.

48. Fisher, M. A., Weiser, G. C., Hunter, D. L., and Ward, A. C. (1999) Use of a polymerase chain reaction method to detect the leukotoxin gene *lkt*A in biogroup and biovariant isolates of *Pasteurella haemolytica* and *P. trehalosi. Am. J. Vet. Res.* **60,** 1402–1406.

49. Schaller, A., Djordjevic, S.P., Eamens, G.J., et al. (2001) Identification and detection of *Actinobacillus pleuropneumoniae* by PCR based on the gene *apx*IVA. *Vet. Microbiol.* **79,** 47–62.

50. Gram, T., Ahrens, P., Andreasen, M., and Nielsen, J.P. (2000) An *Actinobacillus pleuropneumoniae* PCR typing system based on the *apx* and *oml*A genes - evaluation of isolates from lungs and tonsils of pig. *Vet. Microbiol.* **75,** 43–57.

51. de la Puente-Redondo, V. A., del Blanco, N. G., Gutierrez-Martin, C. B., Mendez, J. N., and Rodriquez Ferri, E. F. (2000) Detection and subtyping of *Actinobacillus pleuropneumoniae* strains by PCR-RFLP analysis of the *tbp*A and *tbp*B genes. *Res. Microbiol.* **151,** 669–681.

52. Ooyofo, B. A., Thornten, S. A., Burr, D. H., Trust, T. J., Pavlovskis, O. R., and Guerry, P. (1992) Specific detection of *Campylobacter jejuni* and *Campylobacter coli* by using polymerase chain reaction. *J. Clin. Microbiol.* **30,** 2613–2619.

53. Kaltenböck, B., Schmeer, N., and Schneider, R. (1997) Evidence for numerous *omp*1 alleles of porcine *Chlamydia trachomatis* and novel chlamydial species obtained by PCR *J. Clin. Microbiol.* **35,** 1835–1841.

54. Caron, J., Ouardani, M., and Dea, S. (2000) Diagnosis and differentiation of *Mycoplasma hyopneumoniae* and *Mycoplasma hyorhinis* infection in pigs by PCR amplification of the p36 and p46 genes. *J. Clin. Microbiol.* **38,** 1390–1396.

55. Kasten, R. W., Carpenter, T. E., Snipes, K. P., and Hirsh, D. C. (1997) Detection of *Pasteurella multocida*-specific DNA in turkey flocks by use of the polymerase chain reaction. *Avian. Dis.* **41,** 676–682.

56. Baily, G. G., Krahn, L. B., Drasar, B. S., and Stocker, N. G. (1992) Detection of *Brucella melitensis* and *Brucella abortus* by DNA amplification. *J. Tropical. Med. Hygiene,* **95,** 271–275.

57. Englen, M. D. and Kelley, L. C. (2000) A rapid DNA isolation procedure for the identification of Campylobacter jejuni by the polymerase chain reaction. *Appl. Microbiol.* **31,** 421–426.

58. van Doorn, L. J., Verschuuren-van Haperen, A., Burnens, A., et al. (1999) Rapid identification of thermotolerant *Campylobacter lari,* and *Campylobacter upsaliensis* from various geographic locations by a GTPase-based PCR-reverse hybridization assay. *J. Clin. Microbiol.* **37,** 1790–1796.

59. Tanaka, K., Miyazaki, T., Maesaki, S., Mitsutake, K., Kakeya, H., Yamamoto, Y., Yanagihara, K., Hossain, M. A., Tashiro, T., and Kohno, S. (1996) Detection of *Cryptococcus neoformans* gene in patients with pulmonary cryptococcosis. *J. Clin. Microbiol.* **34,** 2826–2828.

60. Furrer, B., Candrian, U., Hoefelein, Ch., and Luethy, J. (1991) Detection and identification of *Listeria monocytogenes* in cooked sausage products and in milk by in vitro amplification of haemolysin gene fragments. *J. Appl. Bacteriol.* **70,** 372–379.

61. Luk, J. M. C., Kongmuang, U., Reeves, P. R., and Lindberg, A. A. (1993) Selective amplification of abequose and paratose synthase genes (rfb) by polymerase chain reaction for identification of *Salmonella* major serogroups (A, B, C2, and D). *J. Clin. Microbiol.* **31,** 2118–2123.

62. Hoorfar, J., Baggesen, D. L., and Porting, P. H. (1999) A PCR-based strategy for simple and rapid identification of rough presumptive *Salmonella* isolates. *J. Microbiol. Methods* **35,** 77–84.

63. Pinnow, C. C., Butler, J. A., Sachse, K., Hotzel, H., Timms, L. L., and Rosenbusch, R. F. (2001) Detection of *Mycoplasma bovis* in preservative-treated field milk samples. *J. Dairy Sci.* **84,** 1640–1645.

64. Hotzel, H., Heller, M., and Sachse, K. (1999) Enhancement of *Mycoplasma bovis* detection in milk samples by antigen capture prior to PCR. *Mol. Cell. Probes* **13,** 175–178.

65. Subramaniam, S., Bergonier, D., Poumarat, F., Capaul, S., Schlatter, Y., Nicolet, J., and Frey, J. (1998) Species identification of *Mycoplasma bovis* and *Mycoplasma agalactiae* based on the *uvr*C genes by PCR. *Mol. Cell. Probes* **12,** 161–169.

66. Hance, A. J., Grandshamp, B., Levy-Frebault, V., Lecossier, D., Rauzier, J., Bocart, D., and Gisquel, B. (1989) Detection and identification of mycobacteria by amplification of mycobacterial DNA. *Mol. Microbiol.* **3,** 843–849.

67. Steingrube, V. A., Gibson, J. L., Brown, B. A., Zhang, Y., Wilson, R. W., Rajagonpalan, M., and Wallace Jr., R. J. (1995) PCR amplification and restriction endonuclease analysis of a 65-kilodalton heat shock protein gene sequence for taxonomic separation of rapidly growing mycobacteria. *J. Clin. Microbiol.* **33,** 149–153.

68. Taylor, T. B., Patterson, C., Hale, Y., and Safranek, W. W. (1997) Routine use of PCR-restriction fragment lenght polymorphism analysis for identification of mycobacteria growing in liquid media. *J. Clin. Microbiol.* **35,** 79–85.

69. Ericsson, H., and Stalhandske, P. (1997) PCR detection of Listeria monocytogenes in 'gravad' rainbow trout. *Int. J. Food Microbiol.* **35,** 281–285.

70. Rahn, K., De Grandis, S. A., Clarke, R. C., et al. (1992) Amplification of an invA gene sequence of *Salmonella typhimurium* by polymerase chain reaction as a specific method of detection of *Salmonella. Mol. Cell. Probes* **6,** 271–279.

71. Lantz, P. -G., Al-Soud, W. A., Knutsson, R., Hahn-Hägerdal, B., and Radström, P. (2000) Biotechnical use of poymerase chain reaction for microbiological analysis of biological samples. *Biotechnol. Ann. Rev.* **5,** 87–130.

72. Wastling, J. M., Nicoll, S., and Buxton, D. (1993) Comparison of two gene amplification methods for the detection of Toxoplasma gondii in experimentally infected sheep. *J. Med. Microbiol.* **38,** 360–365.

73. Manzano, M., Cocolin, L., Cantoni, C., and Comi, G. (1998) A rapid method for the identification and partial serotyping of *Listeria monocytogenes* in food by PCR and restriction enzyme analysis. *Int. J. Food Microbiol.* **42,** 207–212.

74. Willems, H., Thiele, D., Frölich-Ritter, R., and Krauss, H. (1994) Detection of *Coxiella burnetii* in cow's milk using the polymerase chain reaction (PCR). *J. Vet. Med. B.* **41**, 580–587.

75. Schrader, C., Protz, D., and Süss, J. (2000) *Coxiella burnetii*, in *Molekularbiologische Nachweismethoden ausgewählter Zoonoseerreger* (Sachse, K. and Gallien, P., eds.), *bgvv-Hefte* 02/2000, pp. 63–70.

76. Roring, S., Hughes, M. S., Skuce, R. A., and Neill, S. D. (2000) Simultaneous detection and strain differentiation of *Mycobacterium bovis* directly from bovine tissue specimens by spoligotyping. *Vet. Microbiol.* **74**, 227–236.

77. Thierry, D., Brisson-Noel, A., Vincent-Levy-Frebault, V., Nguyen, S., Guesdon, J. L., and Gicquel, B. (1990) Characterization of a *Mycobacterium tuberculosis* insertion sequence, IS 6110 and its application in diagnosis. *J. Clin. Microbiol.* **28**, 2668–2673.

78. de Lassence, A., Lecossier, D., Piere, C., Cadranel, J., Stern, M., and Hance, A. J. (1992) Detection of mycobacterial DNA in pleural fluid from patients with tuberculosis pleurisy by means of the polymerase chain reaction: comparison of two protocols. *Thorax* **47**, 265–269.

79. Mangiapan, G., Vokurka, M., Schouls, L., Cadranel, J., Lecossier, D., van Embden, J., and Hance, A. J. (1996) Sequence-capture PCR improves detection of mycobacterial DNA in clinical specimens. *J. Clin. Microbiol.* **34**, 1209–1215.

80. Redstone, J. S., and Woodward, M. J. (1996) The development of a ligase mediated PCR with potential for the differentiation of serovars within *Leptospira interrogans*. *Vet. Microbiol.* **51**, 351–362.

81. Appleyard, G. D., Zarlenga, D., Pozio, E., and Gajadhar, A. A. (1999) Differentiation of *Trichinella* genotypes by polymerase chain reaction using sequence-specific primers. *Int. J. Parasitol.* **85**, 556–559.

82. Innis, M.A. and Gelfand, D.H. (1990) Optimization of PCRs, in *PCR Protocols: A Guide to Methods and Applications*, (Innis, M. A., Gelfand, D. H., Sninsky, J. J., and White T. J., eds.), Academic Press, New York, NY, USA, pp. 3–12.

83. Sardelli, A. (1993) Plateau effect - understanding PCR limitations. *Amplifications* **9**, 1–5.

84. Ruano, G., Brash, G. R., and Kidd, K. K. (1991) PCR: The first few cycles. *Amplifications* **7**, 1–4.

85. Schnell, S. and Mendoza, C. (1997) Theoretical description of the polymerase chain reaction. *J. Theor. Biol.* **188**, 313–318.

86. Saiki, R. K., Scharf, S., Faloona, F., Mullis, K. B., Horn, G. T., Erlich, H. A., and Arnheim, N. (1985) Enzymatic amplification of β-globin genomic sequences and restriction site analysis for diagnosis of sickle cell anaemia. *Science* **230**, 1350–1354.

87. Schnell, S., and Mendoza, C. (1997) Enzymological considerations for a theoretical description of the quantitative competitive polymerase chain reaction (QC-PCR). *J. Theor. Biol.* **184**, 433–440.

88. Morrison, C., and Gannon, F. (1994) The impact of the PCR plateau phase on quantitative PCR. *Biochim. Biophys. Acta* **1219**, 493–498.

89. Bloch, W. (1991) A biochemical perspective of the polymerase chain reaction. *Biochem.* **30,** 2735–2747.

90. Chester, N., and Marshak, D. R. (1993) Dimethyl sulfoxide-mediated primer *Tm* reduction: A method for analyzing the role of renaturation temperature in the polymerase chain reaction. *Anal. Biochem.* **209,** 284–290.

91. Compton, T. (1990) Degenerate primers for DNA amplification, in *PCR Protocols: A Guide to Methods and Applications*, (Innis, M. A., Gelfand, D. H., Sninsky, J. J., and White T. J., eds.), Academic Press, New York, pp. 39–45.

92. Viswanathan, V. K., Krcmarik, K., and Cianciotto, N. P. (1999) Template secondary structure promotes polymerase jumping during PCR amplification. *BioTechniques* **27,** 508–511.

93. Loewen, P. C. and Switala, J. (1995) Template secondary structure can increase the error frequency of the DNA polymerase from *Thermus aquaticus*. *Gene* **164,** 56–63.

94. Don, R. H., Cox, P. T., Wainwright, B. J., Baker, K., and Mattick, J. S. (1991) Touchdown PCR to circumvent spurious priming during gene amplification. *Nucl. Acids Res.* **19,** 4008.

95. Watson, R. (1989) The formation of primer artifacts in polymerase chain reactions. *Amplifications* **2,** 5–6.

96. Gelfand, D. H. and White, T. J. (1990) Thermostable DNA polymerases, in *PCR Protocols: A Guide to Methods and Applications*, (Innis, M. A., Gelfand, D. H., Sninsky, J. J., and White T. J., eds.), Academic Press, New York, NY, pp. 129–141.

97. Erlich, A. H., Gelfand, D., and Sninsky, J. J. (1991) Recent advances in the polymerase chain reaction. *Science* **252,** 1643–1651.

98. Chou, Q., Russell, M., Birch, D. E., Raymond, J., and Bloch, W. (1992) Prevention of pre-PCR mis-priming and primer dimerization improves low-copy-number amplifications. *Nucl. Acids Res.* **20,** 1717–1723.

99. Wages, J. M. and Fowler, A. K. (1993) Amplification of low copy number sequences. *Amplifications* **11,** 1–3.

100. Birch, D. E., Kolmodin, J., Wong, J., et al. (1996) The use of a thermally activated DNA polymerase PCR gives improved specificity, sensitivity and product yield without additives or extra process steps. *Nature* **381,** 445–446.

101. Kellogg, D. E., Rabalkin, I., Chen, S., et al. (1994) TaqStart antibody: "hot start" PCR facilitated by neutralizing monoclonal antibody directed against *Taq* DNA polymerase. *BioTechniques* **16,** 1134–1137.

102. Ailenberg, M., and Silverman, M. (2000) Controlled hot start and improved specificity in carrying out PCR utilizing touch-up and loop incorporated primers (TULIPS). *BioTechniques* **29,** 1018–1024.

103. Kainz, P., Schmiedlechner, A., and Strack, H. B. (2000) Specificity-enhanced hot-start PCR: Addition of double-standed DNA fragments adapted to the annealing temperature. *BioTechniques* **28,** 278–282.

104. Albert J., and Fenyö, E. M. (1990) Simple, sensitive, and specific detection of human immunodeficiency virus type 1 clinical specimens by polymerase chain reaction with nested Primer. *J. Clin. Microbiol.* **28,** 1560–1564.

105. Chamberlain, J. S., Gibbs, R. A., Rainier, J. L., and Caskey, C. T. (1990) Multiplex PCR for the diagnosis of Duchenne muscular dystrophy, in *PCR Protocols: A Guide to Methods and Applications* (Innis, M. A., Gelfand, D. H., Sninsky, J. J., and White T. J., eds.), Academic Press, New York, pp. 272–281.

106. Zimmermann, K., and Mannhalter, J. W. (1996) Technical aspects of quantitative competitive PCR. *BioTechniques* **21**, 268–279.

107. Raeymarkers, L. (1993) Quantitative PCR: Theoretical considerations with practical implications. *Anal. Biochem.* **214**, 582–585.

108. Sachse, K., Zagon, J., Rüggeberg, H., Kruse, L., and Broll, H. (2001) Detection of genetic modifications in novel foods. *Food Rev. Intl.* (in press)

109. Becker-André, M., and Hahlbrock, K. (1989) Absolute mRNA quantification using the polymerase chain reaction (PCR). A novel approach by a PCR aided transcript titration assay (PATTY). *Nucleic Acids Res.* **17**, 9437–9446.

110. Piatak, M. Jr., Luk, K. -C., Williams, B., and Lifson, J. D. (1993) Quantitative competitive polymerase chain reaction for accurate quantitation of HIV DNA and RNA species. *BioTechniques* **14**, 70–81.

111. Bercovich, D., Regev, Z., Ratz, T., Luder, A., Plotsky, Y., and Gruenbaum, Y. (1999) Quantitative ratio of primer pairs and annealing temperature affecting PCR products in duplex amplification. *BioTechniques* **27**, 762–770.

112. Henegariu, O. Heerema, N. A., Dlouhy, S. R., Vance, G. H., and Vogt, P. H. (1997) Multiplex PCR: Critical parameters and step-by-step protocol. *BioTechniques* **23**, 504–511.

113. Olcén, P., Lantz, P. -G., Bäckman, A., and Radström, P. (1995) Rapid diagnosis of bacterial meningitis by a seminested PCR strategy. *Scand. J. Infect. Dis.* **27**, 537–539.

114. Radström, P., Bäckman, A., Qian, N., Kragsbjerg, Pahlson, C., and Olcén, P. (1994) Detection of bacterial DNA in cerebrospinal fluid by an assay for simultaneous detection of *Neisseria meningitidis, Haemophilus influenzae*, and streptococci using a seminested PCR strategy. *J. Clin. Microbiol.* **32**, 2738–2744.

115. Tong, C. Y. W., Donnelly, C., Harvey, G., and Sillis, M. (1999) Multiplex polymerase chain reaction for the simultaneous detection of *Mycoplasma pneumoniae, Chlamydia pneumoniae*, and *Chlamydia psittaci* in respiratory samples. *J. Clin. Pathol.* **52**, 257–263.

116. Wang, H., Fadl, A. A., and Khan, M. I. (1997) Multiplex PCR for avian pathogenic mycoplasmas. *Mol. Cell. Probes* **11**, 211–216.

117. Zarlenga, D. S., Chute, M. B., Martin, A., and Kapel, C.M. (1999) A multiplex PCR for unequivocal differentiation of all encapsulated and non-encapsulated genotypes of *Trichinella. Int. J. Parasitol.* **29**, 1859–1867.

118. Burkardt, H. J. (2000) Standardization and quality control of PCR analyses. *Clin. Chem. Lab. Med.* **38**, 87–91.

119. Prariyachatigul, C., Chaiprasert, A., Meevootisom, V., and Pattanakitsakul, S. (1996) Assessment of a PCR technique for the detection and identification of *Cryptococcus neoformans. J. Med. Vet. Mycol.* 34, 251–258.

2

Pre-PCR Processing of Samples

Peter Rådström, Rickard Knutsson, Petra Wolffs, Maria Dahlenborg, and Charlotta Löfström

1. Introduction

Diagnostic polymerase chain reaction (PCR) is an extremely powerful rapid method for diagnosis of microbial infections and genetic diseases, as well as for detecting microorganisms in environmental and food samples. However, the usefulness of diagnostic PCR is limited, in part, by the presence of inhibitory substances in complex biological samples, which reduce or even block the amplification capacity of PCR in comparison with pure solutions of nucleic acids (1). Thus, the presence of substances interfering with amplification will directly influence the performance of diagnostic PCR and, in particular, the assay's sensitivity of detection. Some inhibitors may dramatically interfere with amplification, even at very small amounts. For example, PCR mixtures containing the widely used *Taq* DNA polymerase are totally inhibited in the presence of 0.004% (v/v) human blood (2). Consequently, sample processing prior to PCR is required to enable DNA amplification of the target nucleic acids in the presence of even traces of PCR-inhibitory substances. To improve diagnostic PCR for routine analysis purposes, the processing of the sample is crucial for the robustness and the overall performance of the method. In general, diagnostic PCR may be divided into four steps: (*i*) sampling; (*ii*) sample preparation; (*iii*) nucleic acid amplification; and (*iv*) detection of PCR products (**Fig. 1**). Pre-PCR processing comprises all steps prior to the detection of PCR products. Thus, pre-PCR processing includes the composition of the reaction mixture of PCR and, in particular, the choice of DNA polymerase and amplification facilitators to be used.

This chapter will focus on sample preparation and the use of appropriate DNA polymerases and PCR facilitators for the development of efficient pre-

From: *Methods in Molecular Biology, vol. 216: PCR Detection of Microbial Pathogens: Methods and Protocols*
Edited by: K. Sachse and J. Frey © Humana Press Inc., Totowa, NJ

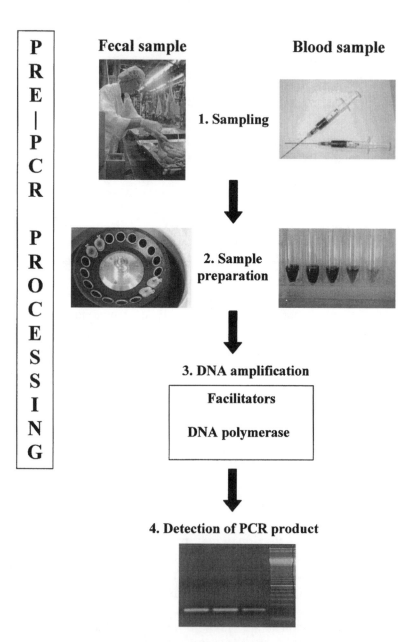

Fig. 1. Illustration of pre-PCR processing. The figure shows the different steps in diagnostic PCR. Pre-PCR processing refers to sampling, sample preparation, and DNA amplification with the addition of PCR facilitators and the use of an appropriate DNA polymerase.

PCR processing strategies for various categories of samples, as well as substances and mechanisms involved in inhibition.

2. PCR Inhibitors

PCR inhibitors originate either from the original sample or from sample preparation prior to PCR, or both *(3)*. In a review by Wilson *(4)*, a systematic list of PCR inhibitors was presented, and the mechanisms by which the inhibitors may act were divided into the following three categories: (*i*) inactivation of the thermostable DNA polymerase; (*ii*) degradation or capture of the nucleic acids, and (*iii*) interference with the cell lysis step. Although many biological samples were reported to inhibit PCR amplification, the identities and biochemical mechanisms of many inhibitors remain unclear.

2.1. Approaches to the Characterization of PCR Inhibitors

The effect of PCR inhibitors can be studied by either increasing the concentration of purified template DNA or adding different concentrations of the PCR-inhibitory samples or by both ways. Increasing the concentration of target DNA may be useful to overcome the effect of inhibitors (interfering with DNA and/or binding reversibly to the DNA-binding domain of the DNA polymerase), whereas adding different concentrations of the inhibitory sample is an alternative approach to evaluate the strength of the inhibitory samples on the amplification capacity of PCR. On the other hand, studying the effect of inhibitors on the polymerization activity of the DNA polymerase can be useful to (*i*) compare the effect of different inhibitors; (*ii*) perform a kinetic analysis of the DNA polymerase in the presence and absence of inhibitors; and (*iii*) evaluate the effect of adding substances that relieve the inhibition, such as bovine serum albumin (BSA). The recent introduction of thermal cyclers with real-time detection of PCR product accumulation offers the possibility to study the quantitative effects of inhibitors more efficiently. These instruments may be used to study the efficiency of the PCR performance and/or to study the DNA polymerase efficiency for the synthesis of DNA in the presence and absence of PCR inhibitors *(5)*.

2.2. Identification of PCR Inhibitors

A limited number of components have been identified as PCR inhibitors, namely, bile salts and complex polysaccharides in feces *(6,7)*, collagen in food samples *(8)*, heme in blood *(9)*, humic substances in soil *(10)*, proteinases in milk *(11)*, and urea in urine *(12)*. The thermostable DNA polymerase is probably the most important target site of PCR-inhibitory substances *(2)*. In a recent study, using various chromatographic procedures, hemoglobin, immunoglobulin G (IgG), and lactoferrin were identified as three major PCR inhibitors in human blood *(5,13)*. The mechanism of PCR inhibition by IgG was found to be dependent on its ability to interact with single-stranded DNA. Furthermore, this interaction was enhanced when DNA was heated with IgG. By testing different specific clones of IgGs, blocking of amplification through the interaction of single-stranded target DNA was found to be a general effect of IgGs. Therefore, in the case of blood specimens, it is not advisable to use boiling as a sample preparation method or to use hot-start PCR protocol.

Hemoglobin and lactoferrin were found to be the major PCR inhibitors in erythrocytes and leukocytes, respectively *(5)*, and both hemoglobin and lactoferrin contain iron. The mechanism of inhibition may be related to the ability of these proteins to release iron ions into the PCR mixture. When the inhibitory effect of iron was investigated, it was found to interfere with DNA synthesis. Furthermore, bilirubin, bile salts and hemin, which are derivatives of hemoglobin, were also found to be PCR inhibitory. It has been suggested that heme regulates DNA polymerase activity and coordinates the synthesis of components in hemoglobin in erythroid cells by feedback inhibition *(14)*. In the same study, it was observed that hemin was a competitive inhibitor with the target DNA and a noncompetitive inhibitor with the nucleotides through direct action against the DNA polymerase. As a result, characterization of PCR inhibitors and detailed knowledge of inhibitory capacities and mechanisms are important prerequisites for the development of more efficient sample preparation methods, which will eliminate the need for extensive processing of biological samples prior to diagnostic PCR.

3. Sample Preparation

The objectives of sample preparation are (*i*) to exclude PCR-inhibitory substances that may reduce the amplification capacity of DNA and the efficiency of amplification (*see* **Chapter 1**); (*ii*) to increase the concentration of the target organism to the practical operating range of a given PCR assay; and (*iii*) to reduce the amount of the heterogeneous bulk sample and produce a homogeneous sample for amplification in order to insure reproducibility and repeatability of the test. All these factors affect the choice of sample preparation method. However, many sample preparation methods are laborious, expensive, and time-consuming or do not provide the desired template quality *(15)*. Since sample preparation is a complex step in diagnostic PCR, a large variety of methods have been developed, and all these methods will affect the PCR analysis differently in terms of specificity and sensitivity *(1)*. The most frequently used sample preparation methods may be divided into four different categories: (*i*) biochemical; (*ii*) immunological; (*iii*) physical; and (*iv*) physiological methods (**Table 1**).

Table 1
Sample Preparation Methods Used for Different Types of Samples[a]

Category of sample preparation method	Subcategory	Sample preparation method	Sample	Reference
Biochemical	Adsorption	Lectin-based separation	Beef meat	*(78)*
		Protein adsorption	Blood	*(9)*
	DNA extraction	DNA purification method	Hemolytic serum	*(79)*
		Lytic methods	Blood anticoagulant	*(80)*
Immunological	Adsorption	Immunomagnetic capture	Blood	*(81)*
Physical		Aqueous two-phase systems	Soft cheese	*(82)*
		Buoyant density centrifugation	Minced meat	*(31)*
		Centrifugation	Urine	*(28)*
		Dilution	Blood	*(30)*
		Filtration	Milk	*(29)*
Physiological		Enrichment	Meat	*(33)*

[a] Modified with permission from **Ref. *1*.**

3.1. Biochemical Methods

The most widely employed biochemical method is DNA extraction. Many different commercial kits are available, such as BAX (Quallcon Inc., Wilmington, DE) *(16)*, PrepMan (Applied BioSystems, Foster City, CA, USA) *(17)*, Purugene (Gentra Systems Inc., Minneapolis, MN, USA) *(18)*, QIAamp® (Qiagen, Valencia, CA, USA) *(19)*, and XTRAX (Gull Laboratories Inc., Salt Lake City, UT, USA) *(20)*. Consequently, several studies have compared and evaluated the quality of the extracted DNA *(18,21,22)*, and a kit that provides the highest yield, concentration, and purity of DNA can be recommended. The advantage of DNA extraction is that a homogeneous sample with high quality is provided for amplification. Most PCR inhibitors are removed, since the template is usually purified and stored in appropriate buffers, such as Tris-EDTA (TE) buffer. The drawback of DNA extraction methods is that the target microorganism usually has to be pre-enriched in medium or on an agar plate prior to extraction. In addition, most DNA extraction methods are laborious and costly. Batch-to-batch variation after DNA extraction may also exist with respect to purity and concentration of the template.

3.2. Immunological Methods

This category is mainly based on the use of magnetic beads coated with antibodies *(23)*. Since antibodies are used, the specificity will be influenced, and the captured cells will be those containing the corresponding antigen. The specificity of the PCR protocol will depend on both the PCR assay used, as well as the specificity of the antibodies. In general, after immunocapture, the sample requires lysis or washing (24), and viruses can then be used directly (25). In most cases, these methods increase the concentration of the target organism. The homogeneity of the PCR sample may differ depending on the processing steps that follow the capture, but usually the template is of appropriate quality after this treatment. Since part of the specificity depends on the antibodies themselves, false negative results can be obtained as a result of cross-reactions. This methodology is quite expensive and also very laborious and time-consuming.

3.3. Physical Methods

Many different physical methods have been used, such as aqueous two-phase systems (26), buoyant-density centrifugation (27), centrifugation (28), filtration (29) and dilution (30). These methods are dependent on the physical properties of the target cells, for example cell density and size. Aqueous two-phase systems provide a gentle way of partitioning PCR inhibitors and target cells between two immiscible phases. For instance, a polyethylene glycol (PEG) 4000 and dextran 40-based system was used in a PCR detection assay for *Helicobacter pylori* in human feces (6). Density centrifugation was shown to be a promising method if fast detection is of importance (31). Density media, such as Percoll (Pharmacia, Uppsala, Sweden) (27) and BactXtractor (Quintessence Research AB, Bålsta, Sweden) (32), were used to concentrate the target organism and remove PCR-inhibitory substances of different density. After this treatment, whole cells were obtained, which could be used as a PCR sample. The homogeneity of the sample may differ depending on the kind of biological sample matrices. If components of the sample matrix have the same density as the cells these may inhibit DNA amplification. The advantage of density centrifugation is that the target organism is being concentrated, which allows rapid detection response. Furthermore, these methods are relatively user friendly.

3.4. Physiological Methods

These methods are based on bacterial growth and biosynthesis of cell components, i.e., genome, cytoplasm, and cell surface constituents. Culture can be carried out in enrichment broth or on agar plates. Again, the aim is to provide detectable concentrations of viable target cells prior to PCR (33). Selective or nonselective agar or enrichment medium can be used, and the specificity will depend partly on the characteristics of the medium. The template quality, as well as the homogeneity of the PCR sample, may differ with respect to the presence of cell components. The advantages of this methodology are its simplicity and low cost. The method provides viable cells to be used in PCR without further lysis steps *(34)*. However, it must be borne in mind that cells contain high concentrations of macromolecules, which might influence and shift the equilibrium in many biochemical reactions *(35)*, for instance the DNA polymerase and its DNA template–primer binding properties *(36)* (*see* **Chapter 1**). Therefore, the DNA polymerase has a key function during DNA amplification in terms of DNA synthesis and resistance to PCR inhibitors.

A comparison of the performance of sample preparation methods described in this section is shown in **Table 2**.

4. DNA Polymerases

The first PCR experiments were carried out with the thermolabile Klenow fragment of *Eschericia coli* DNA polymerase I, which needed to be replenished for every cycle *(37)*. The use of the thermostable DNA polymerase from *Thermus aquaticus* (*Taq*) has greatly simplified PCR and enhanced the specificity *(38)*. With high specific activity, fidelity, and temperature range, *Taq* DNA polymerase and its derivatives became and still are the most widely used enzymes in PCR. Thermostable DNA polymerase is a key component in the amplification reaction, and any factor interfering with the enzymatic activity will affect the amplification capacity. The DNA polymerase can be degraded, denatured, or have its enzymatic activity reduced by a wide variety of compounds present in biological samples *(3,5,9,39)*.

A number of DNA polymerases from other organisms are now commercially available. Examples of commonly used DNA polymerases include r*Tth* and *Tth*, isolated from *Thermus thermophilus*, DyNazyme isolated from *T. brockianus*, as well as Ampli*Taq*® Gold (Applied BioSystems, Foster City, CA, USA) and Platinum *Taq* with built-in hot start, both isolated from *T. aquaticus*. These polymerases exhibit very different properties with regard to resistance to various components in biological samples and performance in the presence of these components. The choice of DNA polymerase was shown to influence the performance of several PCR-based applications, such as genotyping using restriction fragment-length polymorphism (RFLP) *(40)* and random-amplified polymorphic DNA (RAPD) *(41)*, multiplex PCR assays *(42)*, differential display reverse transcription PCR (RT-PCR) *(43)*, and autosticky PCR *(44)*. Recent research indicated that different polymerases have different susceptibilities to PCR inhibitors *(2)*. Therefore, the inhibition of PCR by components of biological samples can be reduced or eliminated by choosing an appropriate thermostable DNA polymerase without the need for extensive sample processing prior to PCR.

The choice of DNA polymerase is determined by several factors related to the application. The level of resistance of DNA polymerase to PCR inhibitors

Table 2
Comparison of the Performance of Different Pre-PCR Sample Preparation Methods

Category of sample preparation method	Product of sample preparation	Homogeneity of product	Concentration of product	Removal of PCR inhibitors	Time required	Cost	Availability
Biochemical: DNA extraction	DNA	Good	Average	Yes	3–6 h	High	Complex
Immunological: Immunomagnetic capture	Cell/DNA	Average	Average	Average	2–4 h	High	Limited
Physical: Buoyant density centrifugation	Cell	Average	Good	Average	30 min	Average	Limited
Physiological: Enrichment	Cell	Low	Good	Low	6–24 h	Low	Good

can be determined by intrinsic factors, such as enzyme purification techniques and reaction buffer composition, as well as its production from native or recombinant strains. Furthermore, the sample preparation protocol and the presence of trace levels of extraction reagents in the purified sample can affect the extraction efficiency and the sensitivity of PCR. *Taq* DNA polymerase from different commercial sources was reported to be inhibited to a different extent by humic substances in soil extracts *(45)*. The source of *Taq* DNA polymerase in the PCR step was also found to affect the banding patterns produced in differential display *(43)*. Variations in the performance of DNA polymerases in co-amplification PCR were also found to be salt-dependent *(46)*. The polymerase *Tth* maintains both DNA- and RNA-dependent DNA polymerase activities in the presence of 5% (v/v) phenol, while a trace amount of phenol was found to be inhibitory to *Taq* DNA polymerase *(39)*. Several studies evaluated the usefulness and characteristics of different DNA polymerases with respect to various PCR samples including clinical samples, blood, feces, and cell material.

4.1. Clinical Samples

It was noted that both *Tfl* and *Tth* DNA polymerases are more resistant to aqueous and vitreous fluids of the eye than the polymerases *Taq*, *Tli*, and the Stoffel fragment *(47)*. *Tth* DNA polymerase was also shown to be less affected by inhibitors present in nasopharyngeal swab samples compared to *Taq* DNA polymerase in an assay detecting influenza A virus *(48)*. The use of hot-start enzymes, such as Ampli*Taq* Gold and Platinum *Taq*, reduces the possibility of carryover contamination. Furthermore, increased specificity using Ampli*Taq* Gold was demonstrated for a multiplex PCR assay detecting middle ear pathogens *(42)*. Amplification of highly degraded DNA from paraffin-embedded tissue using Ampli*Taq* Gold or Platinum *Taq* increased the yield by up to 20 times compared to *Taq*. Improved PCR amplification with less background was observed in the same study for Ampli*Taq* Gold compared to Platinum *Taq* when a time-release PCR protocol was applied *(49)*.

4.2. Blood

When the inhibitory effect of blood on nine thermostable DNA polymerases was studied, Ampli*Taq* Gold and *Taq* DNA polymerases were totally inhibited in the presence of 0.004% (v/v) blood in the PCR mixture, while HotTub, Pwo, rTth, and Tfl DNA polymerases were able to amplify DNA in the presence of at least 20% (v/v) blood without reduced amplification sensitivity *(2)*. Furthermore, it was found that the addition of 1% (v/v) blood was totally inhibitory to *Taq* DNA polymerase, while a target sequence in the presence of up to 4% (v/v) blood was amplified using *Tth* DNA polymerase *(50)*. Different PCR conditions and target DNA concentrations may explain these conflicting results regarding the effect of blood on *Taq* DNA polymerase. The enhancement of amplification yield and specificity using Ampli*Taq* Gold DNA polymerase instead of Ampli*Taq* DNA polymerase in multiplex detection of DNA in blood was also reported *(51,52)*.

4.3. Feces

In a comparison of the amplification efficiency of *Tth* polymerase and *Taq* DNA polymerase in detecting *Helicobacter hepaticus* in mice feces, a 100-fold increase in sensitivity with *Tth* polymerase over *Taq* DNA polymerase was observed *(53)*. Furthermore, it has been reported that *Pwo* and r*Tth* DNA polymerases could amplify DNA in the presence of 0.4% (v/v) feces without reduced sensitivity *(2)*. The inhibitory effect of the microbial flora in pig feces on the amplification capacity of r*Tth* and *Taq* DNA polymerase was observed when detecting *Clostridium botulinum* *(17)*. The results showed a decrease in sensitivity by one log unit when using *Taq* DNA polymerase instead of r*Tth*.

4.4. Cell Material

The DNA polymerases from *T. aquaticus* and *T. flavus* were found to bind to short double-stranded DNA fragments without sequence specificity *(54)*. Furthermore, it was reported that the accumulation of amplification products during later PCR cycles also exerts an inhibitory effect on the DNA polymerases *(55)*. It was indicated that the main factor contributing to the plateau phase in PCR was the binding of DNA polymerase to its amplification products (*see* **Chapter 1**). *Taq* DNA polymerase was replaced with *Tth* DNA polymerase for more sensitive detection of *Staphylococcus aureus* DNA in bovine milk *(8)*. Also, the detection of cells of the poultry pathogen *Mycoplasma iowae* was significantly improved by replacement of *Taq* DNA polymerase with *Tth* DNA polymerase *(56)*.

5. Amplification Facilitators

In the course of the development of PCR methodology the basic master mixture containing DNA polymerase, primers, nucleotides, and a reaction buffer containing Tris-HCl, KCl, and $MgCl_2$, has been extended with numerous compounds to enhance the efficiency of amplification. Such compounds are called amplification enhancers or amplification facilitators *(57)*. They can affect amplification at different stages and under different conditions by (*i*) increasing or decreasing the thermal stability of the DNA template; (*ii*) affecting the error rate of the DNA polymerase; (*iii*) affecting the specificity of the system; and (*iv*) relieving the inhibition of amplification caused by complex biological samples. With the introduction of new DNA polymerases, a number of suppliers have already added amplification facilitators into the

Table 3
Composition of Commercial Buffer Systems for DNA Polymerases

DNA polymerase	Tris-HCl (mM)	KCl (mM)	MgCl$_2$ (mM)	Triton-X (vol%)	Tween 20 (vol%)	BSA (µg/ml)	EGTA (mM)	Glycerol (vol%)
				Buffer components				
DyNazyme	100	500	15	1%	—	—	—	—
Platinum *Taq*	200	500	—	—	—	—	—	—
r*Tth*	100	1000	—	—	0.5%	—	7.5	5%
Taq	100	500	15	—	—	—	—	—
Tth	100	1000	15	—	0.5%	500	—	—

Table 4
Concentration of Facilitators Used in Different Applications

Facilitator	Concentration	Application	Reference
BSA	4.0 g/L	Relief of inhibition by meat, blood, and feces.	*(61)*
BSA	0.4 g/L	Relief of inhibition by bilirubin and humics.	*(59)*
BSA	0.6 g/L	Relief of inhibition by melanin in RT-PCR.	*(83)*
gp32	0.1 g/L	Relief of inhibition by meat, blood, and feces.	*(61)*
gp32	0.15 g/L	Relief of inhibition by bilirubin and humics.	*(59)*
DMSO	5%	Rescue of failed amplification.	*(57)*
DMSO	2–10%	Facilitation of RT-PCR.	*(84)*
Tween 20	2.5 g/L	Relief of inhibition by feces.	*(61)*
Tween 20	0.5%	Relief of inhibition by plant polysaccharides.	*(74)*
Betaine	117.0 g/L	Relief of inhibition by blood and meat.	*(61)*
Betaine	2.5 *M*	PCR of GC-rich sequences.	*(71)*
Glycerol	10–15%	Rescue of failed amplification.	*(57)*
PEG 400	5%	Relief of inhibition by plant polysaccharides.	*(74)*
PEG 400	10–15%	Rescue of failed amplification.	*(57)*
PEG 400	2.0 g/L	Relief of inhibition by blood.	*(61)*

accompanying buffers (**Table 3**). A subdivision of facilitators into five groups was proposed *(58)*: (*i*) proteins; (*ii*) organic solvents; (*iii*) non-ionic detergents; (*iv*) biologically compatible solutes; and (*v*) polymers. These groups will be discussed in more detail, including some of the commonly used compounds within the different groups. Specific amounts of facilitators used by different research groups are listed in **Table 4**.

5.1. Proteins

The two proteins most commonly used to facilitate amplification are BSA *(59–61)* and the single-stranded DNA-binding protein gp32, which is a protein encoded by gene 32 of bacteriophage T4 *(59,61–63)*. The addition of BSA to the amplification mixture was shown to relieve inhibition of amplification by several substances, such as blood, meat, feces *(61)* and heme-containing compounds *(9)*. It has been suggested that BSA can help to overcome PCR inhibition by blood or heme-containing compounds by binding them. Furthermore, it was shown that BSA can bind phenolics and relieve PCR inhibition in this way *(59)*. Inhibition of amplification by fecal samples can be caused by the degradation of DNA polymerase by proteinases. It has been suggested that proteins such as BSA and gp32 can relieve this inhibition effect by serving as a target for the proteinases *(11)*. BSA is often used for the stabilization of proteins in solution, and thus, a possible way of facilitating amplification may consist in stabilization of the DNA polymerase *(64)*. The protein gp32 may facilitate amplification in the same fashion as BSA. However, gp32 can bind single-stranded DNA, protecting it from nuclease digestion *(65)*, and it has been suggested that, in blood, the protein can improve the accessibility of the DNA polymerase when large amounts of coagulated organic material are present in the PCR sample *(50)*.

5.2. Organic Solvents

Examples of frequently used organic solvents as PCR facilitators include dimethyl sulfoxide (DMSO) and formamide. It has been suggested that both solvents affect the thermal stability of the primers and the thermal activity profile of the DNA polymerase *(57)*, thereby increasing the specificity of amplification *(66)*. The effect on thermal stability seems to be caused by the general capability of organic solvents to destabilize DNA in solution *(67,68)*.

5.3. Non-ionic Detergents

The main non-ionic detergents used as PCR facilitators are Tween 20 and Triton-X. It was shown that the addition of Tween 20 stimulates the activity of *Taq* DNA polymerase and reduces false terminations of the enzyme *(69)*. The mechanisms behind these findings are still unclear.

5.4. Biologically Compatible Solutes

Betaine and glycerol are the most common facilitators in the group of biologically compatible solutes. The solutes are used by organisms and cellular systems to maintain biological activity under extreme conditions. For that reason, glycerol is used in the storage buffer of thermostable enzymes. The addition of both betaine and glycerol to amplification reaction mixtures was found to enhance specificity *(66,70)* and to reduce the formation of secondary structures caused by GC-rich regions *(71)*. Also, glycerol may facilitate amplification by enhancing the hydrophobic interactions between protein domains and raising the thermal transition temperature of proteins *(72)*. Glycerol can also lower the strand separation temperature of DNA, thus facilitating amplification *(73)*.

5.5. Polymers

PEG and dextran are polymers that can be used as amplification facilitators. It was shown that PEG can facilitate amplification in similar ways as organic solvents *(57)*. Also, PEG was reported to relieve the inhibition caused by feces *(61)* and dextran sulfate, a plant polysaccharide *(74)*. Furthermore, PEG is known to possess enzyme stabilizing properties comparable to BSA, which serve to maintain enzymatic activity *(64)*. This action could enhance amplification by stabilizing the DNA polymerase.

6. Pre-PCR Processing Strategies

The treatment of complex biological samples prior to amplification is a crucial factor determining the performance of diagnostic PCR assays. The following requirements should be fulfilled to insure optimal conditions *(1)*: (*i*) absence or low concentration of PCR-inhibitory components in the sample; and (*ii*) sufficient concentration of target DNA.

Pre-PCR treatment aims to convert a complex biological sample containing the target microorganisms into PCR-amplifiable samples. Since complex biological samples often contain PCR inhibitors *(4)*, numerous pre-PCR processing protocols have been developed. The reason for the variety in PCR protocols and pre-PCR methods is that the most suitable approach depends on the nature of the sample and the purpose of the PCR analysis. For instance, various sample preparation methods were developed to remove or reduce the effects of PCR inhibitors without knowing the identity of the PCR inhibitors and/or understanding the mechanism of inhibition. Therefore, the characterization of PCR inhibitors represents an important step in the development of efficient sample preparation methods designed to overcome the effects of inhibitory factors. For example, the PCR-inhibitory effect of collagen was partially relieved by adjusting the magnesium ion concentration in the amplification mixture *(75)*.

Once the sample matrix has been characterized regarding PCR inhibitors and concentrations of target DNA, one can predict whether the sample is suitable for PCR analysis or not. Samples can be divided into heterogeneous and homogeneous samples, with most complex biological samples being heterogeneous. Consequently, the conditions for DNA amplification can be optimized through efficient pre-PCR processing. Several different pre-PCR processing strategies can be used: (*i*) optimization of the sample preparation method; (*ii*) optimization of the DNA amplification conditions by the use of alternative DNA polymerases, and/or amplification facilitators; and (*iii*) a combination of both strategies.

Selection and optimization of sample preparation methods is the most frequently used approach to circumvent PCR inhibition *(1)*. Many PCR protocols combine sample preparation methods from different categories. A common strategy for diagnostic PCR consists in the combination of a pre-enrichment method with a biochemical DNA extraction method *(17,76)* or with a physical sample preparation method *(20)*. The enrichment step is usually included to concentrate the target cells to PCR-detectable concentrations *(33)*. The complexity of the various methods must be considered in light of the aim of the PCR analysis, i.e., if the results are to be used for risk assessments or for hazard analysis critical control point (HACCP) purposes.

A summary of the different sample preparation categories is presented in **Table 2**. In general, DNA extraction methods provide templates of high quality, but the method is usually complex. However, automated robust DNA extraction methods have been introduced. Physical methods are favorable, since they do not affect the specificity of the PCR protocol, as may the immunological and physiological methods. The simplest method is to take the PCR sample directly from the enrichment broth and dilute the sample, because of the inhibitory components present in the enrichment broth *(20)*. Recently, a PCR-compatible enrichment medium was developed for detection of *Yersinia enterocolitica*, thus making pre-PCR processing of swab samples unnecessary *(77)*. However, complex matrices present in the culture medium may have a detrimental effect on PCR performance.

The DNA amplification reaction mixture can be optimized by selection of a robust DNA polymerase and by the addition of amplification facilitators, to circumvent the PCR-inhibitory effects of sample components and to maintain the amplification efficiency. This strategy has been employed in the laboratory of the authors for blood samples, and by using the r*Tth* DNA polymerase combined with BSA, it was possible to amplify DNA in the presence of at least 20% (v/v) blood without loss of sensitivity *(2)*. Furthermore, a pre-PCR processing protocol was developed for detection of *Clostridium botulinum* spores in porcine fecal samples, based on inclusion of a sample preparation method and the use of a more robust DNA polymerase *(17)*. After a heat shock (10 min at 70°C: and pre-enrichment for 18 h at 30°C, the feces homogenate was exposed to DNA extraction prior to PCR, and PCR was performed using the more robust r*Tth* DNA polymerase.

In the future development of diagnostic PCR assays, research on pre-PCR processing is most likely to expand in response to the growing demand for rapid, robust and user-friendly PCR protocols. A future challenge for pre-PCR processing strategies is the design of PCR procedures integrating both sampling and DNA amplification as automated operations.

Acknowledgments

This work was supported by grants from VINNOVA, the Swedish Agency for Innovation Systems, and from the Commission of the European Community within the program "Quality of Life and Management of Living Resources," QLRT-1999-00226.

References

1. Lantz, P. G., Abu Al-Soud, W., Knutsson, R., Hahn-Hägerdal, B., and Rådström, P. (2000) Biotechnical use of the polymerase chain reaction for microbiological analysis of biological samples. *Biotechnol. Annu. Rev.* **5,** 87–130.
2. Abu Al-Soud, W., and Rådström, P. (1998) Capacity of nine thermostable DNA polymerases to mediate DNA amplification in the presence of PCR-inhibiting samples. *Appl. Environ. Microbiol.* **64,** 3748–3753.
3. Rossen, L., Nørskov, P., Holmstrøm, K., and Rasmussen, O. F. (1992) Inhibition of PCR by components of food samples, microbial diagnostic assays and DNA-extraction solutions. *Int. J. Food Microbiol.* **17,** 37–45.
4. Wilson, I. G. (1997) Inhibition and facilitation of nucleic acid amplification. *Appl. Environ. Microbiol.* **63,** 3741–3751.
5. Abu Al-Soud, W., and Rådström, P. (2001) Purification and characterization of PCR-inhibitory components in blood cells. *J. Clin. Microbiol.* **39,** 485–493.
6. Lantz, P. -G., Matsson, M., Wadström, T., and Rådström, P. (1997) Removal of PCR inhibitors from human faecal samples through the use of an aqueous two-phase system for sample preparation prior to PCR. *J. Microbiol. Methods* **28,** 159–167.
7. Monteiro, L., Bonnemaison, D., Vekris, A., et al. (1997) Complex polysaccharides as PCR inhibitors in feces: *Helicobacter pylori* model. *J. Clin. Microbiol.* **35,** 995–998.
8. Kim, C. H., Khan, M., Morin, D. E., et al. (2001) Optimization of the PCR for detection of *Staphylococcus aureus nuc* gene in bovine milk. *J. Dairy Sci.* **84,** 74–83.
9. Akane, A., Matsubara, K., Nakamura, H., Takahashi, S. and Kimura, K. (1994) Identification of the heme compound copurified with deoxyribonucleic acid (DNA) from bloodstains, a major inhibitor of polymerase chain reaction (PCR) amplification. *J. Forensic Sci.* **39,** 362–372.
10. Tsai, Y. L., and Olson, B. H. (1992) Rapid method for separation of bacterial DNA from humic substances in sediments for polymerase chain reaction. *Appl. Environ. Microbiol.* **58,** 2292–2295.

11. Powell, H. A., Gooding, C. M., Garret, S. D., Lund, B. M., and McKee, R.A. (1994) Proteinase inhibition of the detection of *Listeria monocytogenes* in milk using the polymerase chain reaction. *J. Clin. Microbiol.* **18,** 59–61.

12. Khan, G., Kangro, H. O., Coates, P. J., and Heath, R. B. (1991) Inhibitory effects of urine on the polymerase chain reaction for cytomegalovirus DNA. *J. Clin. Pathol.* **44,** 360–365.

13. Abu Al-Soud, W. A., Jönsson, L. J., and Rådström, P. (2000) Identification and characterization of immunoglobulin G in blood as a major inhibitor of diagnostic PCR. *J. Clin. Microbiol.* **38,** 345–350.

14. Byrnes, J. J., Downey, K. M., Esserman, L., and So, A.G. (1975) Mechanism of hemin inhibition of erythroid cytoplasmic DNA polymerase. *Biochemistry* **14,** 796–799.

15. Jaffe, R. I., Lane, J. D., and Bates, C.W. (2001) Real-time identification of *Pseudomonas aeruginosa* direct from clinical samples using a rapid extraction method and polymerase chain reaction (PCR). *J. Clin. Lab. Anal.* **15,** 131–137.

16. Bailey, J. S. (1998) Detection of *Salmonella* cells within 24 to 26 hours in poultry samples with the polymerase chain reaction BAX system. *J. Food Prot.* **61,** 792–795.

17. Dahlenborg, M., Borch, E., and Rådström, P. (2001) Development of a combined selection and enrichment PCR procedure for *Clostridium botulinum* types B, E and F and its use to determine prevalence in fecal samples from slaughter pigs. *Appl. Environ. Microbiol.* **67,** 4781–4789.

18. Fahle, G. A., and Fischer, S. H. (2000) Comparison of six commercial DNA extraction kits for recovery of cytomegalovirus DNA from spiked human specimens. *J. Clin. Microbiol.* **38,** 3860–3863.

19. Freise, J., Gerard, H. C., Bunke, T., et al. (2001) Optimised sample DNA preparation for detection of *Chlamydia trachomatis* in synovial tissue by polymerase chain reaction and ligase chain reaction. *Ann. Rheum. Dis.* **60,** 140–145.

20. Lantz, P. G., Knutsson, R., Blixt, Y., Al Soud, W. A., Borch, E., and Rådström, P. (1998) Detection of pathogenic *Yersinia enterocolitica* in enrichment media and pork by a multiplex PCR: a study of sample preparation and PCR- inhibitory components. *Int. J. Food Microbiol.* **45,** 93–105.

21. Shafer, R. W., Levee, D. J., Winters, M. A., Richmond, K. L., Huang, D. and Merigan, T. C. (1997) Comparison of QIAamp HCV kit spin columns, silica beads, and phenol- chloroform for recovering human immunodeficiency virus type 1 RNA from plasma. *J. Clin. Microbiol.* **35,** 520–522.

22. Kramvis, A., Bukofzer, S., and Kew, M. C. (1996) Comparison of hepatitis B virus DNA extractions from serum by the QIAamp blood kit, GeneReleaser, and the phenol-chloroform method. *J. Clin. Microbiol.* **34,** 2731–2733.

23. Hallier-Soulier, S., and Guillot, E. (1999) An immunomagnetic separation polymerase chain reaction assay for rapid and ultra-sensitive detection of *Cryptosporidium parvum* in drinking water. *FEMS Microbiol. Lett.* **176,** 285–289.

24. Antognoli, M. C., Salman, M. D., Triantis, J., Hernandez, J., and Keefe, T. (2001) A one-tube nested polymerase chain reaction for the detection of *Mycobacterium*

bovis in spiked milk samples: an evaluation of concentration and lytic techniques. *J. Vet. Diagn. Invest.* **13**, 111–116.

25. Jothikumar, N., Cliver, D. O., and Mariam, T.W. (1998) Immunomagnetic capture PCR for rapid concentration and detection of hepatitis A virus from environmental samples. *Appl. Environ. Microbiol.* **64**, 504–508.

26. Lantz, P. -G., Tjerneld, F., Hahn-Hägerdal, B., and Rådström, P. (1996) Use of aqueous two-phase systems in sample preparation for polymerase chain reaction-based detection of microorganisms. *J. Chromat. B.* **680**, 165–170.

27. Lindqvist, R., Norling, B., and Lambertz, S.T. (1997) A rapid sample preparation method for PCR detection of food pathogens based on buoyant density centrifugation. *Lett. Appl. Microbiol.* **24**, 306–310.

28. Gerritsen, M. J., Olyhoek, T., Smits, M. A., and Bokhout, B. A. (1991) Sample preparation method for polymerase chain reaction-based semiquantitative detection of *Leptospira interrogans* serovar *hardjo* subtype *hardjobovis* in bovine urine. *J. Clin. Microbiol.* **29**, 2805–2808.

29. Starbuck, M. A., Hill, P. J., and Stewart, G.S. (1992) Ultra sensitive detection of *Listeria monocytogenes* in milk by the polymerase chain reaction (PCR). *Lett. Appl. Microbiol.* **15**, 248–252.

30. Abu Al-Soud, W., Lantz, P. -G., Bäckman, A., Olcén, P., and Rådström, P. (1998) A sample preparation method which facilitates detection of bacteria in blood cultures by the polymerase chain reaction. *J. Microbiol. Meth.* **32**, 217–224.

31. Lindqvist, R. (1997) Preparation of PCR samples from food by a rapid and simple centrifugation technique evaluated by detection of *Escherichia coli* O157:H7. *Int. J. Food Microbiol.* **37**, 73–82.

32. Thisted Lambertz, S., Lindqvist, R., Ballagi-Pordány, A. and Danielsson-Tham, M.-L. (2000) A combined culture and PCR method for detection of pathogenic *Yersinia enterocolitica* in food. *Int. J. Food Microbiol.* **57**, 63–73.

33. Sharma, V. K., and Carlson, S.A. (2000) Simultaneous detection of *Salmonella* strains and *Escherichia coli* O157:H7 with fluorogenic PCR and single-enrichment-broth culture. *Appl. Environ. Microbiol.* **66**, 5472–5476.

34. Knutsson, R., Blixt, Y., Grage, H., Borch, E., and Rådström, P. (2002) Evaluation of selective enrichment PCR procedures for *Yersinia enterocolitica*. *Int. J. Food Microbiol.* **73**, 35–46.

35. Minton, A. P., and Wilf, J. (1981) Effect of macromolecular crowding upon the structure and function of an enzyme: glyceraldehyde-3-phosphate dehydrogenase. *Biochemistry* **20**, 4821–4826.

36. Zimmerman, S. B., and Trach, S. O. (1988) Macromolecular crowding extends the range of conditions under which DNA polymerase is functional. *Biochim. Biophys. Acta.* **949**, 297–304.

37. Saiki, R. K., Scharf, S., Faloona, F., et al. (1985) Enzymatic amplification of beta-globin genomic sequences and restriction site analysis for diagnosis of sickle cell anemia. *Science* **230**, 1350–1354.

38. Saiki, R. K., Gelfand, D. H., Stoffel, S., et al. (1988) Primer-directed enzymatic amplification of DNA with a thermostable DNA polymerase. *Science* **239**, 487–491.

39. Katcher, H. L., and Schwartz, I. (1994) A distinctive property of *Tth* DNA polymerase: enzymatic amplification in the presence of phenol. *Biotechniques* **16**, 84–92.
40. Zsolnai, A., and Fesus, L. (1997) Enhancement of PCR-RFLP typing of bovine leukocyte adhesion deficiency. *Biotechniques* **23**, 380–382.
41. Diakou, A., and Dovas, C.I. (2001) Optimization of random-amplified polymorphic DNA producing amplicons up to 8500 bp and revealing intraspecies polymorphism in *Leishmania infantum* isolates. *Anal. Biochem.* **288**, 195–200.
42. Hendolin, P. H., Paulin, L., and Ylikoski, J. (2000) Clinically applicable multiplex PCR for four middle ear pathogens. *J. Clin. Microbiol.* **38**, 125–132.
43. Haag, E., and Raman, V. (1994) Effects of primer choice and source of *Taq* DNA polymerase on the banding patterns of differential display RT-PCR. *Biotechniques* **17**, 226–228.
44. Gál, J., Schnell, R. and Kálmán, M. (2000) Polymerase dependence of autosticky polymerase chain reaction. *Anal. Biochem.* **282**, 156–158.
45. Tebbe, C. C., and Vahjen, W. (1993) Interference of humic acids and DNA extracted directly from soil in detection and transformation of recombinant DNA from bacteria and a yeast. *Appl. Environ. Microbiol.* **59**, 2657–2665.
46. Favre, N., and Rudin, W. (1996) Salt-dependent performance variation of DNA polymerases in co- amplification PCR. *Biotechniques* **21**, 28–30.
47. Wiedbrauk, D. L., Werner, J. C., and Drevon, A. M. (1995) Inhibition of PCR by aqueous and vitreous fluids. *J. Clin. Microbiol.* **33**, 2643–2646.
48. Poddar, S. K., Sawyer, M. H., and Connor, J. D. (1998) Effect of inhibitors in clinical specimens on *Taq* and *Tth* DNA polymerase-based PCR amplification of influenza A virus. *J. Med. Microbiol.* **47**, 1131–1135.
49. Akalu, A., and Reichardt, J. K. (1999) A reliable PCR amplification method for microdissected tumor cells obtained from paraffin-embedded tissue. *Genet. Anal.* **15**, 229–233.
50. Panaccio, M., and Lew, A. (1991) PCR based diagnosis in the presence of 8% (v/v) blood. *Nucleic Acids Res.* **19**, 1151.
51. Kebelmann-Betzing, C., Seeger, K., Dragon, S., et al. (1998) Advantages of a new *Taq* DNA polymerase in multiplex PCR and time-release PCR. *Biotechniques* **24**, 154–158.
52. Moretti, T., Koons, B., and Budowle, B. (1998) Enhancement of PCR amplification yield and specificity using Ampli*Taq* Gold DNA polymerase. *Biotechniques* **25**, 716–722.
53. Shames, B., Fox, J. G., Dewhirst, F., Yan, L., Shen, Z., and Taylor, N. S. (1995) Identification of widespread *Helicobacter hepaticus* infection in feces in commercial mouse colonies by culture and PCR assay. *J. Clin. Microbiol.* **33**, 2968–2972.
54. Kainz, P., Schmiedlechner, A. and Strack, H. B. (2000) Specificity-enhanced hotstart PCR: addition of double-stranded DNA fragments adapted to the annealing temperature. *Biotechniques* **28**, 278–282.
55. Kainz, P. (2000) The PCR plateau phase - towards an understanding of its limitations. *Biochim. Biophys. Acta.* **1494**, 23–27.

56. Laigret, F., Deaville, J., Bove, J. M., and Bradbury, J. M. (1996) Specific detection of *Mycoplasma iowae* using polymerase chain reaction. *Mol. Cell. Probes* **10**, 23–29.

57. Pomp, D., and Medrano, J. F. (1991) Organic solvents as facilitators of polymerase chain reaction. *Biotechniques* **10**, 58–59.

58. Abu Al-Soud, W. (2000) Optimisation of diagnostic PCR: A study of PCR inhibitors in blood and sample pretreatment. *Doctoral thesis.* Department of Applied Microbiology, Lund University, Lund, Sweden.

59. Kreader, C. A. (1996) Relief of amplification inhibition in PCR with bovine serum albumin or T4 gene 32 protein. *Appl. Environ. Microbiol.* **62**, 1102–1106.

60. Tsutsui, K., and Mueller, G.C. (1987) Hemin inhibits virion-associated reverse transcriptase of murine leukemia virus. *Biochem. Biophys. Res. Commun.* **149**, 628–634.

61. Abu Al-Soud, W., and Rådström, P. (2000) Effects of amplification facilitators on diagnostic PCR in the presence of blood, feces, and meat. *J. Clin. Microbiol.* **38**, 4463–4470.

62. Topal, M. D., and Sinha, N. K. (1983) Products of bacteriophage T4 genes 32 and 45 improve the accuracy of DNA replication in vitro. *J. Biol. Chem.* **258**, 12,274–12,279.

63. Chandler, D. P., Wagnon, C. A., and Bolton, H., Jr. (1998) Reverse transcriptase (RT) inhibition of PCR at low concentrations of template and its implications for quantitative RT-PCR. *Appl. Environ. Microbiol.* **64**, 669–677.

64. Jordan, S. P., Zugay, J., Darke, P. L. and Kuo, L. C. (1992) Activity and dimerization of human immunodeficiency virus protease as a function of solvent composition and enzyme concentration. *J. Biol. Chem.* **267**, 20,028–20,032.

65. Wu, J. R., and Yeh, Y. C. (1973) Requirement of a functional gene 32 product of bacteriophage T4 in UV. *J. Virol.* **12**, 758–765.

66. Varadaraj, K., and Skinner, D .M. (1994) Denaturants or cosolvents improve the specificity of PCR amplification of a G + C-rich DNA using genetically engineered DNA polymerases. *Gene* **140**, 1–5.

67. Lee, C. H., Mizusawa, H., and Kakefuda, T. (1981) Unwinding of double-stranded DNA helix by dehydration. *Proc. Natl. Acad. Sci. USA* **78**, 2838–2842.

68. Dutton, C. M., Paynton, C., and Sommer, S. S. (1993) General method for amplifying regions of very high G+C content. *Nucleic Acids Res.* **21**, 2953–2954.

69. Innis, M. A., Myambo, K. B., Gelfand, D. H., and Brow, M. A. D. (1988) DNA sequencing with *Thermus aquaticus* DNA polymerase and direct sequencing of polymerase chain reaction-amplified DNA. *Proc. Nat. Acad. Sci. USA* **85**, 9436–9440.

70. Frackman, S., Kobs, G., Simpson, D., and Storts, D. (1998) Betaine and DMSO: Enhancing agents for PCR. *Promega Notes* **27**.

71. Henke, W., Herdel, K., Jung, K., Schnorr, D., and Loening, S.A. (1997) Betaine improves the PCR amplification of GC-rich DNA sequences. *Nucleic Acids Res.* **25**, 3957–3958.

72. Back, J. F., Oakenfull, D., and Smith, M. B. (1979) Increased thermal stability of proteins in the presence of sugars and polyols. *Biochemistry* **18**, 5191–5196.

73. Nagai, M., Yoshida, A., and Sato, N. (1998) Additive effects of bovine serum albumin, dithiothreitol, and glycerol on PCR. *Biochem. Mol. Biol. Int.* **44,** 157–163.
74. Demeke, T., and Adams, R.P. (1992) The effects of plant polysaccharides and buffer additives on PCR. *Biotechniques* **12,** 332–334.
75. Kim, S., Labbe, R. G., and Ryu, S. (2000) Inhibitory effects of collagen on the PCR for detection of *Clostridium perfringens. Appl. Environ. Microbiol.* **66,** 1213–1215.
76. Chen, S., Yee, A., Griffiths, M., et al. (1997) The evaluation of a fluorogenic polymerase chain reaction assay for the detection of *Salmonella* species in food commodities. *Int. J. Food. Microbiol.* **35,** 239–250.
77. Knutsson, R., Fontanesi, F., Grage, H., and Rådström, P. (2001) Development of a PCR-compatible enrichment medium for *Yersinia enterocolitica*: Amplification precision and dynamic detection range during cultivation. *Int. J. Food Microbiol.* **72,** 185–201.
78. Grant, K. A., Dickinson, J. H., Payne, M. J., Campbell, S., Collins, M. D., and Kroll, R. G. (1993) Use of the polymerase chain reaction and 16S rRNA sequences for the rapid detection of *Brochothrix* spp. in foods. *J. Appl. Bacteriol.* **74,** 260–267.
79. Klein, P. G., and Juneja, V. K. (1997) Sensitive detection of viable *Listeria monocytogenes* by reverse transcription-PCR. *Appl. Environ. Microbiol.* **63,** 4441–4448.
80. Nordvåg, B., Riise, H., Husby, G., Nilsen, I., and El-Gewely, M. R. (1995) Direct use of blood in PCR. *Methods Neurosci.* **26,** 15–25.
81. Seesod, N., Lundeberg, J., Hedrum, A., et al. (1993) Immunomagnetic purification to facilitate DNA diagnosis of *Plasmodium falciparum. J. Clin. Microbiol.* **31,** 2715–2719.
82. Lantz, P. -G., Tjerneld, F., Borch, E., Hahn-Hägerdal, B., and Rådström, P. (1994) Enhanced sensitivity in PCR detection of *Listeria monocytogenes* in soft cheese through use of an aqueous two-phase system as a sample preparation method. *Appl. Environ. Microbiol.* **60,** 3416–3418.
83. Giambernardi, T. A., Rodeck, U., and Klebe, R. J. (1998) Bovine serum albumin reverses inhibition of RT-PCR by melanin. *Biotechniques* **25,** 564–566.
84. Sidhu, M. K., Liao, M. J., and Rashidbaigi, A. (1996) Dimethyl sulfoxide improves RNA amplification. *Biotechniques* **21,** 44–47.

3

Critical Aspects of Standardization of PCR

Jeffrey Hoorfar and Nigel Cook

1. Introduction

Food-borne zoonotic bacteria, transferred from animals to food products, are a major concern in modern food production and, consequently, for human health *(1)*. Accordingly, there is a need to control the entire food chain, from infections at the herd level to the consumer, through screening and certification programs, which apply highly sensitive and cost-effective methods for detection of food-borne pathogens. Such tests would also be useful to the food industry and clinical diagnostic laboratories. Molecular methods, using nucleic acid diagnostics, are receiving increasing attention for testing the microbiological safety of food *(2)*. There are several advantages of nucleic acid diagnostics compared to bacteriological culturing and immunochemical methods. Nucleic acid molecules can, via sequence combinations, precisely prescribe the phenotypic characteristics of a microorganism, and detection of specific sequences unique to a particular species can obviate any requirement for confirmatory tests. Furthermore, detection of nucleic acids is very rapid compared to conventional culture-based analysis. The most widely used nucleic acid diagnostic test is based on the powerful PCR technology *(3)*.

The enthusiasms of scientists, and the enormous bulk of publications presenting convincing data, have encouraged many diagnostic laboratories to implement PCR-based methods for pathogen detection *(4)*. However, the results of tests developed or published by one laboratory can sometimes be difficult to reproduce by other laboratories. Although this relates to most laboratory techniques, lack of reproducibility is more pronounced in molecular techniques due to sensitive reagents, complex equipment, and the need for personnel with specific skills. Proper validation based on consensus criteria is an absolute prerequisite for successful adoption of PCR-based diagnostic methodology.

From: *Methods in Molecular Biology, vol. 216: PCR Detection of Microbial Pathogens: Methods and Protocols*
Edited by: K. Sachse and J. Frey © Humana Press Inc., Totowa, NJ

2. FOOD-PCR: A European Effort

Recognizing this, in 1999, the European Commission approved the research project FOOD-PCR (www.PCR.dk) with the aim of validating and standardizing the use of diagnostic PCR for detection of bacterial pathogens in foods. A consortium of 35 institutes, companies, and universities from 14 EU countries and 7 associate and applicant states was formed to work on this project.

An intention of FOOD-PCR was to devise standardized PCR-based detection methods for 5 major pathogens: *Salmonella enterica*, thermophilic *Campylobacter* spp., enterohemorrhagic *Eschericia coli* (EHEC), *Listeria monocytogenes* and *Yersinia enterocolitica*. The methods would focus on three sample types from primary food production: poultry-carcass rinse, pig-carcass swab, and milk.

The 3-yr project comprises 6 work packages and 20 tasks (*see* **Fig. 1**), including production of certified DNA material, preparation of a thermal cycler performance guideline, and performance of PCR ring trials. Another important area would be automated detection, including enzymer-linked immunosorbent assay (ELISA)-format and real-time PCRs.

The development of the standardized methods was designed to proceed in three phases (*see* **Fig. 2**). In Phase 1, researchers working in expert laboratories prepared defined DNA material, selected promising candidate PCR methods, and tested them for efficiency and selectivity against comprehensive collections of reference strain DNA. The final selected PCRs were then optimized and taken forward into Phases 2 and 3 (*see* **Subheading 9.**). This would provide a thorough evaluation of the efficiency and robustness of the PCRs.

The project also had a work package devoted to sample pretreatment. Here, methods would be developed based on current International Organization for Standardization (ISO) pre-enrichment procedures *(5,6)*, adapting then-existing procedures, where necessary, to allow PCR to replace conventional postenrichment and/or detection. In Phase 3, the complete procedure comprising sample pretreatment and PCR would be subjected to intelaboratory trials, to provide validated PCR-based pathogen detection protocols.

Among the intended outcomes of the project was the production of a guideline and a biochemical kit for proficiency testing of different types and brands of cyclers, a simple method for purifying DNA from bacterial cultures, production of reference DNA material, an on-line database containing validated PCR protocols, organized workshops for end users, and preparation of standardized guidelines in collaboration with European Committee on Standardization (CEN), Working Group 6 (WG6).

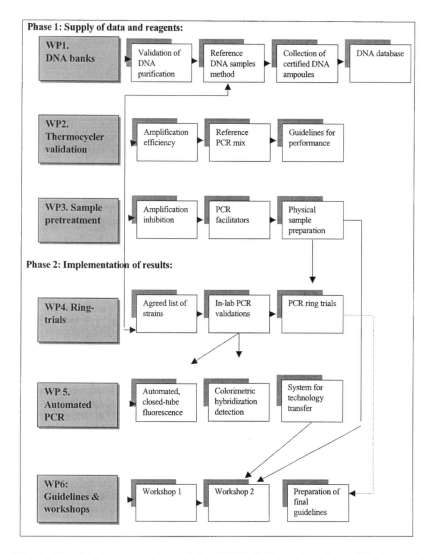

Fig. 1. Graphical presentation of the FOOD-PCR project for validation and standardization of PCR for detection of important food-borne pathogens.

3. A Common Vision

The experience gained through the FOOD-PCR project and activities of various international working groups suggest that a very basic aspect of PCR standardization is elaboration of a vision, or far sighted goals. As with any other major effort, the work should be based on a clear vision, a focused strategy,

Phase 1:
Evaluation of several PCR primer sets for each
pathogen
•One set adopted
•PCR optimized
•Detection limit defined

Phase 2:
Validation of accuracy and robustness of the PCRs
a - Partners use standardized reagents
b - Partners use reagents of choice
•Reference strain list prepared for each pathogen
•DNA extracted from reference strains
•Specificity of PCR evaluated by 10 -12 laboratories

Phase 3:
Validation of complete PCR-based detection
methods
•Using spiked samples at high, medium, low, and
zero cell number
•10 - 12 participants
•Partners perform pre-enrichment, extraction, and
PCR detection, according to SOPs
•Spiked samples analysed by existing approved
methods
•Performance of PCR-based methods compared to
approved methods

Fig. 2. Practical steps when performing ring trials for validation of PCR.

and a few basic principles shared by the organizations, industries, scientists
and authorities involved. The vision adopted by the PCR working group
(TAG3/WG6) of CEN is that diagnostic PCR will have the same status as
conventional bacteriological culture techniques by the year 2010 (*see* **Fig. 3**).
The strategy or long-term plan to achieve that vision is rapid publication of a
few basic guidelines. The principles of the work will be based on noncommer-
cial and international effort, including extensive multicenter ring trials, with
the aim of providing end-users with nonexclusive protocols. The emphasis in
the strategy is the speed and simplicity of the publications, thus avoiding protocols outdated by the amazing pace of technology development in the field of
molecular diagnostics.

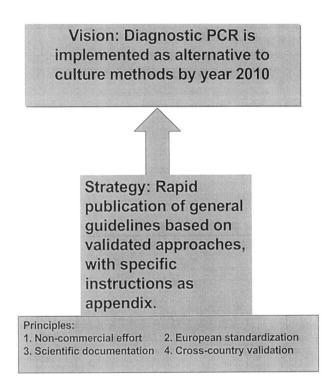

Fig. 3. Proposed vision, strategy and principles for standardization of diagnostic PCR for detection of food-borne pathogens.

4. Focused Strategy

Preparation of international standards can be notoriously lengthy and cumbersome and can take up to several yr. This has at least been the case with bacteriological culture techniques and, interestingly, also with PCR-based detection of genetically modified organisms (GMO), creating a technological vacuum for legislators.

In addition, most standards were initially prepared without available validation data, which has had to be provided later with some considerable effort. This can be avoided for diagnostic PCR if the strategy is based on rapid publication of timely standards prepared as guidelines, rather than detailed laboratory protocols, which in some cases may even be outdated by the time they have been through the usual international voting procedure. A European working group (TC275/WG6/TAG3) has planned preparation of overall guidelines on pre-PCR sample preparation, thermal cycler efficiency testing, and amplification and detection (*see* **Fig. 4**).

```
1. An overall PCR guideline, including the
necessary definitions.

2. A general document on sample treatment and
DNA extraction.

3. A guideline on validation of thermal cyclers.

4. A general document for amplification and
detection, including three examples for gel
electrophoresis, PCR-ELISA and real-time PCR.

5. A certified, updated website for available
primers and test conditions.
```

Fig. 4. The strategy for preparation of standards for PCR testing.

5. Validation Criteria

Often the terms "sensitivity" and "detection limit" are used interchangeably in the literature and in the scientific community. The confusion can be worse for PCR laboratories, where sensitivity can also refer to the lack of robustness or the well-described fragility of the polymerase enzyme to inhibitors in the sample. Fortunately, guidelines laid in the draft of the MicroVal protocol (7) provide us with some general definitions for interpretation and communication of the results of validation of alternative microbiological methods.

Some national or regional validation organizations, e.g., Nordic Validation Organ (NordVal), Association Francaise de Mormalisation (AFNOR), Deutsches Institute für Normung (DIN), etc., have also published definitions and guidelines for validation of alternative microbiological methods. Based on the literature and international working papers available, we propose a set of specific definitions for validation parameters to be used for evaluation of PCR testing (see Tables 1 and 2)

6. Interpretation of Results

Due to the risk of false positive PCR cases, the TAG3 European working group is currently considering inclusion of a solution hybridization step for confirmation of amplicons (8). However, in our experience, if primer sequences that amplify genetic regions unique to a species have been carefully chosen, tested, and validated, post-PCR confirmation is unnecessary and will negate the advantage of rapidity, although it may look convincing on a paper for standard protocol. In addition, it could be too complicated for many small laboratories and exclude commercial kits based solely on gel electrophoresis.

Table 1
Definition of Terms Proposed to Be Used in Validation of PCR Testing[a]

Validation:	Demonstration that adequate confidence is provided that results obtained by PCR are comparable to those obtained by the reference method.
Qualitative PCR:	The test response is either the presence or absence of PCR product (amplicon), detected either by observation or in equipment.
Quantitative PCR:	The test response is the copy number of amplicon, detected indirectly and related to the number of target microorganisms.
Detection limit (DL):	The smallest number of culturable target microorganisms necessary to create a PCR positive response.
Selectivity:	A measure of inclusivity of target strains (from a wide range of strains) and exclusivity (the lack of amplicon from a relevant range of closely related nontarget strains).
Positive deviation (PD):	A PCR positive case when the reference method gives a negative result (false positive).
Negative deviation (ND):	A PCR negative case when the reference method gives a positive result (false negative).
Positive agreement (PA):	A sample being positive in both PCR and the reference method.
Negative agreement (NA):	A sample being negative in both PCR and the reference method.
Diagnostic accuracy (AC):	The degree of correspondence between the response obtained by PCR method and the response obtained by reference method on identical natural samples [AC = (PA+NA) / Total number of samples].
Diagnostic sensitivity (SE):	The ability of PCR to detect the microorganism when it is detected by the reference method [(PA/N+) × 100].
Diagnostic specificity (SP):	The ability of PCR to **not** detect the microorganism when it is **not** detected by the reference [(NA/N−) ×100].
Robustness:	Reproducibility by other laboratories using different batches and brands of reagents and validated thermal cyclers and equipment.

[a] Adapted from the MicroVal protocol (7).
N− is the total number of negative results with the reference method.
N+ is the total number of positive results with the reference method.

Table 2
Paired Results of Reference and PCR Methods (7)

PCR response	Reference method positive (R+)	Reference method negative (R−)
Alternative method (A+)	+ / + positive agreement (PA)	− / + positive deviation (PD) (R−/A+)
Alternative method negative (A−)	+ / − negative deviation (ND) (A− / R+)	−/ − negative agreement (NA)

Table 3
Test Controls Necessary for Performance of Diagnostic PCR

Internal amplification control (IAC):	Containing chimeric nonrelevant DNA added to master mixture and to be amplified by the same primer set as the target DNA but with an amplicon size visually distinguishable from the target amplicon.
Processing positive control (PPC):	A negative sample spiked with sufficient amount of the target pathogen and processed throughout the entire protocol.
Processing negative control (PNC):	A negative sample spiked with sufficient amount of a closely related, but nontarget strain processed throughout the entire protocol.
Reagent control (Blank):	Containing all reagents, but no nucleic acid apart from the primers.
Premise control:	A tube containing the master mixture left open in the PCR setup room to detect possible contaminating DNA in the environment (to be done in certain intervals as part of the quality assurance program).
Standard concentrations:	3 to 4- samples containing 10-fold dilution series of known number of target DNA copies in a range above the detection limit (necessary only for quantitative PCR).

Possible false positive results can be revealed through further culturing of the enriched microorganisms from PCR positive cases, although we should have in mind the suboptimal diagnostic sensitivity of most culture methods. Just to mention one: for *Salmonella* culture, it was found necessary to pre-enrich some samples in two different pre-enrichment broths in order to obtain an optimal diagnostic sensitivity *(9,10)*.

7. Test Controls

This brings us back to the well-known dilemma of PCR vs culture methods, where we actually compare "apples" with "pears". In PCR, we are amplifying DNA, while culture methods isolate live bacteria, and in some cases, leaving the injured target bacteria behind. Many workers have addressed this issue by spike-in experiments that demonstrate a detection limit of one target bacterium in a 25-g sample. However, the applicability to real life situations of spiked studies using fresh cultures of "healthy" inoculates can be justifiably questioned.

One of the strengths of diagnostic PCR is its ability to be used as an identification tool, particularly using the 16S sequence regions of pathogenic bacteria (*see* **Chapters 1** and **4**). Although this application appears to be quite straightforward, the importance of including proper test controls may be easily overseen due to the excitement of workers and the elegance of the technique.

A PCR cannot be given diagnostic status, no matter how limited the application, before it includes, as a minimum, an amplification control, a processing positive control, a reagent control (blank), and a processing negative control (*see* **Table 3**), as suggested in several working documents.

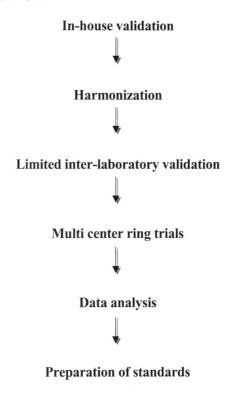

Fig. 5. The proper sequence of events in preparation of PCR standards.

8. Sequence of Events

The occurrence of an incorrect sequence of events is seen with some of the traditional culture techniques. When many workers begin to face the challenge of quality control requirements it is realized that we actually have to go several steps back and begin with proper validations (*see* **Fig. 5**). The writing of standards must be based on validation studies and experience gained through ring trials. Harmonization and validation are the prerequisites of standardization, in that sequence, and the writing of standards must be based on successful validation studies and laboratory experience gained through ring trials. Implementation of quality assurance programs is often substantially facilitated if relevant standards are available.

9. Development and Validation of Standard PCR-Based Methods to Detect Food-Borne Pathogens

We propose that development and validation of a standard PCR-based method to detect a food-borne pathogen should take place in three phases.

The choosing and testing of primers should be the first phase in the production of a standard PCR. Here, laboratories, which are expert in PCR methodology for a particular pathogen, should decide collectively which primer sets will be selected for evaluation. This will generally be done on the basis of those laboratories' experience of the primers in use. Next, laboratories with particular expertise in working with the pathogen concerned should draw up a specificity strain list. This list should contain strains that comprehensively represent the pathogen and representatives of cross-reacting species. These other nontarget species should be ones that are closely related to the target species, and the list may also contain more distantly related or nonrelated species, if it is considered that they may be encountered with the target species in the food matrix ultimately to be examined. DNA should be extracted from each strain and analyzed in a series of PCRs containing the primer sets to be evaluated. Each PCR should contain reagents that are as identical as possible (supplier, batch, etc.) and use thermal cyclers that are routinely calibrated and checked. The initial evaluation should be performed through a limited interlaboratory trial, including at least three expert laboratories as partners (*see* **Fig. 2**). The criteria for successful evaluation should be strong, with specific amplification of correct target sequences and no others. One primer set should be chosen to take forward into the next rounds of the standardization process. The PCR should then be optimized (*see* **Chapter 4**). The detection limit of the PCR, in terms of the number of cells it can detect with 99 % probability *(11)*, should be established thereafter.

The second phase should take the form of a large-scale interlaboratory trial to confirm the specificity of the PCR. This trial should involve 10–12 partners *(7)*, in addition to the organizing laboratory. Each participating laboratory receives a Standard Operating Procedure (SOP), samples of DNA from the strain list established previously, and sufficient reagents (once again as identical as possible, e.g., same supplier and batch) to perform PCRs in duplicate upon the DNA. The DNA samples should be blind, i.e., their identity known only to the organizing laboratory, and coded. The participants perform the PCRs and report the results to the organizing laboratory. The percentage ratio of true positive results to false positive and true negative results to false negatives should be recorded (*see* **Table 2**). With a robust and accurate PCR that is correctly performed, this ratio should be 100% in each case.

A robust PCR could in addition work as well with different reagents from different suppliers. The Phase 2 ring trial participants, this time using nonidentical reagents, should evaluate this. This is done by repeating the specificity evaluation, but this time using enzymes and reagents of their choice (while still conforming to the reaction concentrations and conditions as specified in the SOP). This will provide a demonstration of the ultimate robustness of the PCR. If it works as well with other reagents as with those used in the originating laboratory, this should reduce the logistic burden of the next phase of the validation, as the organizing laboratory need not supply all the reagents to each participant.

The final PCR-based detection protocol should contain a simple and universal sample treatment of the matrix in combination with a PCR-compatible (pre)enrichment broth prior to amplification. This will give a degree of familiarity to the method, which should encourage end users to replace conventional methodology with it. Pre-enrichment is commonly employed in several microbial detection techniques associated with foods and is used for several reasons. It can ensure the presence of target cells in numbers above the detection limit of the technique and either dilutes out food-derived inhibitory substances or allows their digestion through bacterial growth *(12,13)*. The pre-enrichment step of the PCR-based method should be similar, if not identical, to the conventional method, and in the case of food-borne pathogens, there are several standard pre-enrichment methods currently available *(14–16)*. The simplest way of using a pre-enrichment culture would be to put an aliquot directly into the PCR as target. This, however, may result in inhibition of the PCR, which should be carefully ascertained. Some form of secondary treatment may be necessary, ranging in complexity from simple dilution of the enrichment culture, to DNA extraction, to the use of PCR facilitators *(17)* (*see* **Chapter 2**). The most straightforward method should be chosen, as it should always be remembered that the final method is intended for use by routine high-throughput laboratories and general technical staff.

The next and final phase, therefore, involves validation of the complete PCR-based method as a comparison with the equivalent conventional method. Again, this is done as an inter-laboratory trial involving 10–12 partners. Samples of matrix spiked at various target cell densities are sent to each participant by the organizing laboratory. In the Nordval guideline *(18)*, these levels should be zero, 1–10 cells, and 10–100 cells /25-g sample. The participants should then incubate (pre-enrich) the samples, apply the PCR according to the SOP, and record the results. Concurrently, they perform a conventional detection procedure (plating, etc.) after the pre-enrichment and record the results. The organizing laboratory should compare the results, or responses, of each method. Results from at least eight laboratories with valid results must be available, if the comparison is to be thorough *(7)*. The diagnostic specificity, diagnostic sensitivity, and the overall accuracy of each method should be determined and compared. A robust and efficient PCR-based method should be at least as accurate as the conventional method.

10. Demanding Logistics

The experience from FOOD-PCR and those of others shows that ring trials can be very costly and that they require a specialized infrastructure for packaging, shipment, and data collection *(19)*. If the participating laboratories are not all accredited, it is then necessary to provide detailed SOPs for basic methods and equipment involved, and the originating laboratory, as well as organizing the shipments, must centralize as much of the handling techniques as possible.

The ideal situation would be to assign one specialized company to take care of the centralized purchase of reagents, the packaging of chemicals and hazardous pathogens according to international regulations, include and evaluate temperature markers in packages, and assess the overall logistics performance of ring trials. This will minimize the risk of sample and reagent variation and allow the scientist's time to be focused on the test performance rather than organizing the logistics. Major projects can then easily subcontract the practical work with sufficient funding to such a company. In addition, this kind of logistics infrastructure would be beneficial to other standardization programs once established.

11. Transparency of Information

What most scientists engaged in basic research may naturally overlook, when working in multilateral research consortia involving both companies and control authorities, is agreement on access rights and level of information flow, which can turn any standardization project into a nightmare.

In our opinion, transparency is so crucial to public acceptance of international standards, that the issue of protection of knowledge must come second. Naturally, any entrepreneur should be welcome to take advantage of available standards for producing ready-to-go kits and automated platforms. This is already the case, e.g., with culture media used for traditional microbiological techniques. End users who would prefer to procure kits due to lack of space or skilled personnel, but would still wish to conform to international standards, should be able to do so through commercial kits. However, the commercial aspects of the knowledge produced through ring trials and exchange of technology must not come in the way of standardization for the benefit of public health.

Here, it is important to emphasize that leadership of various standardization working groups involved can play a major role in creating an atmosphere of trust and credibility. Specifically, we recommend assigning persons with no affiliation to commercial interests to chair various working groups. A level of awareness would always overshadow the work of an ad hoc group, if a company member chairs it, regardless of how hard they try to exercise impartiality.

References

1. Tauxe, R. V. (1991) *Salmonella*: A postmodern pathogen. *J. Food Protection* **54,** 563–568.
2. Vaneechoutte, M., and Van Eldere, J. (1997) The possibilities and limitations of nucleic acid amplification technology in diagnostic microbiology. *J. Med. Microbiol.* **46,** 188–194.
3. Mullis, K., Faloona, F., Scharf, S., Siaki, R., Horn, G., Erlich, H. (1986) Specific enzymatic amplification of DNA in vitro: The polymerase chain reaction. *Cold Spring Harb. Symp. Quant. Biol.* **51,** 263–273.
4. Scheu P. M., Berghof K., and Stahl U. (1998) Detection of pathogenic and spoilage micro-organisms in food with the polymerase chain reaction. *Food Micorbiol.* **15,** 13–31.
5. Anon. (1999A) Microbiology of food and animal feeding stuff—carcass sampling for microbiological analysis. ISO/CD 17604. International Organization for Standardization, Geneva, Switzerland.
6. Anon. (2000) Microbiology of food and animal feeding stuff—Preparation of test samples, initial suspension and decimal dilution for microbiological examination. Doc CEN/TC 275/ WG 6 N 114. Published by AFNORE, Paris, France.
7. Anon. (1999) Microbiology of food and feeding stuffs— Protocol for the validation of alternative methods. Document established by the Joint Group of MicroVal (WG 6) and CEN (WG 6/TAG 2). ISO/DIS 16140, Geneva, Switzerland.
8. Anon. (2001) General requirement relating to methods of detecting foodborne pathogens using the polymerase chain reaction (PCR). CEN/TC 275/WG6/TAG3 N 10. Published by DIN. Berlin, Germany.
9. D'Aoust, J.-Y., Sewell, A., and Jean, A. (1990) Limited sensitivity of short (6h) selective enrichment for detection of foodborne *Salmonella*. *J. Food Protection* **53,** 562–265.
10. Hoorfar, J. and Baggesen, D. L. (1998) Importance of preenrichment media for isolation of *Salmonella* spp. from swine and poultry. *FEMS Microbiol. Lett.* **169,** 125–130.
11. Knutsson, R., Blixt, Y., Grage, H., Borch, E., and Rådström, P. (2002) Evaluation of selective enrichment PCR procedures for *Yersinia enterocolitica. Int. J. Food Microbiol.* **73,** 35–46.
12. Uyttendaele, M., Shukkink, R., van Gemen, B., and Debevere, J. (1995) Detection of *Campylobacter jejuni* added to foods by using a combined selective enrichment and nucleic acid sequence-based amplification (NASBA). *Appl. Environ. Microbiol.* **61,** 1341–1347.
13. Szabo, E. A., and Mackey, B. M. (1999) Detection of *Salmonella enteritidis* by reverse transcription-polymerase chain reaction (PCR). *Int. J. Food Microbiol.* **51,** 113–122.
14. Anon. (1993) Microbiology—general guidance on methods for the detection of Salmonella. International Organization for Standardization, ISO 6579, 3rd edition. Geneva, Switzerland.

15. Anon. (1995) Microbiology of food and animal feeding stuffs - Horizontal method for detection of thermotolerant *Campylobacter*. International Standard Organisation, ISO 10272, 1st edition. Geneva, Switzerland.
16. Anon. (1996) *Yersinia enterocolitica*. Detection in food. Nordic Committee on Food Analysis, Method no. 117, 3rd Edition. Oslo, Norway.
17. Simon, M . C., Gray, D. I., and Cook, N. (1996) DNA extraction and PCR methods for the detection of Listeria monocytogenes in cold - smoked salmon. *Appl. Environ. Microbiol.* **62,** 822–824.
18. Anon. (2001) Protocol for the validation of alternative microbiological methods. NV-DOC.D-2001-04-25. Published by NordVal. Copenhagen, Denmark.
19. Bohnert, M.., Humbert, F. and Lombard, B. (2001) Validation of ISO microbiological methods. Detection of *Salmonella* according to draft standard pr EN ISO/ DIS 6579: 2000. Final Report. Ploufragan, France.

4

PCR Technology and Applications
to Zoonotic Food-Borne Bacterial Pathogens

Peter Stephensen Lübeck and Jeffrey Hoorfar

1. Introduction

Among molecular biological methods targeting nucleic acids, the polymerase chain reaction (PCR) has become the most popular diagnostic method in human and veterinary medicine, as well as in microbiological food testing *(1,2)*.

The first publications on PCR came from the group of R. K. Saiki and colleagues in 1985 *(3)* and K. B. Mullis and F. A. Faloona in 1987 *(4)*. Initially, the Klenow fragment from *Eschericia coli* DNA polymerase I was used by manual addition of the enzyme to the reaction after each individual cycle about 40 times. The introduction of *Thermus aquaticus* (*Taq*) DNA polymerase *(5)* with its relative stability at temperatures above 90°C eliminated the need for adding enzyme after each cycle. This simplified the PCR technology significantly and contributed to its rapid and widespread acceptance, not only in basic, but also in applied research. As a result of amplification, it is now possible to study genetic material that is present in such tiny amounts that it cannot be detected or analyzed by other methods directly. In little more than a decade, PCR has evolved from being a technique used by a few molecular biologists to a universally accepted tool for rapid diagnosis of pathogens.

2. Detection of Food-Borne Bacteria:
DNA-Based Methods vs Culture

Classical methods of detecting bacteria often involve a pre-enrichment culture and a selective enrichment step followed by subsequent identification through morphological, biochemical and/or immunochemical tests. Cultural methods tend to be labor-intensive and time-consuming. Therefore, many research efforts were devoted to the development of rapid and sensitive meth-

From: *Methods in Molecular Biology, vol. 216: PCR Detection of Microbial Pathogens: Methods and Protocols*
Edited by: K. Sachse and J. Frey © Humana Press Inc., Totowa, NJ

ods for the detection of microorganisms *(6,7)*. Application of DNA and RNA probes, immunochemical methods and detection of bioluminescence are some of the tools currently used *(8–12)*.

PCR testing offers the possibility to improve detection and characterization of pathogenic bacteria, since one can target species-specific DNA regions and specific traits of pathogenicity, especially genes coding for toxins, virulence factors, or major antigens. The PCR technique has several advantages over classical bacteriology with respect to detection limit, speed, and potential for automation *(13,14)*. The latter capability is indeed necessary for application of the test in extensive screening programs. Currently, probes and PCR methods are available for many important food-borne pathogens, such as *Salmonella*, enterohemorrhagic *E. coli*, *Yersinia enterocolitica*, *Campylobacter* spp., and *Listeria monocytogenes*.

Salmonella: Salmonellosis is an infectious disease of humans and animals caused by organisms of two species of salmonella, *S. enterica* and *S. bongori*. In order to increase the sensitivity, specificity, and speed of detection, several different DNA methods have been developed *(15,16)*. However, due to the lack of common genes for toxins or other virulence factors, the approach for isolation of specific DNA probes has been to select randomly cloned chromosomal fragments. Furthermore, ribosomal RNA-based oligonucleotide probes have been used successfully in a single-phase hybridization assay to detect a large number of serovars of *Salmonella*. (For PCR-based detection of *Salmonella*, see **Chapter 19**.)

E. coli: The pathogenic and nonpathogenic strains of *E. coli* are difficult to distinguish using selective cultivation methods, whereas pathogenic properties are reliably identified through detection of toxin genes *(17,18)*. Cloned oligonucleotide probes have been designed for detection of genes encoding enterotoxins of both the heat-stable (ST), the heat-labile (LT), and shiga-like cytotoxin (SLTs) families *(19)*. Today, probes and oligonucleotides are mostly used for characterization of *E. coli* which has been isolated by traditional culture methods *(20)*. (For PCR-based detection of shiga toxin-producing *E. coli* [STEC]), *see* **Chapter 11**.)

Campylobacter spp.: Most species of *Campylobacter* are recognized as pathogens in humans *(21)*, with the thermophilic species *C. coli* and *C. jejuni* being the most important pathogenic species. Animals are the main reservoir of the microorganisms, and infections of humans are mainly of food-related origin. Detection of *Campylobacter* by culture has to cope with several problems *(22)*. First, in samples such as food, the agent is present in very low numbers. Second, due to their extreme sensitivity to environmental factors, such as atmospheric oxygen, low pH, dryness, and ambient temperature, the number of viable *Campylobacter* cells can be substantially reduced during transportation.

Third, in clinical samples, antibiotics may inhibit growth of bacteria. Finally, despite being viable, the organism may still not grow in culture media. Application of alternative methods can, therefore, help to overcome the aforementioned problems. A large number of reports were published on the application of oligonucleotides, cloned DNA fragments, and chromosomal DNA as hybridization probes, as well as PCR *(23–25)*. (For PCR-based detection of *Campylobacter, see* **Chapter 7**.)

Y. enterocolitica: The pathogenic serotypes of *Y. enterocolitica* are important food-borne pathogens frequently found in pig herds and pig products *(26,27)*. Conventional culture methods for detection of the organism can take up to a month to complete and are rather laborious *(28)*. In addition, nonpathogenic strains of *Y. enterocolitica* may be present in diagnostic samples to the same extent as the pathogenic ones, and a main goal for nucleic acid-based methods is the exclusion of this group *(29)*. The first PCR method used for the detection of pathogenic strains was developed in 1990 *(30)*. Since then, various other amplification protocols have been used to detect the bacteria in culture *(31)*, feces *(32)*, water, and spiked food samples *(33)*, as well as in tonsils from swine *(34)*. (For PCR-based detection of *Y. enterocolitica, see* **Chapter 22**.)

L. monocytogenes: This microorganism is recognized as an important food-borne pathogen because of its widespread distribution in the environment and its ability to grow at lower temperature *(35)*. However, only the hemolytic species *L. monocytogenes, L. seeligeri, L. ivanovii* are associated with pathogenicity *(36)*. PCR has shown great promise as tool for detection of the hemolysin (*hly*A) gene of *L. monocytogenes*, which is found in the pathogenic strains *(37–39)*. (For PCR-based detection of *L. monocytogenes, see* **Chapter 12**.)

Generally speaking, nucleic acid amplification methods are extremely sensitive and, therefore, require just minute amounts of the target organism *(40)*. PCR possesses all the virtues to make it a versatile tool for the detection of pathogens; it is very specific if suitable primer sequences are chosen. In addition, amplification of DNA copies can be accomplished rapidly, and the method is of particular value in situations where the target microbes are viable but nonculturable, or when isolation of the microorganism is difficult *(41)*. While other DNA-based methods require careful handling and processing of samples to provide good quality target DNA, PCR can still detect sheared or partially degraded DNA *(42)*.

The specificity of DNA hybridization techniques depends primarily on the sequence of the probe. This is similar to detection by PCR, where specificity primarily depends on the sequence of the primers and the target sequence flanked by them. In this connection, it should be mentioned that the distinction between primers and probes is blurred, since specific primers can also be used as probes *(43)*, and probes can be used for selection of specific primers for amplification. The possibility to design both narrow and broad ranges of specificity allows differentiation at different taxonomic levels and the monitoring of specific populations or strains *(44,45)*.

3. Fundamentals of PCR

PCR has the ability to amplify specific DNA sequences in an exponential fashion by in vitro DNA synthesis *(4,5)*. Starting from a few molecules of template DNA, it is possible to produce millions of copies of a characteristic genomic segment. In microbiological diagnosis, the technique is used to detect, identify and differentiate germs present in clinical and environmental samples.

3.1. Principle of the Reaction

PCR represents a cyclic reaction where target DNA is amplified in vitro by a series of polymerization cycles. Each cycle includes three steps: a heating step at 91°–97°C, where the DNA template duplex is denatured (melted) to single strands, an annealing step usually at 40°–65°C where short oligonucleotide primers bind to the single-stranded DNA template, and an extension step at 68°–73°C where thermostable DNA polymerase catalyzes the synthesis of a new DNA strand by elongation of the primed strand.

The reaction requires two short oligonucleotides (primers) flanking the target region to be amplified, which are present in large molar excess and hybridize to complementary segments of DNA. During the reaction, deoxynucleotide triphosphates (dNTP), i.e., dATP, dCTP, dGTP and dTTP, are bound to the free 3'-hydroxyl end of the new strand. Only deoxynucleotide monophosphate is incorporated in the DNA chain, cleaving off a pyrophasphate group.

Ideally, the number of DNA copies is doubled in each cycle. Consequently, a single copy of target DNA should theoretically be multiplied to 2^{30}, i.e., 1.074×10^9, copies after 30 cycles. In practice, however, the number of copies in the final reaction product is lower, mainly due to inhibitory effects (*see* **Chapter 2**), the influence of structural and methodological parameters (*see* **Chapter 1**), as well as the exhaustion of reagents.

3.2. Thermostable DNA Polymerases

The key reagent of PCR is the thermostable DNA polymerase, which was originally extracted from thermophilic bacteria. The various sorts of today's commercially available polymerases exhibit different degrees of resilience towards inhibitory components in clinical, environmental, and food samples *(46)* and are often designed for specific applications, e.g., hot start PCR, reverse transcription PCR (RT-PCR), long-range PCR, high-fidelity amplification (proofreading enzymes), or all-purpose use. This makes the choice of polymerase an important task that should be considered before a specific PCR assay is set up (*see* **Chapter 2**).

PCR facilitators, compounds such as bovine serum albumin (BSA) and the single-stranded DNA-binding T4 gene 32 protein, can be used as additives in the PCR buffer to prevent inhibition of amplification *(47)*. Thus, by selecting an appropriate combination of thermostable enzyme and PCR facilitator(s), it may be possible to improve the amplification in the presence of background biological matrix.

3.3. Buffers and Master Mixtures

The composition of the polymerase reaction buffer should be optimized for the chosen enzyme. It consists at least of Tris-HCl, pH 7.3–8.9, KCl, and $MgCl_2$. Moreover, EDTA, BSA, and detergents such as Tween® and Triton® X-100 may be required for some enzymes to inactivate inhibitors and insure maximum activity. As a cofactor for the DNA polymerase, Mg^{2+}-ions are crucial for successful polymerization. In general, excess Mg^{2+} will result in accumulation of nonspecific amplification products, and insufficient Mg^{2+} will reduce the yield of amplicon *(48)*. Also, too high concentrations of dNTP may promote misincorporations by the polymerase *(48)*. The dNTPs bind to free Mg^{2+} in a 1:1 molar ratio, and efficient binding of different primers to template DNA is dependent on Mg^{2+}-concentration. Therefore, each PCR assay with a given set of primers has an optimal concentration of free Mg^{2+}, which should be known to the user.

A master mixture contains all the components for a PCR except sample (template) DNA. Thus, all compounds except one are mixed and distributed in a number of tubes followed by addition of the sample DNA, after which the amplification can be started. The idea of making a master mixture is to insure that all components in the reaction are identical in all tubes, so that only the component to be investigated is variable. This minimizes the risk of contamination caused by repeated pipeting. Moreover, the master mixture can contain PCR facilitators and/or components for the elimination of PCR inhibitors.

3.4. Optimization

The enormous variety of applications of PCR makes it impossible to describe a single set of conditions for a successful reaction. As described in the previous paragraphs, many parameters influence the efficiency of amplification, such as the buffer composition, the choice of thermostable DNA polymerase, the length and sequence of the primers, concentration of $MgCl_2$ and dNTP, amount of template and nontarget background DNA, other cellular components, and polymerase inhibitors from sample matrix.

It may be necessary to change the composition of the reaction buffer provided with the DNA polymerase in order to improve the performance of the assay, especially when high sensitivity is a requirement.

Furthermore, methods optimized in one laboratory may lack reproducibility in another laboratory, since the efficiency of amplification is very sensitive to various factors, as for example the quality of the sample with regards to the presence of inhibitors and thermocycler performance. A major problem with PCR detection is, therefore, the possibility of obtaining false negative results. One way of checking whether the DNA is of suitable quality and quantity for PCR is the inclusion of an internal control in the reaction (duplex PCR) *(49)*.

In order to optimize a PCR assay, it is necessary to start with a general reaction and subsequently change the components according to a defined scheme. After optimization, the method should be assessed for its analytical accuracy with a relevant set of target and nontarget strains and species. If the optimized protocol is valid for different thermal cyclers in different laboratories using compounds from different companies, the method is regarded as robust and, therefore, reproducible among laboratories. This means that the PCR protocol has a buffer capacity tolerating certain variations. If the method works only on a certain thermal cycler using certain compounds in the laboratory where it was developed, the method has limited or no buffer capacity and must be optimized each time when introduced into a new laboratory on a new thermal cycler. Unfortunately, most PCR protocols are without buffer capacity and, therefore, have to be optimized when transferred to another laboratory.

4. Oligonucleotide Primers

Primers are synthetic sequences of single-stranded DNA between 6 and 30 nucleotides in length, most often between 16 and 24. Several parameters are important to consider when designing a primer set for a successful PCR. The complementary binding of primer and template DNA, also called annealing, is due to hydrogen bonding between the bases in primer and template DNA. Guanine (G) and cytosine (C) are linked via three hydrogen bonds, and adenine (A) and thymine (T) form two hydrogen linkages. Therefore, the annealing temperature of a primer of a given length is dependent on its base composition. A typical primer should have a molar G + C content of 40–60%. The length and composition of a primer determine the melting temperature (T_m) of the duplex DNA, which can be roughly calculated by the following simple formula *(50)*:

$$T_m = 2(A + T)^\circ + 4(G + C)^\circ \qquad\qquad \text{[Eq. 1]}$$

The formula is not exact, but provides a temperature range for optimal annealing.

4.1. Primer Design

A critical aspect of any PCR assay is the design and selection of optimal primers. Poorly designed primers are among the main causes for nonspecific amplification of nontarget DNA or complete failure of amplification of target DNA. Commercially available primer design software, e.g., OLIGO *(51)*, can facilitate a systematic search for suitable primers.

An important feature of an efficient primer is its complete match with the target DNA. When primers have to be designed on the basis of a single or a few available sequences of the target organism, it may turn out later that they do not match completely to all strains because of sequence variation. In these circumstances, the primers must match the target sequence at least at the 3' end, covering 5 to 6 nucleotides towards the 3' end. A few mismatches of the primer at the 5' end can be tolerated. The amount of tolerable mismatch is dependent on the nucleobases facing each other. While GC and AT pairing are perfect, and AC pairs are acceptable, GT, CT and GA pairs are unacceptable because of weak hybridization or steric hindrance. The consequences of poor primer annealing due to mismatches can be false negative results of the PCR assay.

Another important factor impairing the performance of an amplification assay is the formation of primer–dimers as a result of complementary base pairing between primer molecules. Complementary stretches of 2 to 3 bases in the two primers are sufficient to cause this phenomenon. Moreover, primers forming hairpin structures by internal hybridization will reduce the efficiency of amplification. In both cases, high annealing temperatures are necessary to prevent the unwanted hybridization, or new primers should be designed to insure optimal amplification.

Following the design of specific primers, it is necessary to check their selectivity in diagnostic PCR (*see* **Chapter 3**). There should be a positive PCR signal for all members of the target taxon and no amplification of DNA from organisms outside the taxon. Specificity in PCR is dependent on the annealing temperature. For specific detection with a completely matching primer set, it is recommended to use the highest possible annealing temperature to insure high stringency *(45,49,52)*. Optimal annealing temperatures are typically at 55–60°C. In contrast, primers not matching 100 % will anneal only at lower temperatures, which leads to lower sensitivity and specificity. This can be compensated by nested amplification *(53,54)* (*see* **Chapters 1** and **3**).

4.2. Diagnostic Specificity

When developing a technique for detection and identification of organisms in a complex environment, one of the most central aspects to be considered is its diagnostic specificity (*see* **Chapters 1** and **3**). In most cases, the criteria of specificity can only be fulfilled on the basis of the knowledge of relationships at the genetic level between the targeted organism(s) and related organisms. Therefore it is necessary to characterize a number of isolates genetically, including individuals from the target group as well as individuals from related groups *(55)*.

Specific detection by PCR requires knowledge of sequences of the target gene and of corresponding genes in related organisms. These sequences can be identified using specific probes that already exist for detection of the organism *(51)*, by studying specific genes with unique sequences *(56)*, or by sequencing amplified fragments from fingerprint methods *(57)*. Other potential targets for specific amplification can be repetitive extragenic sequences present in bacteria *(58)* or species-specific tandem DNA repeats in many eukaryotes *(44,59,60)*. In some instances, published sequences can be used solely as the basis for primer design from highly variable regions, but often only few sequences are available from such target genes. One strategy is to align known sequences *(61,62)*, design primers (amplimers) from the conserved regions, and use these to amplify the corresponding region of related organisms, including the target organism *(63)*. Direct sequencing and subsequent alignment of the sequences may then facilitate the design of specific primers for the target organism in variable regions.

If no DNA sequence data of a potential target gene are available, degenerate primers based on protein sequences *(64)* or alignments of DNA sequences from related genes of other organisms can be designed. Degenerate primers represent mixtures of primers with different nucleotides at certain positions. They can be used to amplify a variety of target sequences that are related to each other, but share a lower degree of homology.

By using two or more primer pairs in the same reaction, i.e., multiplex PCR (*see* **Chapter 1**), it is possible to amplify several target sequences simultaneously *(65,66)*. For instance, Lee Lang et al. *(67)* used a triplex PCR to screen *E. coli* for three different toxins in marine waters. Multiplex PCR can also be used with internal controls to identify potential false negative results *(49,68,69)*. In bacteria this procedure can be performed by adding universal 16S rDNA primers to each specific amplification *(68)*. An example of multiplex PCR in combination with nested primers and an internal control was reported for *Listeria* spp. *(70)*.

4.3. Diagnostic Sensitivity and Detection Limit

Good detection limit (*see* **Chapter 3**) is crucial when a pathogen has to be identified directly from the sample without cultural enrichment. There are several approaches to increase the detection limit, such as double PCR, nested PCR, and RT-PCR. In double PCR, a portion of the amplification products from the first reaction is used in a second reaction using the same primers *(66)*. In nested PCR, however, different primers are used in the second reaction, which are internal to the first primer set *(4,54,71)*. In semi-nested PCR, only one of the two primers is internal, the other being the same as in the first round of amplification *(72–74)*. The use of nested or semi-nested primers in the second amplification step furthermore enhances the specificity by eliminating possible false positive products from the first round *(66)*.

As detection by PCR does not discriminate between live or dead bacteria, it may be necessary to combine PCR with other methods to evaluate the presence of viable cells. One way to overcome this problem may involve methods like BIO-PCR (combined biological and enzymatic amplification), in which only living cells seem to be detected *(51)*, or RT-PCR where the target for detection are specific mRNAs. RT-PCR is reverse transcription of RNA to DNA followed by a PCR amplification. RT-PCR is primarily used in studies of gene expression *(75)* and in detection of virus with RNA genomes *(76–78)*. Conventional RT-PCR includes only one specific set of primers and can detect only one type of target RNA sequence in a sample. As in conventional PCR, this approach can be used with several sets of primers, i.e., multiplex RT-PCR *(77)*.

Detection of small numbers of target organisms requires a low detection limit. There are reports that additional cycles of amplification enhance the detection limit *(66,79)*, but according to Henson and French *(44)* and Arai et al. *(80)*, too many cycles can increase the risk of the accumulation of nonspecific PCR-amplified products. Another strategy to increase detection limit includes the use of labeled primers or probes to detect amplification products that are not visible on agarose gels *(53,81)*.

5. Identification and Typing

5.1. Ribosomal Approaches for Identification

The presence of highly conserved regions in the ribosomal DNA has facilitated the development of primers to be used in a broad range of organisms (*see* **Chapter 1**). Because of the special features of rDNA, this region represents an attractive target for genotyping and also the analysis of phylogenetic relatedness by PCR *(82)*. The existence of an extensive catalog of bacterial 16S rDNA sequences in databanks such as Ribosomal Database Project (RDP), (http://rdp.cme.msu.edu/html/) permits rapid comparison of 16S sequences and development of primers for PCR assays at different phylogenetic levels.

5.2. Fingerprinting Methods for Typing

Quite a number of PCR fingerprinting methods have been developed since 1989. Generally, the methods can be subdivided into whole genome fingerprinting and fingerprinting of genomic regions containing repetitive elements.

The latter include enterobacterial repetitive intergenic consensus PCR (ERIC-PCR) *(83)*, repetitive extragenic palindromic PCR (REP-PCR) *(83)*, denaturing gradient gel electrophoresis (DGGE), *(84)*, terminal restriction fragment length polymorphism (T-RFLP) *(85)*, and amplified ribosomal DNA restriction analysis (ARDRA) *(86)*. Among whole genome fingerprinting PCR methods, random amplified polymorphic DNA (RAPD) *(87)*, and amplified fragment length polymorphism (AFLP) *(88)* are most common, but also arbitrarily primed-PCR (AP-PCR) *(89)* and universally primed PCR (UP-PCR) *(90)* are used.

RAPD, AP-PCR, and UP-PCR can all be used for typing of organisms without previous knowledge of DNA sequences. The use of one single primer leads to amplification of several DNA fragments randomly distributed throughout the genome. AFLP makes use of restriction enzyme digestion and addition of adapters for primer annealing before amplification of the DNA for differentiation. In all these methods, multiple bands are produced, and the fragments are separated by gel electrophoresis, providing direct analysis of polymorphism of different isolates. UP-PCR differs from the more well-known RAPD and AP-PCR techniques in the design of primers. The primers used in RAPD are short, usually 8–12-mers, with random sequence composition. The amplification is very sensitive to reaction conditions, in particular the annealing temperature. Primers in UP-PCR are usually 15–21-mers and have a unique design targeting mainly evolutionary younger intergenic segments of the genome. Based on the UP-PCR technique, different derivative methods have been developed. One of these, the species identification method, is a cross-blot hybridization variant, in which hybridization of UP-PCR products obtained from isolates of one species with the same primer reveals DNA homology *(91,92)*. The cross-blot hybridization variant has a potential as a DNA array-based typing method *(93)*. Another application is the development of diagnostics of specific strains by identification of unique markers that can be detected selectively by conversion of the marker into sequence characterized amplified method (SCAR) and using pairwise combinations of selected primers (SCAR primers) for amplification of a specific product *(94)*.

Fingerprinting methods are not suitable for direct detection of target organisms in complex diagnostic samples. These methods require pure cultures due to the nature of the primers that allow amplification from almost any organisms. The main advantage of the methods is their potential to differentiate very similar isolates of the same species, which makes them suitable for studies of microbial populations in molecular epidemiology.

6. Critical Parameters in PCR

6.1. Inhibition of DNA Amplification

Failure of DNA amplification due to the presence of inhibitory substances is a common problem in PCR and may, in some cases, be the main cause of false negative reactions (*see* **Chapter 2**) *(94)*. Inhibition influences the outcome of PCR by lowering the amplification efficiency or complete prevention of amplification. The inhibition acts at one or more essential stages of the reaction, the most important of which is inhibition of polymerase activity. Inhibition of PCR encountered in complex samples, such as blood, stool, serum, chicken carcass, soil, cheese, bean sprout, and oyster have been described *(95,96)*. It has been shown that a number of substances, frequently found in enrichment media, DNA preparation solutions and food samples, inhibit PCR *(97)*. Moreover, it was shown that different sample preparation procedures have different efficiency depending on the organism to be detected *(98,99)*. Therefore, research in sample preparation and standardization is still needed in order to exploit the full potential of PCR in microbiological diagnosis.

Not only the substances influence the outcome of the PCR, but the DNA polymerase itself can be a limiting factor because different brands of the enzyme can be inhibited by different inhibitors *(46,100)*. Therefore, inhibition of PCR in different biological samples can be reduced or eliminated by choosing an appropriate thermostable DNA polymerase without the need for extensive sample processing prior to PCR *(100)*. Although optimization and refinement of the PCR technique have been given much attention, the problems associated with the presence of PCR inhibitors in complex samples have not been overcome. This makes sample preparation steps prior to PCR a bottleneck in routine application and quality assurance of PCR-based methods.

6.2. Carry-over Contamination

Nucleic acid amplification techniques in general, and PCR in particular, are notoriously susceptible to contamination *(101,102)*. Once a reagent becomes contaminated with target DNA, the only solution is to discard it. It is often impossible to accurately determine which reagent is contaminated and when the contamination occurred. However, one of the greatest risks of contamination is the possibility of introducing a small amount of previously amplified DNA into a new reaction. This is referred to as "the carryover problem" *(103)*.

Enzymatic and chemical methods have been developed to control this phenomenon *(104,105)*. One approach is based on the use of the enzyme uracil-*N*-glycosylase (UNG), which preferentially cleaves uracil-containing DNA. As dTTP is replaced with dUTP in the master mixture *(104)*, UNG does not degrade native DNA templates or primers, but it cleaves the DNA products of previous amplifications at the uracil residues, rendering the contamination source non-amplifiable. UNG, which is present in the mixture from the beginning, is not active during the amplification reaction, so newly synthesized molecules are unaffected. The use of this technology is essentially invisible to the user, and no additional manipulations or equipment are necessary.

A simple chemical method for carryover control is based on the inactivation of amplified DNA using a photochemical technique based on isopsoralen derivatives *(105)*. Amplification is carried out in the presence of isopsoralen. At the end, the PCR tube is irradiated with UV light, and the isopsoralens form cyclobutane monoadducts with pyrimidine bases of the DNA. If this modified DNA is carried over into subsequent amplification reactions, it will not serve as a template for amplification. However, double PCR or nested PCR cannot be controlled by either of these procedures.

7. Further Prospects

7.1. Quantitation

There have been many efforts to use PCR for quantitation *(106–109)*. There will be a further increase in the popularity of real-time PCR as commercially available equipment has become more reliable and affordable. However, great care must be taken with optimization of reaction conditions and interpretation of quantitative PCR results. Minor differences in the efficiency of amplification among samples can give rise to markedly different amounts of the final product, due to exponential nature of amplification. Ferre *(110)* showed, for the first time that results obtained by quantitative PCR correlated well with results from hybridization experiments. Besnard and Andre *(111)* developed a quantitative method in which a competitor DNA was added to the sample at known concentrations before nucleic acid extraction, so that both targets would be co-processed and quantitative results be more reproducible.

The inclusion of known amounts of internal control DNA (competitors) is one of the most frequently used methodologies in quantitative PCR *(75,107,109)*. By using this strategy, it is possible to reproducibly quantify the amount of target DNA in an environmental sample by titrating unknown amounts of target DNA against a dilution series of competitor DNA *(66,75)*.

Quantitative competitive PCR (QC-PCR) is a method consisting of co-amplification, in the same tube, of two different templates of similar length carrying the same primer binding sites. This insures both templates being amplified at equal efficiency and the ratio of products remaining constant throughout the reaction. The relative amount of each product can be determined by ethidium bromide-stained gels or by counting bands radiolabeled with ^{32}P *(112)*.

7.2. Automation

Since the invention of thermal cyclers in connection with thermostable DNA polymerases, real-time detection equipment has introduced the second major automation event into PCR technology. The real-time mode of amplification has basically abolished the need to open PCR tubes following amplification, which is the main source of carryover.

In addition, application of solution hybridization probes in combination with fluorescence dyes can increase diagnostic specificity and sensitivity of PCR testing, making the use of nested PCR or double PCR unnecessary.

Finally, automation has increased the throughput capacity of PCR laboratories substantially, providing a real opportunity for cost-effective testing of large number of samples, with minimum requirements for skilled labor and large dedicated work spaces. The major automation effort of the future will be directed at pre-PCR sample treatment.

References

1. D'Aoust, J.-Y. (1994) *Salmonella* and the international food trade. *Int. J. Food Microbiol.* **24,** 11–31.
2. Hill, W. E., and Keasler, S. P. (1991) Identification of food borne pathogens by nucleic acid hybridization. *Int. J. Food Microbiol.* **12,** 67–76.
3. Saiki, R. K., Scharf, S. J, Faloona, F. A., et al. (1985) Enzymatic amplification of β-globin genomic sequences and restriction site analysis for diagnosis of sickle cell anemia. *Science* **230,** 1350–1354.
4. Mullis, K.B. and Faloona, F.A. (1987) Specific synthesis of DNA in vitro via a polymerase-catalyzed chain reaction. *Methods Enzymol.* **155,** 335–351.
5. Saiki, R. K., Gelfand, D. H., Stoffel, S., et al. (1988) Primer-directed enzymatic amplification of DNA with a thermostable DNA polymerase. *Science* **239,** 487–491.
6. Swaminathan, B., and Feng, P. (1994) Rapid detection of food-borne pathogenic bacteria. *Ann. Rev. Microbiol.* **48,** 401–426.
7. Scheu, P. M., Berghof, K., and Stahl, U. (1998). Detection of pathogenic and spoilage micro-organisms in food with the polymerase chain reaction. *Food Microbiol.* **15,** 13–31.
8. Easter, M. C. (1985) Rapid and automated detection of *Salmonella* by electrical measurements. *J. Hyg. Camb.* **94,** 245–262.
9. Ibrahim, G. F., and Fleet, G.,H. (1985) Detection of salmonellae using accelerated methods. *Int. J. Food Microbiol.* **2,** 259–272.
10. Beumer, R. R., Brinkman, E., and Rombouts F. M. (1991) Enzyme-linked immunoassays for the detection of *Salmonella* spp.: a comparison with other methods. *Int. J. Food Microbiol.* **12,** 363–374.
11. Ellison, A., Perry, S.F. and Stewart, G.S.A.B. (1991) Bioluminescence as a real-time monitor of injury and recovery in *Salmonella typhimurium*. *Int. J. Food Microbiol.* **12,** 323–332.
12. Blackburn, C. de W. (1993) Rapid and alternative methods for the detection of salmonellas. *J. Appl. Bacteriol.* **75,** 199–214.

13. Hanai, K., Satake, M., Naakanishi, H., and Venkateswaran, K. (1997) Comparison of commercially available kits for detection of Salmonella strains in foods. *Appl. Environ. Microbiol.* **63**, 775–778.

14. Vaneechoutte, M. and Van Eldere, J. (1997) The possibilities and limitations of nucleic acid amplification technology in diagnostic microbiology. *J. Med. Microbiol.* **46**, 188–194.

15. Olsen, J. E., Aabo, S., Hill, W., et al. (1995) Probes and polymerase chain reaction for detection of food-borne bacterial pathogens. *Int. J. Food Microbiol.* **28**, 1–78.

16. Hoorfar, J., Ahrens, P., and Rådström, P. (2000) Automated 5' nuclease PCR assay for identification of *Salmonella enterica*. *J. Clin. Microbiol.* **38**, 3429–3435.

17. Kuhnert, P., Boerlin, P. and Frey, J. (2000) Target genes for virulence assessment of Escherichia coli isolates from water, food and the environment. *FEMS Microbiol. Rev.* **24**, 107–117.

18. Gyles, C.L. (1992) *Escherichia coli* cytotoxins and enterotoxins. *Can. J. Microbiol.* **38**, 734–746.

19. Newland, J. W., and Neil, R.J. (1988) DNA probes for shiga-like toxins I and II and for toxin converting bacteriophages. *J. Clin. Microbiol.* **26**, 1292–1297.

20. Ojeniyi, B., Ahrens, P., and Meyling, A. (1994) Detection of fimbrial and toxin genes in *Escherichia coli* and their prevalence in piglets with diarrhoea. The application of colony hybridization assay, polymerase chain reaction and phenotypic assays. *Zentralbl. Veterinarmed* **41**, 49–59.

21. On, S. L. W. (1997) Identification methods for Campylobacter, Helicobacters, and related organisms. *Clin. Microbiol. Rev.* **9**, 405–422.

22. Goossens, H., and Butzler, J. P. (1992) Isolation and identification of *Campylobacter* spp., in *Campylobacter jejuni. Current status and future trends.* (Nachamkin, I., Blaser M.J., Tompkins, L.S., eds.), ASM Press, Washington DC, USA, pp. 93–109.

23. Romaniuk, P. J., and Trust, T.J. (1987) Rapid identification of *Campylobacter* species using oligonucleotide probes to 16S ribosomal RNA. *Mol. Cell. Probes.* **3**, 133–142.

24. Eyers, M., Chapeeli, S., van Gamp, G., Goosens, H., and DeWachter, R. (1993) Discrimination among thermophilic *Campylobacter* species by polymerase chain reaction amplification of 23s rRNA gene fragments. *J. Clin. Microbiol.* **31**, 3340–3343.

25. Lübeck, P. S., On, S. L. W., and Hoorfar, J. (2001) Development of a PCR detection method for *Campylobacter jejuni*, *C. coli* and *C. lari* in foods. *Int. J. Med. Microbiol.* **291**, 110.

26. Pedersen, K. B. (1979) Occurrence of *Yersinia enterocolitica* in the throat of swine. *Contr. Microbiol. Immunol.* **5**, 253–256.

27. Nesbakken, T., and Kapperud, G. (1985) *Yersinia enterocolitica* and *Yersinia enterocolitica*-like bacteria in Norwegian slaughter pigs. *Int. J. Food Microbiol.* **1**, 301–309.

28. Hoorfar, J. and Holmvig, C.B.F. (1999) Evaluation of culture methods for rapid screening of swine faecal samples for *Yersinia enterocolitica* O:3 / biotype 4. *Vet. Med. B* **46**, 189–198.

29. Kapperud, G., Dommersnes, K., Skurnik, M., and Hornes, E. (1990) A synthetic oligonucleotide probe and a cloned polynucleotide probe based on the yopA gene for detection and enumeration of virulent *Yersinia enterocolitica*. *Appl. Environ. Microbiol.* **56**, 17–23.

30. Wren, B. W., and Tabaqchali, S. (1990) Detection of pathogenic *Yersinia enterocolitica* by the polymerase chain reaction. *Lancet* **336**, 693.

31. Fenwick, S. G., and Murray, A. (1990) Detection of pathogenic *Yersinia enterocolitica* by polymerase chain reaction. *Lancet* **337**, 496–497.

32. Rasmussen, H. N., Rasmussen, O. F., Christensen, H., and Olsen, J.E. (1995) Detection of *Yersinia enterocolitica* O:3 in faecal samples and tonsil swabs from pigs using IMS and PCR. *J. Appl. Bacteriol.* **78**, 563–568.

33. Kapperud, G., Vardund, T., Skjerve, E., Hornes, E., and Michaelsen, T. E. (1993) Detection of pathogenic *Yersinia enterocolitica* in foods and water by immunomagnetic separation, nested polymerase chain reactions, and colorimetric detection of amplified DNA. *Appl. Environ. Microbiol.* **59**, 2938–2944.

34. Lambertz, S. T., Ballagi-Pordány, A., Nilsson, A., Norberg, P., and Tham, M. L. M. (1996) A comparison between PCR method and a conventional culture method for detecting pathogenic *Yersinia enterocolitica* in food. *J. Appl. Bacteriol.* **81**, 303–308.

35. Gelling, B. G., and Broome C.V. (1989) Listeriosis. *JAMA* **261**, 1313–1320.

36. Vazquez-Bolan, J. A., Kuhn, M., Berche, P., Chakraborty, T., Domínguez-Bernal, G., Goebel, W., Gonzalez-Zorn, B., Wehland, J., and Kreft, J. (2001) *Listeria* pathogenesis and molecular virulence determinants. *Clin. Microbiol. Rev.* **14**, 584–640.

37. Graham, T., Golsteyn-Thomas, E. J., Gannon, V. P., and Thomas, J. E. (1996) Genus- and species-specific detection of *Listeria monocytogenes* using polymerase chain reaction assays targeting the 16S/23S intergenic spacer region of the rRNA operon. *Can. J. Microbiol.* **42**, 1155–1162.

38. Ericsson, H. and Stalhandske, P. (1997) PCR detection of *Listeria monocytogenes* in 'gravad' rainbow trout. *Int. J. Food Microbiol.* **35**, 281–285.

39. Manzano, M., Cocolin, L., Cantoni, C., and Comi, G. (1997) Detection and identification of Listeria monocytogenes from milk and cheese by a single-step PCR. *Mol. Biotech.* **7**, 85–88.

40. Wolcott, M. J. (1992) Advances in nucleic acid-based detection methods. *Clin. Microbiol. Rev.* **5**, 370–386.

41. Gingeras, T. R., Dichman, D. D., Kwoth, D. Y., and Guatelli, J. C. (1990) Methodologies for in vitro nucleic acid amplification and their application. *Vet. Microbiol.* **24**, 235–251.

42. Taylor, G. R. (1991) Polymerase chain reaction: basic principles and automation, in *PCR—A practical approach* (McPherson, M. J., Quirke, P., and Taylor, G. R., eds.), Oxford University Press, Oxford, UK, pp. 1–14.

43. Mitchell, T. G., Freedman, E. Z., White, T. J., and Taylor, J. W. (1994) Unique oligonucleotide primers in PCR for identification of *Cryptococcus neoformans*. *J. Clin. Microbiol.* **32,** 253–255.

44. Henson, J. M., and French, R. (1993) The polymerase chain reaction and plant disease diagnosis. *Ann. Rev. Phytopathol.* **31,** 81–109.

45. Ward, E. (1994) Use of the polymerase chain reaction for identifying plant pathogens, in *Ecology of Plant Pathology* (Blakeman, J. P., and Williamsen, B. eds.), CAB International, USA, 143–160.

46. Abu Al-Soud, W. A., and Rådström, P. (1998) Capacity of nine thermostable DNA polymerases to mediate DNA amplification in the presence of PCR-inhibiting samples. *Appl. Environ. Microbiol.* **64,** 3748–3753.

47. Kreader, C.A. (1996) Relief of amplification inhibition in PCR with bovine serum albumin or T4 gene 32 protein. *Appl. Environ. Microbiol.* **62,** 1102–1106.

48. Saiki, R. K. (1989) *The Design and optimization of the PCR*, (Erlich, H. A., ed.), Stockton Press, New York, NY, USA, pp 7–16.

49. Ward, E. (1995) Improved polymerase chain reaction (PCR) detection of *Gaumannomyces graminis* including a safeguard against false negatives. *European J. Plant Pathol.* **101,** 561–566.

50. Suggs, S. V., Wallace, R. B., Hirose, T., Kawashima, E. H., and Itakura, K. (1981) Use of synthetic oligonucleotides as hybridization probes: isolation of cloned cDNA sequences for human beta 2-microglobulin. *Proc. Natl. Acad. Sci. USA* **78,** 6613–6617.

51. Schaad, N. W., Cheong, S. S., Tamaki, S., Hatziloukas, and Panopoulos, N. J. (1995) A combined biological and enzymatic amplification (BIO-PCR) technique to detect *Pseudomonas syringae* pv. *phaseolicola* in bean seed extracts. *Phytopathology* **85,** 243–248.

52. Innis, M. A. and Gelfand, D. H. (1990) Optimization of PCRs, in *PCR Protocols. A Guide to Methods and Applications* (Innis, M. A., Gelfand, D. H., Sninsky, J. J., and White, T.J ., eds.), Academic Press, San Diego, CA, USA, pp. 3–12.

53. Steffan, R. J. and Atlas, R. M. (1988) DNA amplification to enhance detection of genetically engineered bacteria in environmental samples. *Appl. Environ. Microbiol.* **54,** 2185–2191.

54. McManus, P. S., and Jones, A. L. (1995) Detection of *Erwinia amylovora* by nested PCR and PCR-dot-blot and reverse-blot hybridizations. *Phytopathology* **85,** 618–623.

55. Lübeck, M., and Lübeck, P.S. (1996) PCR - a promising tool for detection and identification of fungi in soil, in *Developments in Plant Pathology: Monitoring Antagonistic Fungi Deliberately Released into the Environment* (Jensen, D. F., Jansson, H. -B., and Tronsmo, A., eds.), Kluwer Academic Publishers, Norwell, MA, USA, pp. 113–122.

56. Zolg, J.W., and Philippi-Schulz, S. (1994) The superoxide dismutase gene, a target for detection and identification of mycobacteria by PCR. *J. Clin. Microbiol.* **32,** 2801–2812.

57. Bulat, S. A., Lübeck, M., Alekhina, I. A., Knudsen, I. M. B., Jensen, D. F. and Lübeck, P. S. (2000) Identification of an UP-PCR derived SCAR marker for an antagonistic strain of *Clonostachys rosea* and development of a strain-specific PCR detection assay. *Appl. Environ. Microbiol.* **66**, 4758–4763.

58. Louws, F. J., Fulbright, D. W., Stephens, C. T., and de Bruijn, F.J. (1995) Differentiation of genomic structure by rep-PCR fingerprinting to rapidly classify *Xanthomonas campestris* pv. *vesicatoria. Phytopathology* **85**, 528–536.

59. Meyer, W., Mitchell, T. G., Freedman, E. Z. and Vilgalys, R. (1993) Hybridization probes for conventional DNA fingerprinting used as single primers in the polymerase chain reaction to distinguish strains of *Cryptococcus neoformans. J. Clin. Microbiol* **31**, 2274–2280.

60. Edel, V., Steinberg, C., Avelange, I., Laguerre, G., and Alabouvette, C. (1995) Comparison of three molecular methods for the characterization of *Fusarium oxysporum* strains. *Phytopathology* **85**, 579–585.

61. Doolittle, R. F. (1990) Molecular evolution: computer analysis of protein and nucleic acid sequences, *Methods Enzymol.* **183**

62. Wheeler, W. C. (1994) Sources of ambiguity in nucleic acid sequence alignment, in *Molecular Ecology and Evolution: Approaches and Applications* (Schierwater, B., Streit, B., Wagner, G. P., and DeSalle, R., eds.), Birkhäuser Verlag, Basel, pp. 323–352.

63. McPherson, M. J., Jones, K. M., and Gurr, S. J. (1994) PCR with highly degenerate primers, in *PCR—A Practical Approach* (McPherson, M. J., Quirke, P., and Taylor, G. R., eds.), The Practical Approach Series Vol. 1, Oxford University Press, UK, pp. 171–186.

64. Lübeck, P. S., Paulin, L., Degefu, Y., Lübeck, M., Alekhina, I. A., Bulat, S. A., and Collinge, D. B. (1997) PCR cloning, DNA sequencing and phylogenetic analysis of a xylanase gene from the phytopathogenic fungus *Ascochyta pisi* Lib. *Physiol. Mol. Plant Pathol.* **51**, 377–389.

65. Bej, A. K., McCarty, S. C., and Atlas, R. M. (1991) Detection of coliform bacteria and *Escherichia coli* by multiplex polymerase chain reaction: comparison with defined substrate and plating methods for water quality monitoring. *Appl. Environ. Microbiol.* **57**, 2429–2432.

66. Steffan, R. J., and Atlas, R. M. (1991) Polymerase chain reaction: applications in environmental microbiology. *Ann. Rev. Microbiol.* **45**, 137–162.

67. Lee Lang, A., Tsai, Y. -L., Mayer, C. L., Patton, K. C. and Palmer, C. J. (1994) Multiplex PCR for detection of the heat-labile toxin gene and Shiga-like toxin I and II genes in *Escherichia coli* isolated from natural waters. *Appl. Environ. Microbiol.* **60**, 3145–3149.

68. Geha, D. J., Uhl, J. R., Gustaferro, C. A., and Persing, D. H. (1994) Multiplex PCR for identification of methicillin-resistant Staphylococci in the clinical laboratory. *J. Clin. Microbiol.* **32**, 1768–1772.

69. Liébana, E., Aranaz, A., Mateos, A., et al. (1995) Simple and rapid detection of Mycobacterium tuberculosis complex organisms in bovine tissue samples by PCR. *J. Clin. Microbiol.* **33**, 33–36.

70. Lawrence, L. M., and Gilmour, A. (1994) Incidence of *Listeria* spp. and *Listeria monocytogenes* in a poultry processing environment and in poultry products and their rapid confirmation by multiplex PCR. *Appl. Environ. Microbiol.* **60**, 4600–4604.

71. Tsushima, S., Hasebe, A., Komoto, Y., et al. (1995) Detection of genetically engineered microorganisms in paddy soil using a simple and rapid "nested" polymerase chain reaction method. *Soil Biol. Biochem.* **27**, 219–227.

72. Rådström, P., Bäckman, A., Qian, N., Kragsbjerg, P., Påhlson, C., and Olcén, P. (1994) Detection of bacterial DNA in cerebrospinal fluid by an assay for simultaneous detection of *Neisseria meningitidis, Haemophilus influenzae*, and streptococci using a seminested PCR strategy. *J. Clin. Microbiol.* **32**, 2738–2744.

73. Straub, T.M., Pepper, I.L., Abbaszadegan, M. and Gerba, C.P. (1994) A method to detect enteroviruses in sewage sludge-amended soil using the PCR. *Appl. Environ. Microbiol.* 60, 1014–1017.

74. Catalan, V., Moreno, C., Dasi, M.A., Munoz, C. and Apraiz, D. (1994) Nested polymerase chain reaction for detection of *Legionella pneumophila* in water. *Res. Microbiol.* **145**, 603–610.

75. Gilliland, G., Perrin, S., and Bunn, H. (1990) Competitive PCR for quantitation of mRNA, in *PCR Protocols. A Guide to Methods and Applications*, (Innis, M. A., Gelfand, D. H., Sninsky, J. J., and White, T.J., eds.), Academic Press, San Diego, CA, USA, pp. 60–69.

76. Robinson, D. J. (1992) Detection of tobacco rattle virus by reverse transcription and polymerase chain reaction. *J. Virol. Methods* **40**, 57–66.

77. Tsai, Y. -L., Tran, B., Sangermano, L. R., and Palmer, C. J. (1994) Detection of poliovirus, hepatitis A virus, and rotavirus from sewage and ocean water by triplex reverse transcriptase PCR. *Appl. Environ. Microbiol.* **60**, 2400–2407.

78. Gilgen, M., Wegmüller, B., Burkhalter, P., et al. (1995) Reverse transcription PCR to detect enteroviruses in surface water. *Appl. Environ. Microbiol.* **61**, 1226–1231.

79. Bej, A. K., and Mahbubani, M.H. (1992) Applications of the polymerase chain reaction in environmental microbiology. *PCR Methods Applications* **1**, 151–159.

80. Arai, M., Mizukoshi, C., Kubochi, F., Kakutani, T., and Wataya, Y. (1994) Detection of *Plasmodium falciparum* in human blood by a nested polymerase reaction. *Am. J. Trop. Med. Hyg.* **51**, 617–626.

81. Schraft, H., and Griffiths, M.W. (1995) Specific oligonucleotide primers for detection of lecithinase-positive *Bacillus* spp. by PCR. *Appl. Environ. Microbiol.* **61**, 98–102.

82. Anderson, B. (1998) Identifying noval bacteria using a broad-range polymerase chain reaction, in *Rapid Detection of Infectious Agents* (Specter, S., Bendinelli, M., and Friedman, H., eds.), Plenum Press, New York and London, pp. 117–129.

83. de Bruijn, F. J. (1992) Use of repetitive (repetitive extragenic palindromic and enterobacterial repetitive intergenic consensus) sequences and the polymerase chain reaction to fingerprint the genomes of Rhizobium meliloti isolates and other soil bacteria. *Appl. Environ. Microbiol.* **58**, 2180–2187.

84. Fischer, S. G., and Lerman, L. S. (1983) DNA fragments differing by single basepair substitutions are separated in denaturing gradient gels: correspondence with melting theory. *Proc. Natl. Acad. Sci. USA* **80,** 1579–1583.

85. Liu, W. -T., Marsh, L. T., Cheng, H., and Forney, L. J. (1997) Characterization of microbial diversity by determining terminal restriction fragment length polymorphisms of genes encoding 16S rRNA. *App. Environ Microbiol.* **63,** 4516–4522.

86. Moyer, C. L., Dobbs, F. C., and Karl, D. M. (1994) Estimation of diversity and community structure through restriction fragment length polymorphism distribution analysis of bacterial 16S rRNA genes from a microbial mat at an active, hydrothermal vent system, Loihi Seamount, Hawaii. *Appl. Environ. Microbiol.* **60,** 871–879.

87. Williams, J. G. K., Kubelik, A. R., Livak, K. J., Rafalski, J. A., and Tingey, S. V. (1990) DNA polymorphisms amplified by arbitrary primers are useful as genetic markers. *Nucleic Acids Res.* **18,** 6531–6535.

88. Vos, P., Hogers, R., Bleeker, M., et al. (1995) AFLP: A new technique for DNA fingerprinting. *Nucleic Acids Res.* **23,** 4407–4414

89. Welsh, J., and McClelland, M. (1990) Fingerprinting genomes using PCR with arbitrary primers. *Nucleic Acids Res.* **18,** 7213–7218.

90. Bulat, S. A., and Mironenko, N.V. (1990) Species identity of the phytopathogenic fungi *Pyrenophora teres* Dreschler and *P. graminea* Ito and Kuribayashi. *Mikol. Fitopatol.* **24,** 435–441. (In Russian)

91. Bulat, S. A., Lübeck, M., Mironenko, N. V., Jensen, D. F., and Lübeck, P.S. (1998) UP-PCR analysis and ITS1 ribotyping of *Trichoderma* and *Gliocladium* fungi. *Mycol. Res.* **102,** 933–943.

92. Lübeck, M., Alekhina, I. A., Lübeck, P. S., Jensen, D.F., and Bulat, S. A. (1999) Delineation of *Trichoderma harzianum* Rifai into two different genotypic groups by a highly robust fingerprinting method, UP-PCR, and UP-PCR product cross-hybridisation. *Mycol. Res.* **103,** 289–298.

93. Lübeck, M., and Poulsen, H. (2001) UP-PCR and UP-PCR cross blot hybridization as a tool for studying anastomosis group relationship in the *Rhizoctonia solani* complex. *FEMS Microbiol. Lett.* **201,** 83–89.

94. Wilson, I.G. (1997) Inhibition and facilitation of nucleic acid amplification. *Appl. Environ. Microbiol.* **63,** 3741–3751.

95. Wernars, K,. Heuvelman, C. J., Chakraborty, T., and Notermans S. H. W. (1991) Use of the polymerase chain reaction for direct detection of *Listeria monocytogenes* in soft cheese. *J. Appl. Bacteriol.* **70,** 121–126.

96. Greenfield, L., and White, T. J. (1993) Sample preparation methods, in *Diagnostic Molecular Microbiology* (Persing, D. H., Smith, T. F., Tenover, F. C., and White, T. J. eds.), ASM Press, Washington DC, USA.

97. Rossen, L., Nørskov, P., Holmstrøm, K., and Rasmussen, O. F. (1992) Inhibition of PCR by components of food samples, microbial diagnostic assays and DNA extraction solutions. *Int. J. Food Microbiol.* **15,** 37–45.

98. Lantz, P.G., Hahn-Hägerdal, B., and Rådström, P. (1994) Sample preparation methods in PCR-based detection of food pathogens. *Trends Food Sci. Technol.* **5**, 384–389.

99. Lantz, P. G., Matsson, M., Wadström, T., and Rådström, P. (1997) Removal of PCR inhibitors from human fecal samples through the use of an aqueous two-phase system for sample preparation prior to PCR. *J. Microbiol. Methods* **28**, 159–167.

100. Lantz, P. G., Knutsson, R., Blixt, Y., Abu Al-Soud, W., Borch, E, and Rådström P. (1998) Detection of pathogenic *Yersinia enterocolitica* in enrichment media and pork by a multiplex PCR: a study of sample preparation and PCR-inhibitory components. *Int. J. Food Microbiol.* **45**, 93–105.

101. Dragon, E. A., Spadoro, J. P., and Madej R. (1993) Quality control of polymerase chain reaction. in *Diagnostic Molecular Microbiology*, (Persing, D. H., Smith, T. F., Tenover, F. C,. and White, T.J. eds.), ASM Press, Washington DC, USA.

102. Kitchin, P. A., and Bootman, J.S. (1993) Quality assurance of the polymerase chain reaction. *Med. Virol.* **3**, 107–114.

103. Ivinson, A. J., and Taylor, G. R. (1991) PCR in genetic diagnosis, in *PCR— A Practical Approach*, (McPherson, M. J., Quirke, P., and Taylor, G. R. eds.), Oxford University Press, USA, pp 15–27.

104. Longo, M., Berninger, M. S., and Hartley, J.L. (1990) Use of uracil DNA glycosylase to control carry-over contamination in polymerase chain reaction. *Gene* **93**, 125–128.

105. Cimino, G. D., Metchette, K. C., Tessman, J. W., Hearst, J. E., and Isaacs, S. T. (1991) Post-PCR sterilization: a method to control carryover contamination for the polymerase chain reaction. *Nucleic Acids Res.* **19**, 99–107.

106. Picard, C., Ponsonnet, C., Paget, E., Nesme, X., and Simonet, P. (1992) Detection and enumeration of bacteria in soil by direct DNA extraction and polymerase chain reaction. *Appl. Environ. Microbiol.* **58**, 2717–2722.

107. Moukhamedov, R., Hu, X., Nazar, N. and Robb, J. (1994) Use of polymerase chain reaction-amplified ribosomal intergenic sequences for the diagnosis of *Verticillium tricorpus*. *Phytopathology* **84**, 256–259.

108. Degrange, V., and Bardin, R. (1995) Detection and counting of *Nitrobacter* populations in soil by PCR. *Appl. Environ. Microbiol.* **61**, 2093–2098.

109. Lamar, R. T., Schoenike, B., Vanden Wymelenberg, A., Stewart, P., Dietrich, D. M., and Cullen, D. (1995) Quantitation of fungal mRNAs in complex substrates by reverse transcription PCR and its application to *Phanerochaete chrysosporium*-colonized soil. *Appl. Environ. Microbiol.* **61**, 2122–2126.

110. Ferre, F. (1992) Quantitative or semi-quantitative PCR: reality versus myth. *PCR Methods Applications* **2**, 1–9.

111. Besnard, N. C., and Andre, P. M. (1994) Automated quantitative determination of hepatitis C virus viremia by reverse transcription-PCR. *J. Clinical. Microbiol.* **32**, 1887–1893.

112. Merzouki, A., Mo, T., Vellani, N., et al. (1994) Accurate and differential quantification of HIV-1 *tat*, *rev* and *nef* mRNAs by competitive PCR. *J. Virol. Methods* **50**, 115–128.

II

PROTOCOLS

5

Detection, Identification, and Subtyping of *Actinobacillus pleuropneumoniae*

Joachim Frey

1. Introduction

Actinobacillus pleuropneumoniae is the etiological agent of porcine pleuropneumonia, which causes significant losses in industrialized swine production worldwide. Bacterial diagnosis of contagious porcine pleuropneumonia is generally done by bacteriological isolation and cultivation of *A. pleuropneumoniae*, followed by serological typing which differentiates 15 serotypes *(1–3)*. There are significant differences in virulence among the 15 serotypes. Serotypes 1, 5, and also 9 and 11 are involved in particularly severe outbreaks of disease with high mortality and severe pulmonary lesions. Serotypes 2–4, 6–8, 12, and 15 are generally less virulent, causing moderate mortality but relatively strong lung lesions. Serotype 3 seems to be mostly of low epidemiological importance, while the remaining serotypes are isolated only very rarely. The degree of virulence seems to be mainly due to the presence of one or two of the pore-forming RTX toxins ApxI, ApxII, and ApxIII, found in *A. pleuropneumoniae*, which characterize the different serotypes of *A. pleuropneumoniae* *(4,5)* (*see* **Table 1**). Toxin ApxI is encoded by the genes *apxICABD*, with gene *A* specifying the structural protein toxin, *C* coding for its activator, and *B* and *D* coding for the corresponding type I secretion pathway. ApxI is produced and secreted by the highly pathogenic serotypes 1, 5, 9, and 11, and also in serotype 10 and 14 of *A. pleuropneumoniae*. ApxII, encoded by *apxIICA*, is produced by all serotypes except 10 and 14. However, serotype 3 is unable to actively secrete ApxII due to the lack of the *apxIBD* genes, which specify the type I secretion pathway for the export of the hemolytic and cytotoxic ApxI and ApxII. The cytotoxic ApxIII, encoded by the operon *apxIIICABD*, is secreted by serotypes 2, 3, 4, 6, 8, and 15. ApxIII seems to be involved in virulence together with ApxII. Consequently,

From: *Methods in Molecular Biology, vol. 216: PCR Detection of Microbial Pathogens: Methods and Protocols*
Edited by: K. Sachse and J. Frey © Humana Press Inc., Totowa, NJ

Table 1
Presence of the Various *apx* genes in Different Serotypes
of *A. pleuropneumoniae* and Apx Toxin that is Actively Secreted

Serotype	Reference strain	*apxI* operon	*apxII* operon	*pxIII* operon	Secreted Apx toxin
1	4074	C A B D	C A		I + II
2	S1536	B D	C A	C A B D	II + III
3	S1421	C A		C A B D	III
4	M62	B D	C A	C A B D	II + III
5a	K17	C A B D	C A		I + II
5b	L20	C A B D	C A		I + II
6	Femø	B D	C A	C A B D	II + III
7	WF83	B D	C A		II
8	405	B D	C A	C A B D	II + III
9	CVI 13261	C A B D	C A		I + II
10	13039	C A B D			I
11	56153	C A B D	C A		I + II
12	8329	B D	C A		II
13	N273	B D	C A		II
14	3906	C A B D			I
15	HS143	B D	C A	C A B D	II + III

strains secreting both ApxII and ApxIII are more pathogenic than those secreting ApxII alone, such as serotypes 7 and 12. Hence, typing of *A. pleuropneumoniae* on the basis of their ability to secrete the three main toxins, or on the basis of the presence of the genes encoding the corresponding structural toxins A, their activators C, and their secretion proteins B and D, respectively, is a useful method to subtype *A. pleuropneumoniae*. This method can, to a certain extent, substitute serotyping of *A. pleuropneumoniae* strains.

Variants encoding the toxins ApxI, ApxII, and ApxIII are also found in a few other *Actinobacillus* species, such as *A. suis* and *A. rossii*, as well as yet unidentified *Actinobacillus* species, which are occasionally isolated from swine (*6*). A fourth toxin, ApxIV, encoded by *apxIVA*, was shown to be specific to the species *A. pleuropneumoniae* (*7*) and was not found in other *Actinobacillus* species. Hence, *apxIVA* was proposed as a genetic target for the identification of the species *A. pleuropneumoniae*. The current report describes a polymerase chain reaction (PCR) method for the identification of the species *A. pleuropneumoniae* and its subtyping into major toxin groups (*5,8*). A second method consisting of a nested PCR, based on the *apxIV* gene, is devoted to the detection of *A. pleuropneumoniae* in lung tissue and in nasal secretory fluid (*9*).

Table 2
Sequences of Oligonucleotide Primers^a Used for PCR and Amplicon Sizes

Amplified gene	Oligonucleotide	Sequence (5'–3')	Fragment (bp)
ApxICA	XICA-L	TTGCCTCGCTAGTTGCGGAT	2420
	XICA-R	TCCCAAGTTCGAATGGGCTT	
apxIICA	XIICA-L	CCATACGATATTGGAAGGGCAAT	2088
	XIICA-R	TCCCCGCCATCAATAACGGT	
apxIIICA	XIIICA-L	CCTGGTTCTACAGAAGCGAAAATC	1755
	XIIICA-R	TTTCGCCCTTAGTTGGATCGA	
apxIBD	XIBD-L	GTATCGGCGGGATTCCGT	1447
	XIBD-R	ATCCGCATCGGCTCCCAA	
apxIIIBD	XIIIBD-L	TCCAAGCATGTCTATGGAACG	968
	XIIIBD-R	AATTAAATGACGTCGGCCAGTC	
ApxIVA v	APXIVA-1L	TGGCACTGACGGTGATGAT	441
	APXIVA-1R	GGCCATCGACTCAACCAT	
apxIVA nested	APXIVAN-1L	GGGGACGTAACTCGGTGATT	377
	APXIVAN-1R	GCTCACCAACGTTTGCTCAT	

^{a} The extension L is used for the forward primer, and R for the reverse primer (*see* **Note 1**).
^{b} *See* **Note 2**.

2. Materials

1. *A. pleuropneumoniae* serotype 1 reference strain 4074 and serotype 2 reference strain S 1536 as positive control samples.
2. Heating block for Eppendorf® tubes at 56°C, 60°C, and at 95°C.
3. Vortex blender.
4. Safety cabinet for PCR preparations ("Clean Spot"; Coy Laboratories, Grass Lake, MI, USA) .
5. Eppendorf centrifuge (relative centrifugal force [RCF] ≥ 13000 *g*) and Eppendorf tubes.
6. Agarose gel equipment including agarose, TBE running buffer (90 m*M* Tris-base, 90 m*M* boric acid, 2 m*M* EDTA, pH 8.3), gel loading buffer, ethidium bromide, DNA size marker as well as photographic equipment for documentation (*see* **ref. *[10]***).
7. Cotton swabs.
8. Thermal cycler and corresponding thin-wall tubes for reactions.
9. Phosphate-buffered saline (PBS) buffer: 50 m*M* Na-phosphate, pH 7.5, 750 m*M* NaCl.
10. Instagene matrix (Bio-Rad, Hercules, CA, USA).
11. Oligonucleotide primers, each 10 μM (*see* **Table 2**).
12. 10X PCR buffer: 10 m*M* Tris-HCl, pH 8.3, 1.5 m*M* MgCl$_2$, 50 m*M* KCl, 0.005 % Tween® 20.

13. Lysis buffer: 100 m*M* Tris-HCl, 0.05 % Tween 20, 0.2 mg/mL proteinase K, pH 8.5.
14. dATP 100 m*M*, dCTP 100 m*M*, dGTP 100 m*M*, dTTP 100 m*M*.
15. *Taq* DNA polymerase 5 U/μl.

3. Methods

The methods described below are aimed at (*i*) the identification and toxin gene typing of *A. pleuropneumoniae* isolates from cultures and (*ii*) the direct detection of *A. pleuropneumoniae* by PCR in lung tissue and nasal secretory fluid.

3.1. Identification and Toxin Gene Typing

Identification of the species *A. pleuropneumoniae* and toxin gene typing of an isolate or a strain from culture on solid medium (e.g., blood agar, chocolate agar, or Columbia agar-βNAD) is performed by preparing a lysate of the culture followed by six individual PCRs using the lysed bacteria as template. As positive controls, lysates from *A. pleuropneumoniae* cultures of serotype 1 strain 4074 and serotype 2 strain S1536 are used.

3.1.1. Preparation of Lysates

Lysates are made from all strains to be tested and from each of the two control strains.

1. Add 450 μL lysis buffer to an Eppendorf tube.
2. Resuspend 4 to 5 bacterial colonies.
3. Vortex mix strongly, to obtain a good suspension.
4. Incubate tube at 60°C for 60 min (heating block).
5. Vortex mix strongly.
6. Incubate at 95°–97°C for 15 min (inactivate proteinase).
7. Use lysates directly as templates for PCR, or store at −20°C.

3.1.2. Preparation of PCR Premix

For each specific PCR (*see* **Table 2**), 1 mL PCR premixture is first prepared in the safety cabinet, taking all necessary precautions to prevent contamination. Then, PCRs are made for each of the genes *apxICA*, *apxIICA*, *apxIIICA*, *apxIBD*, *apxIIIBD*, and for *apxIVA* (*see* **Note 1**). The PCR based on *apxIV* is made to confirm the species *A. pleuropneumoniae* (*see* **Notes 2** and **3**).

PCR Premixture (label tube with corresponding gene pairs to be amplified):

H₂O bidistilled	843 μL
10X PCR buffer	100 μL
dATP 100 m*M*	1.7 μL
dCTP 100 m*M*	1.7 μL
dGTP 100 m*M*	1.7 μL
dTTP 100 m*M*	1.7 μL
Primer-L 10 m*M* (*see* **Table 2**)	25 μL
Primer-R 10 m*M*(*see* **Table 2**)	25 μL

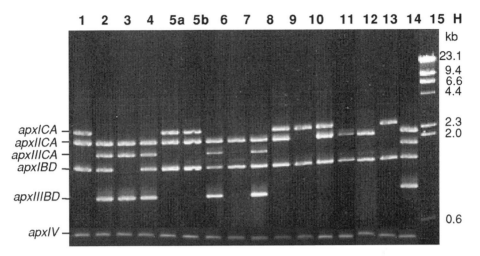

Fig. 1. Agarose gel electrophoresis of the PCR products obtained with the specific primers for genes *apxICA*, *apxIBD*, *apxIICA*, *apxIIICA* and *apxIIIBD* for the 15 different serotypes and subtypes 5a and 5b. Note that the PCR amplification products of the different individual reactions are united and analyzed in the same slot of the gel. Lanes 1–15, PCR products obtained with the reference strains of the different serotypes, respectively. H, bacteriophage λ DNA digested with *Hin*dIII and used as molecular size standard.

3.1.3. PCR and Electrophoretic Analysis

1. Use thin-walled thermal cycler tubes cooled on ice, or cooling block (4°C).
2. Add for each reaction: corresponding PCR premixture (47.0 µL), *Taq* DNA polymerase (0.25 µL), and lysate (2.0 µL).
3. Pre-heat thermal cycler to 95°C.
4. Place the cooled tubes directly into the thermal cycler.
5. Run 35 cycles with the parameters 30 s at 94°C, 30 s at 54°C, 1.5 min at 72°C.
6. Take the tubes to a different room for analysis of the amplicons.
7. Analyze 5 µL of each PCR amplicons on a 0.7 % agarose gel.
8. For a given strain, the PCR amplicons (5 µL of each) can be united and run together on the electrophoresis, since the products have a different length (*see* **Fig. 1**).
9. Stain the gel and photograph.
10. Analyze the results and determine the toxin type according to the patterns of **Fig. 1.** and toxin data of **Table 1**.

3.2. Detection of A. pleuropneumoniae in Lung Tissue and Nasal Secretory Fluid

For the detection of *A. pleuropneumoniae* in lung tissue and nasal secretory fluid, DNA of the samples is recovered, using Instagene matrix prior to the use as template for PCR. In addition, a nested PCR based on the *apxIVA* gene is required in order to attain an appropriate sensitivity of the test. The method allows the detection of *A. pleuropneumoniae* bacteria in lung tissue and nasal secretory fluid. For subtyping, however, it is necessary to isolate *A. pleuropneumoniae* by culture *(9)*.

3.2.1. Preparation of Template from Lung Tissue (see **Note 4**)

A. *pleuropneumoniae* can be detected directly from pathological lung lesions by PCR. The preparation of the template DNA is done using the following method:

1. Excise 0.25 g lung tissue from the edge of a lesion.
2. Place in a sterile tube and homogenize by hand using a sterile pestle.
3. Add 0.5 mL PBS to facilitate homogenization.
4. Centrifuge 5 s at 10,000*g* to remove large debris.
5. Retain 200 μL of the supernatant, discard the rest.
6. Centrifuge the supernatant 5 min at 13,000*g*.
7. Keep the pellet.
8. Add 200 μL Instagene matrix and resuspend pellet.
9. Heat to 56°C for 30 min.
10. Vortex mix briefly.
11. Heat to 95°–97°C (or boil) for 8 min.
12. Vortex mix briefly.
13. Centrifuge 5 min at 13,000 *g*.
14. Keep the supernatant as template for PCR.

3.2.2. Preparation of Template from Nasal Secretory Fluid (see **Note 4**)

A. *pleuropneumoniae* can be detected by PCR directly from nasal secretory fluid taken by simple cotton-tipped swabs:

1. Insert a cotton-tipped swab 2 cm into the nasal cavity of the pig.
2. Return swab immediately to its sterile plastic housing.
3. Keep on ice until further processing.
4. Immerse swab in 1 mL PBS for 30 min on ice.
5. Twirl for 1 min, then remove swab, and discard.
6. Vortex mix suspension.
7. Centrifuge 5 min at 13,000*g*.
8. Discard supernatant.
9. Add 200 μL Instagene matrix and resuspend pellet.
10. Heat to 56°C for 30 min.
11. Vortex mix briefly.
12. Heat to 95–97°C (or boil) for 8 min.
13. Vortex mix briefly.
14. Centrifuge 5 min at 13,000*g*.
15. Keep the supernatant as template .

3.2.3. Nested PCR

The detection of *A. pleuropneumoniae* DNA in the samples prepared from lung tissues (*see* **Subheading 3.2.1.**) and nasal secretory fluid (*see* **Subheading 3.2.2.**) will be performed by a highly sensitive 2-step nested PCR. Particular care has to be taken in order to avoid any cross-contamination by spill-over and by aerosols originating from samples containing PCR amplicons, especially from the first amplification step. Care must be taken when opening PCR tubes (avoid aerosol formation). The preparation of the second step, which involves the pipeting of the PCR amplicon of the first step, should be done, if ever possible, in a separate safety cabinet for PCR preparations, which should be treated by UV light after each use. It is highly recommended to run several negative control samples.

1. Use thin-walled thermal cycler tubes pre-cooled at 4°C.
2. Add for each first-step reaction: PCR premixture *apxIVA* (40.0 µL), *Taq* DNA polymerase (0.25 µL), template (*see* **Subheadings 3.2.2.** or **3.2.3.**) (10.0 µL).
3. Preheat thermal cycler to 95°C.
4. Place the cooled tubes directly into preheated thermal cycler.
5. Run 35 cycles with: 30 s at 94°C, 30 s at 54°C, 30 s at 72°C.
6. Prepare the second step in a precooled thin-walled tube by mixing: PCR premixture *apxIVA*-nested (49 µL), *Taq* DNA polymerase(0.25 µL), and Amplicon of the first step (1 µL)
7. Preheat thermal cycler to 95°C.
8. Directly place the cooled tubes into the preheated thermal cycler.
9. Run 35 cycles with: 30 s at 94°C, 30 s at 54°C, and 30 s at 72°C.
10. Take the tubes to a different room for analysis.
11. Analyze 10 µL of each PCR amplicon on a 0.7 % agarose gel.
12. Stain the gel and photograph.
13. Analyze the results (377-bp band) and verify all negative controls.

4. Notes

1. The oligonucleotide primers (**Table 2**) were designed in order to get the same annealing temperature of 54°C for all seven PCRs. However, we do not recommend the use of several primer pairs in the same reaction (multiplex PCR), since our experience showed that, in some cases, certain genes are only weakly amplified in the presence of several primer pairs. The primer pairs for the genes of the toxins ApxI, II and III were developed from DNA sequence data obtained from the corresponding genes of the different serotype reference strains and field strains. They were tested in over 300 different serotype reference and field strains including at least 10 field strains from each serotype (**ref. 8**). However, the resulting PCR products are relatively large. In case of redesigning new primer pairs, variability of the *apx* toxin genes must be considered.
2. Identification of the species *A. pleuropneumoniae*, using primer pair APXIVA-1L/APXIVA-1R, was performed on a large number of *A. pleuropneumoniae*

strains, including all serotype reference strains and 200 field strains collected from laboratories worldwide *(9)*. In addition, strains from a large collection of *Actinobacillus* species and other *Pasteurellacea*, as well as yet unknown species were tested as negative controls. PCR findings were verified by *rrs* (16S rRNA) sequence determination and phenotypic identification. The *apxIVA* PCR fully matched with phenotypic and phylogenetic identification.

3. Alternative methods for the detection of *A. pleuropneumoniae* were developed based on 16S rRNA genes *(11)*, genes of outer membrane proteins *(12,13)*, housekeeping genes *(14)*, or a random by chosen specific DNA segment of *A. pleuropneumoniae (15)*. Furthermore, a serotype 5-specific PCR test based on capsular biosynthesis genes was developed *(16)*.

4. Direct detection of *A. pleuropneumoniae* from pathological lung tissue revealed a high recovery of *A. pleuropneumoniae* reaching 94% of all lungs with lesions indicative of pleuropneumonia. Culture technique was less sensitive and recovered *A. pleuropneumoniae* in 76% of the cases *(9)*. The sensitivity of the test from nasal swabs seemed to be lower in this study, ranging from 43–64% of the potentially infected animals *(9)*. However, *A. pleuropneumoniae* was not cultivated from nasal swabs in this study *(9)*.

Acknowledgments

The author thanks Yvonne Schlatter, Edy Viley, and Peter Kuhnert for their valuable help with the preparation of this manuscript. This work was supported in part by grant no. 5002-578117/5002-045027 of the Priority Programme "Biotechnology" of the Swiss Science Foundation NF, the Swiss Federal Veterinary Office, and by the research fund of the Institute for Veterinary Bacteriology.

References

1. Taylor, D. J. (1999) *Actinobacillus pleuropneumoniae*, in *Diseases of swine*, (Straw, B. E., D'Allaire, S., Mengeling, W. L., and Taylor, D. J., eds.), Iowa State University Press, Ames, IA, USA, pp. 343–354.

2. Nielsen, R. Andresen, L. O., Plambeck, T., Nielsen, J. P., Krarup, L. Y., Jorsal, S. E. (1997) Serological characterization of *Actinobacillus pleuropneumoniae* biotype 2 strains isolated from pigs in two Danish herds. *Vet. Microbiol.* **54,** 35–46.

3. Blackall, P. J., Klaasen, H. B. L. M., van den Bosch, H., Kuhnert, P., and Frey, J. (2001) Proposal of a new serovar of *Actinobacillus pleuropneumoniae*: serovar 15. *Vet.Microbiol.* **84,** 47–52.

4. Frey, J. (1995) Virulence in *Actinobacillus pleuropneumoniae* and RTX toxins. *Trends Microbiol.* **3,** 257–261.

5. Beck, M., Vandenbosch, J. F., Jongenelen, I. M. C. A., Loeffen, P. L. W., Nielsen, R., Nicolet, J., and Frey, J. (1994) RTX toxin genotypes and phenotypes in *Actinobacillus pleuropneumoniae* field strains. *J. Clin. Microbiol.* **32,** 2749–2754.

6. Schaller, A., Kuhnert, P., De la Puente-Redondo, V. A., Nicolet, J., and Frey, J. (2000) Apx toxins in *Pasteurellaceae* species from animals. *Vet. Microbiol.* **74,** 365–376.

7. Schaller, A., Kuhn,R., Kuhnert, P., Nicolet, J., et al. (1999) Characterization of *apxIVA*, a new RTX determinant of *Actinobacillus pleuropneumoniae. Microbiology* **145,** 2105–2116.

8. Frey, J., Beck, M., Vandenbosch, J. F., Segers, R. P. A. M., and Nicolet, J. (1995) Development of an efficient PCR method for toxin typing of *Actinobacillus pleuropneumoniae* strains. *Mol. Cell. Probes* **9,** 277–282.

9. Schaller, A., Djordjevic, S. P., Eamens, G. J., et al. (2001) Identification and detection of *Actinobacillus pleuropneumoniae* by PCR based on the gene *apxIVA. Vet. Microbiol.* **79,** 47–62.

10. Ausubel, F. M., Brent, R., Kingston, R. E., et al. (1999) *Current Protocols in Molecular Biology.* John Wiley & Sons, Inc., New York, NY, USA.

11. Fussing, V., Paster, B. J., Dewhirst, F. E., and Poulsen., L. K. (1998) Differentiation of *Actinobacillus pleuropneumoniae* strains by sequence analysis of 16S rDNA and ribosomal intergenic regions, and development of a species specific oligonucleotide for *in situ* detection. *Syst. Appl. Microbiol.* **21,** 408–418.

12. Gram, T., Ahrens, P., and Nielsen, J. P. (1996) Evaluation of a PCR for detection of *Actinobacillus pleuropneumoniae* in mixed bacterial cultures from tonsils. *Vet. Microbiol.* **51,** 95–104.

13. Gram, T., and Ahrens, P. (1998) Improved diagnostic PCR assay for *Actinobacillus pleuropneumoniae* based on the nucleotide sequence of an outer membrane lipoprotein. *J. Clin. Microbiol.* **36,** 443–448.

14. Hernanz, M. C., Cascon, S. A., Sanchez, S. M., Yugueros, J., Suarez, R. S., and Naharro, C. G. (1999) Molecular cloning and sequencing of the aroA gene from *Actinobacillus pleuropneumoniae* and its use in a PCR assay for rapid identification. *J. Clin. Microbiol.* **37,** 1575–1578.

15. Sirois, M., Lemire, E. G., and Levesque, R. C. (1991) Construction of a DNA probe and detection of *Actinobacillus pleuropneumoniae* by using polymerase chain reaction. *J. Clin. Microbiol.* **29,** 1183–1187.

16. Lo, T. M., Ward, C. K., and Inzana T. J. (1998) Detection and identification of *Actinobacillus pleuropneumoniae* serotype 5 by multiplex PCR. *J. Clin. Microbiol.* **36,** 1704–1710.

6

Identification and Differentiation of *Brucella abortus* Field and Vaccine Strains by BaSS-PCR

Darla R. Ewalt and Betsy J. Bricker

1. Introduction

Brucellosis is a bacterial disease affecting livestock worldwide. Historically, at least seven species of pathogenic *Brucella* have been described, based primarily on host preference. Genetically, it appears that there is only a single species with host-adapted strains (*1*). *Brucella abortus* typically infects cattle and causes abortion. It is easily spread within a herd and can cause significant economic loss. It is also zoonotic, causing a range of chronic symptoms in humans. Many countries have developed a brucellosis eradication program and have significantly reduced or eliminated the disease, thanks in part to the use of two live-vaccines administered in early calfhood, *B. abortus* strain 19 and strain RB51. These are the most commonly used vaccines for cattle.

Countries that have successfully eliminated the disease maintain active surveillance measures necessary to prevent reintroduction of the disease. The initial screening protocol is based on serological reaction to *Brucella* antigens. However, cross-reacting bacterial infections and other factors can cause false positive reactions. Isolation and identification of the *Brucella* bacteria is the definitive diagnostic tool.

The extreme level of genetic conservation among Brucella species and strains has made differential identification difficult. The conventional "gold standard" identification consists of a panel of traits based on cultural characteristics, substrate-dependent metabolic rates, phage susceptibility, and antibiotic resistance. Twenty-five separate characteristics must be considered in making a specific identification, a process that typically takes 1 to 2 wk.

To expedite the process, numerous polymerase chain reaction (PCR) assays have been developed and published. However, only one assay, the AMOS assay

From: *Methods in Molecular Biology, vol. 216: PCR Detection of Microbial Pathogens: Methods and Protocols*
Edited by: K. Sachse and J. Frey © Humana Press Inc., Totowa, NJ

(named for the species it identifies: *B. abortus*, *B. melitensis*, *B. ovis*, and *B. suis*), has been developed to identify and differentiate the major *Brucella* species and also to differentiate the *B. abortus* vaccine strains from field isolates *(2,3)*. The AMOS assay is a single-tube multiplex PCR assay designed to amplify up to three independent targets differing in size. Identification is based on the pattern of DNA products amplified from specific DNA targets located within the unknown isolate's genome. In the field, most cases involve the specific identification of *B. abortus* and differentiation of field isolates from the vaccine strains. For this reason, the assay was abbreviated to include only those primers necessary to identify *B. abortus* and differentiate field isolates from the vaccine strains and from other *Brucella* species. In the laboratory setting, this abridged assay was evaluated and found to be in good agreement with the conventional tests *(4)*.

Over time, the original protocol has been modified. Primers have been added or changed, and the assay conditions have been optimized. An internal control was included to detect inhibitors or insufficient target concentration. In its current form, the assay involves amplification of up to four different loci by seven unique primers (**Fig. 1**). The specific loci selected were: (*i*) a region of the 16S rRNA gene conserved in most bacteria (internal, positive control); (*ii*) the DNA sequence in which the *Brucella*-specific element IS*711* is inserted adjacent to the *alk*B gene (an arrangement present in all *B. abortus* biovars found in North America *[2]*); (*iii*) the DNA sequence in which the *Brucella*-specific element IS*711* is inserted within the *wboA* gene (an arrangement that appears to be found only in *B. abortus* vaccine strain RB51 *[5]*); and (*iv*) a segment of the erythritol (*eri*) catabolic operon that includes a 702-bp sequence absent from the *B. abortus* vaccine strain S19 genome *(6,7)*. Identification is based on the presence or absence of these loci as indicated by PCR amplification. This paper describes the improved protocol that, for convenience, is now referred to as the *Brucella abortus* strain-specific (BaSS)-PCR assay.

2. Materials

2.1. Equipment

1. A dedicated work area completely free of contaminating *Brucella* or *Brucella* DNA.
2. A set of adjustable pipets dedicated to PCR setup only (e.g., P-2, P-10, P-200, and P-1000; Rainin Instruments, Woburn, MA, USA).
3. Additional pipets (e.g., P-10 and P-200) for dispensing template and detection procedures.
4. Disposable pipet tips with aerosol-preventing filters (e.g., RT-10F [or RT-10FG, *see* item 5], RT-200F, and RT-1000F; Rainin Instruments).
5. (Optional) Disposable pipet tips 0.5–10 μL with ShaftGard™ (*see* **Note 1**) and aerosol-preventing filters (e.g., RT-10GF in place of RT-10F).
6. Test tube racks designed for 0.2-mL microtubes which are dedicated to PCR setup only.

2.2. Sample Preparation

1. Nutrient agar plates such as tryptose agar (Difco no. 264300, Becton Dickinson, Sparks, MD, USA), trypticase soy agar (BBL no. 211043; Becton Dickinson), or *Brucella* agar (BBL no. 211086), containing 5% serum (bovine, calf, or fetal calf).
2. Saline: 0.85% (w/v) NaCl in sterile water.
3. CO_2 (10%) incubator or jar.
4. Inoculating loops.
5. Spectrophotometer (e.g., DU 650; Beckman Instruments, Fullerton, CA, USA).
6. Disinfectant, e.g., 1% Lysol IC™, (Reckitt & Colman, Wayne, NJ, USA).
   The following items are needed for the methanol preservation of cells (optional):
7. Methanol.
8. Screw cap tubes, 20 × 125 mm.
9. Trypticase soy broth with 5% bovine serum (30 mL per 250-mL screw cap flask).
10. Shaking (rotating) water bath set at 37°C.
11. Screw cap centrifuge tubes (50 mL).

2.3. PCR Amplification

1. Thermal cycler (*see* **Note 2**).
2. Disposable 200-µL thin-walled tubes.
3. (Optional) A repeating pipet such as the Eppendorf® Repeater™ Model 4780 (Cat. no. 22-26-000-6; Eppendorf AG, Westbury, NY, USA) fitted with a 1.25-mL combitip set to dispense 25 µL (*see* **Note 3**).
4. PCR-grade water (*see* **Note 4**).
5. FastStart™ *Taq* DNA Polymerase (Cat. no. 2-032-902; Roche Molecular Biochemicals, Indianapolis, IN, USA) (*see* **Note 5**).
6. 10X Reaction buffer without $MgCl_2$ (included with FastStart *Taq* DNA Polymerase) (*see* **Note 6**).
7. 25 m*M* $MgCl_2$ (*see* **Note 6**).
8. 10 m*M* dNTP mixture: 2.5 m*M* each of dATP, dCTP, dGTP, and dTTP (e.g., Cat. no. R725-01; Invitrogen, Carlsbad, CA, USA).
9. Oligonucleotide primers (*see* **Table 1** and **Subheading 3.1.1.**).
10. (Optional) GC Rich Enhancer (included with FastStart *Taq* DNA Polymerase), (*see* **Note 7**).
11. TE: 10 m*M* Tris-HCl, pH 8.0, 1 m*M* EDTA; can be stored for at least 2 yr at room temperature.

2.4. Detection of Amplification

1. Electrophoresis equipment with power supply (e.g., Horizon 11-14 Gel Electrophoresis Apparatus, cat. no. 11068-012; Life Technologies/Gibco-BRL, Rockville, MD, USA).
2. Electrophoresis grade agarose (e.g., SeaKem® GTG agarose, cat. no. 50071; FMC/BioWhittaker, Rockland, ME, USA).
3. 6X Loading dye (e.g., cat. no. E190; AMRESCO, Solon, OH, USA).

4. Ethidium bromide 10 mg/mL (e.g., cat. no X328; AMRESCO).
   CAUTION: ethidium bromide is a mutagen and potential carcinogen.
5. 0.5X TBE: 44.6 mM Tris, 44.5 mM boric acid, 1 mM EDTA, pH 8.3 (*see* **Note 8**).
6. DNA size standard, preferably a 100-bp ladder in the range of 100–1000 bp (e.g., cat. no. 15628-019; LifeTechnologies/Gibco-BRL).

3. Methods

3.1. Precautions

The extreme sensitivity of the PCR makes this technique highly vulnerable to contamination artifacts. Since as little as a single copy of template will amplify the corresponding products, significant care and planning must be implemented before performing any PCR assays for diagnostic purposes. Preparation of the bulk quantities of the reaction mixture (referred to as the Master Mixture) should be done in a dedicated room or hood that is free of bacteria or DNA. It is not sufficient to simply decontaminate an area with bactericidal solutions, since the DNA of the dead bacteria will still be amplified by PCR. The treatment of surfaces with 10% bleach (v/v) *(8)* will destroy the DNA as well as kill the bacteria, as long as all surfaces are exposed. Pipets, tube racks, and stock solutions should be dedicated to PCR. Addition of the sample template, amplification, and detection should be performed in another area. The use of disposable tips containing aerosol-preventing filters is highly recommended for all stages of the procedure.

3.2. Sample Preparation

Biohazard Warning: *Brucella* is a Class III pathogen; all steps involving the use of live *Brucella* should be done in a BL2 approved Biological Safety Cabinet with appropriate precautions (*9*; this reference is an excellent resource for techniques in handling, isolating, culturing and identifying *Brucella*) (*see* **Note 9**).

1. The *B. abortus* from abortion material, lymph nodes, milk, and reproductive organs is cultivated on a primary isolation plate (a basic nutrient agar plate, such as tryptose agar, trypticase-soy agar, or brucella agar, containing 5% serum [bovine, calf or fetal calf serum] and antibiotics) (*see* **Note 9**). The plate is incubated in a 10% CO_2 atmosphere at 37°C for 24–72 hrs. If there are numerous pure colonies on the primary isolation plate, the isolate can be set up immediately for PCR (*see* **step 4**).
2. If pure colonies of *Brucella* cannot be obtained from the primary isolation plate, the suspected *Brucella* colonies are reinoculated onto a secondary isolation plate and incubated at 37°C in a 10% CO_2 atmosphere for 24–48 h.
3. (Optional) Culture suspensions may be preserved in methanol for later use (*see* **Note 10**). Using a sterile inoculation loop, transfer bacteria from a single colony to a flask containing 30 mL of trypticase soy broth with 5% bovine serum. Incubate in a shaking 37°C water bath for 24–72 hrs until a heavy growth is achieved

Table 1
PCR Primer Sequences and Stock Concentrations

Primer	Nucleotide sequence 5' to 3'	Concentration of 100X Stock
IS*711*-specific	TGC-CGA-TCA-CTT-AAG-GGC-CTT-CAT-TGC-CAG	1.90 µg/µL
Abortus-specific	GAC-GAA-CGG-AAT-TTT-TCC-AAT-CCC	1.55 µg/µL
16S-universal-F	GTG-CCA-GCA-GCC-GCC-GTA-ATA-C	1.40 µg/µL
16S-universal-R	TGG-TGT-GAC-GGG-CGG-TGT-GTA-CAA-G	1.60 µg/µL
eri-F	GCG-CCG-CGA-AGA-ACT-TAT-CAA	1.35 µg/µL
eri-R	CGC-CAT-GTT-AGC-GGC-GGT-GA	1.30 µg/µL
RB51-3	GCC-AAC-CAA-CCC-AAA-TGC-TCA-CAA	1.55 µg/µL

(the broth will be deeply opaque). Transfer the bacteria-broth suspension to a screw top centrifuge tube that has been weighed. Centrifuge the sample at 7000*g* for 15 min. Discard the supernatant into a bactericidal disinfectant (e.g., 1% Lysol IC), and determine the wet weight of the pellet. Resuspend the pellet in 100 µL methanol and 50 µL saline for each mg wet weight of the pellet. Incubate the sample at 4°C for at least 1 wk, thoroughly mixing by inversion daily, to assure complete killing of the bacteria.

4. Standardize each suspension to a density of 1.5–2.0 U of absorbance at 600 nm, (DU650 Spectrophotometer) by removing supernatant or adding additional 66.6% methanol/33.3% saline.
5. To prepare bacteria from agar plates (primary or secondary) for PCR, use a sterile inoculating loop to transfer bacteria from several colonies to 500 µL of saline. Adjust the concentration of bacteria to the specifications given in **step 3** with saline.
6. Immediately before use, remix the culture suspension and dilute an aliquot 1/10 in PCR-grade water (e.g., 5-mL suspension in 45-µL water). Mix gently but thoroughly. The diluted material should be appropriately discarded after use. (*see* **Note 11**).

3.3. PCR Amplification

3.3.1. Preparation of the Master Mixture (100 Assays; see **Note 12**)

1. Synthetic oligonucleotides should be dissolved in TE buffer to a concentration of 100X (*see* **Table 1**). The 100X stock (20 µ*M* of each primer) is stable at 4°C for at least 2 yr as long as care is taken not to contaminate the solution.
2. Prepare the Primer Cocktail by dispensing the following 100X concentrates into a 1.5-mL microfuge tube: 233 µL PCR-grade water, 2.5 µL IS*711*-specific primer, 2.5 µL *B. abortus*-specific primer, 2.5 µL 16S universal primer-F, 2.5 µL 16S universal primer-R, 2.5 µL *eri* primer-F, 2.5 µL *eri* primer-R, and 2.5 µL RB51-primer.
3. Prepare the Master Mixture by dispensing the following into a 3- or 5-mL disposable tube: 1130 µL PCR-grade water, 250 µL 10X reaction buffer without MgCl$_2$ (*see* **Note 6**), 150 µL 25 m*M* MgCl$_2$, 200 µL 10 m*M* dNTP mixture, 250 µL

Primer Cocktail from **step 2**, 500 µL GC Rich Enhancer (optional) (*see* **Note 7**). (If an enhancer is not used, then 500 µl PCR-grade water should be substituted), and 20 µL FastStart *Taq* DNA Polymerase.

4. Mix the solution thoroughly, but gently, by pipeting up and down (*see* **Note 13**).
5. Aliquot the Master Mixture in 25-µL quantities into 0.2-µL thin-walled PCR tubes (or alternatively, a PCR-certified 96-well plate) (*see* **Note 14**). Store the assay tubes at –20° ±2°C.
6. Prior to use, thaw enough Master Mixture tubes for unknowns and controls and mix thoroughly but gently by finger tapping.

3.3.2. Amplification of Products by PCR

1. Add between 1.0 and 2.5 µL of unknown sample or control to each assay tube (*see* **Note 15**). Be sure to mix each sample thoroughly, just before removing the aliquot, since *Brucella* tends to settle out quickly.
2. Amplify the PCR products by using the following parameters: 1 cycle of 95°C for 5.0 min (*see* **Note 16**), then 40 cycles of 95°C for 15 s, 52°C for 30 s, 72°C for 1 min 30 s, and then 4°C indefinitely. The choice of ramp-time does not appear to be critical.
3. After amplification, the unopened samples can be stored at 4°C until ready for detection.

3.4. Detection of Amplified Products

1. Prepare a 5-mm thick 2.0% agarose gel (in 0.5X TBE) with an appropriate number of wells (*see* **Note 17**).
2. Combine 1 µL of 6X loading dye with 8 µL amplified sample and mix well before loading into the gel well (*see* **Note 18**).
3. Run the gel in 0.5X TBE until the bromophenol blue marker is at least 5 cm from the well to achieve good separation of the bands. For the equipment described here, 80–85 V for 2.5 h maximizes resolution without significant diffusion of the amplified DNA bands. Adjustments in voltage and time may be needed for other brands of equipment.
4. Stain the gel for 45 min in ethidium bromide solution (250 µg/500-mL of 0.5X TBE). Alternatively, the gel can be stained before electrophoresis or during electrophoresis by adding ethidium bromide to the running buffer (*see* **Note 19**). CAUTION: ethidium bromide is a mutagen and potential carcinogen (*see* **Note 20**).

3.5. Interpretation of Data

Identification is based on the number and the sizes of the products amplified by PCR (*see* **Fig. 1**). All samples except the negative controls should amplify at least 1 DNA product, the 800-bp 16S sequence. If this band is not present, then the sample may contain PCR inhibitors, the DNA was degraded, or the sample was not dispensed into the Master Mixture. It may be necessary to dilute the original sample to decrease the level of inhibitors in the reaction, repeat the assay with a fresh sample, or simply repeat the assay with the original sample.

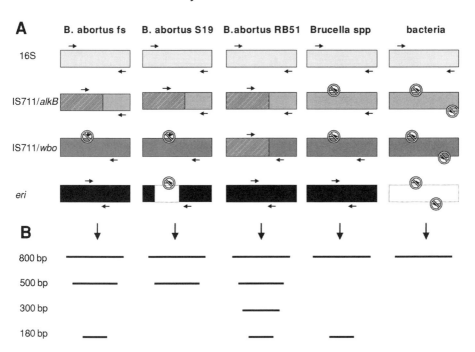

Fig. 1. Predicted amplified loci (rows) for various categories of unknowns (columns). (**A**) The four loci for each category are shown with their hybridizing primers; (**B**) the predicted products resulting from successful amplification. ▢, 16S locus; ▨, *alkB locus;* ▨, IS711; ▨, *wboA* locus; ■, *eri* locus; ☐, DNA absent; →, hybridizing amplification primer; ⊘, nonhybridizing primer.

A 500-bp product from the IS*711-alkB* locus is amplified with all *B. abortus* (biovar 1, 2, and 4) templates including those from the vaccine strains. Other *Brucella* species and bacteria do not have this DNA sequence, and no product will be amplified. A 300-bp product will be amplified from the IS*711-wboA* locus found only in the vaccine strain RB51. A 180-bp product is amplified from the *eri* gene present in all *Brucella* species and strains except *B. abortus* vaccine strain S19. Other bacterial genomes lack this locus and will not amplify the 180-bp product. Sample results are shown in **Fig. 2**.

3.6. Troubleshooting

The *B. abortus* RB51 positive control should amplify all four products. If some or all of the products are missing, then there was a problem with the Master Mixture or the cycling parameters. Repeat the assay with fresh Master Mixture.

The presence of amplification in the negative controls indicates a contamination problem. If both negative controls are contaminated, then the entire batch of Master Mixture was probably contaminated and should be discarded.

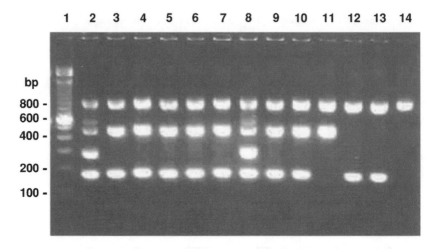

Fig. 2. Typical patterns amplified from bacterial bovine isolates as detected by agarose gel electrophoresis. Lane 1, 100-bp ladder; lane 2, *B. abortus* RB51; lane 3, *B. abortus* field strain; lane 4, *B. abortus* field strain; lane 5, *B. abortus* field strain; lane 6, *B. abortus* field strain; lane 7, *B. abortus* field strain; lane 8, *B. abortus* RB51; lane 9, *B. abortus* field strain; lane 10, *B. abortus* field strain; lane 11, *B. abortus* strain S19; lane 12, *Brucella* species (not *B. abortus*); lane 13, *Brucella* species (not *abortus*); and lane 14, non-*Brucella* bacteria. A 2% agarose gel was loaded with 8-µL amplified product and 1-µL loading dye per well, electrophoresed for 2.5 h at 70 V, stained with ethidium bromide, and visualized with UV light.

If only the water control is contaminated, then it means the problem probably occurred during the current assay, and the rest of the unused batch is probably usable. Repeat the assay with fresh water and extra care to prevent contamination. If the contamination of either control amplifies only the large 16S gene product, then any exposure to bacteria during the process could have caused the contamination. This may be difficult to avoid since even commercial enzymes and buffers may contain trace levels of nonspecific bacterial DNA, and absolute sterility in the larger laboratory environment is difficult to maintain.

The appearance of weak bands on a gel can complicate the analysis. Weak bands can result from cross-well contamination, improper amplification parameters or sample contamination. Only a few bacteria contaminating an isolate can result in visible amplified product. Multiplex PCR is more sensitive to variations in the amplification parameters than is traditional PCR. The annealing temperature is a good example. The addition of more primers increases the range of individual optimal melting temperatures (T_ms) and also the potential for mispriming events. The optimal temperature for each primer is a compromise between a lower temperature for stable annealing to the correct locus (but also possible mispriming) and a higher temperature to discourage mispriming (but also weaker binding to the correct locus). The BaSS-PCR assay was optimized at 52°C but may be run with a few degrees variation without compromising the results. The *eri* primers have the lowest T_m, and this product is the most likely to be affected by high temperature, inhibitors, and other destabilizing agents.

Amplification of smaller products is favored over larger products synthesized at the same time. Certain PCR enhancers may even prevent larger products from being amplified (*see* **Note 6**). Weak amplification of the 800-bp 16S amplicon across an entire group of samples would suggest that the operating parameters were not optimal.

4. Notes

1. Disposable pipet tips with ShaftGard protection (an extension of the top of the tip that covers a larger area of the pipet barrel) are particularly useful for taking aliquots (e.g., templates) from test tubes deeper than 35 mm (the size of a standard 1.5-mL microfuge tube) without contaminating the pipet barrel.
2. The protocol is written for a thermal cycler with a heated lid to prevent evaporation. If the machine to be used does not have a heated lid, the sample must be overlaid with an equal volume of bacteria-free DNA-free mineral oil or paraffin wax.
3. If a repeating pipet is used to dispense the Master Mixture, it should be used only for PCR. The use of individually wrapped disposable combitips is particularly helpful in preventing contamination. The tip should be replaced when making each batch of Master Mixture.
4. All stocks and solutions need to be made with PCR-grade water (i.e., sterile DNA-free RNA-free nuclease-free water).
5. Because of the complexity of the assay components, the choice of polymerase is critical to the success of this assay. Numerous polymerases have been examined with variable success. Proofreading polymerase formulations such as the Expand™ Long Polymerase mixture (Cat. no. 1 681 834; Roche Molecular Biochemicals) performed the best and most consistently, provided that the assay mixture was prepared fresh (*see* **Note 9**). The FastStart *Taq* DNA Polymerase also performs well and was selected as the polymerase of choice. *Taq* DNA polymerase from other commercial sources should be tested for suitability. AmpliTaq Gold® (Cat. no. N808-2041; Applied Biosystems, Foster City, CA, USA) was consistently a poor performer and is not recommended for this assay.
6. The success of the assay is also dependent on the pH and the $MgCl_2$ concentration of the reaction buffer. In general, PCR of *Brucella* DNA works best at pH 9.0 and a final $MgCl_2$ concentration of 1.5 mM. However, for proper activation of the modified FastStart Taq DNA Polymerase, the reaction buffer needs to be pH 8.3. This was taken into consideration during optimization of the assay. If other polymerases are used, the pH and $MgCl_2$ concentration will need to be optimized.
7. Since *Brucella* DNA is moderately GC rich (56%), we examined a number of PCR-enhancing products on the market. Both MasterAmp™ 10X PCR Enhancer

with betaine (Cat. no. ME81201; Epicenter Technologies, Madison, WI, USA), and Perfect Match™ (Cat. no. 600129, Stratagene, La Jolla, CA, USA) enhanced amplification of the smaller products at the expense of the larger products to the point that often the larger products were not synthesized. These products are not recommended for use in this assay. Addition of 10% dimethyl sulfoxide (DMSO) appeared to have no effect on the assay. The GC Rich Enhancer that comes with the FastStart Taq DNA Polymerase didn't appear to affect amplification of products but did seem to reduce primer–dimer formation. Its inclusion in the Master Mixture is optional.

8. TBE can be conveniently prepared as a 5X solution: 54 g Tris base, 27.5 g boric acid, and 4.16 g EDTA-tetrasodium salt dihydrate/L. Store the stock in a glass bottle at room temperature. If the solution starts to precipitate during storage, it can be redissolved by autoclaving. Dilute the stock 1/10 to make a 0.5X solution for preparing gels, for staining, and for use as the running buffer.

9. A comprehensive description of the primary isolation of *Brucella* is beyond the scope of this paper. Detailed information on the isolation procedures is located in other sources *(10,11)*.

10. Preservation of samples in methanol is recommended for samples that will not be tested within 7–10 d. The preserved cells are a little more difficult to lyse during the 95°C precycle step, but serve as an adequate source of target DNA when used at the recommended concentration this protocol. When preserved as described, samples can be stored at 4°C for at least 5 yr without noticeable deterioration. As an additional benefit, the bacteria are killed during preservation, reducing the risk of infection and eliminating further need for BL3 containment.

11. Once it has been diluted in water, the DNA inside the bacteria is not as stable, and so, the diluted samples should not be stored for later use (longer than about 12 h).

12. The individual assay tubes can be stored at –20° C for at least 6 mo without a significant change in activity. Preparing large batches of Master Mixture that are stored frozen in individual aliquots has several advantages: (*i*) there will be fewer variations between experiments; (*ii*) less opportunity for contamination since the stock solutions are opened and used fewer times; (iii) assays can be performed immediately as samples trickle in; and (iv) repeated freeze-thaw cycles are avoided. However if the Master Mixture is stored frozen in its complete formulation with polymerase, dNTPs, and primers, it is necessary to use an artificial hot start component to prevent premature polymerization of primers into hairpin and primer–dimer structures. This can be accomplished by physical separation of components with a paraffin wax barrier (such as Ampliwax Gems®, Cat. no. N808-0150; Applied Biosystems), by using a chemically modified polymerase (such as FastStart *Taq* DNA Polymerase, Roche Molecular Biochemicals) designed to be inactive until heated to a high temperature for several min, or by inactivating the polymerase active site with an antibody (such as TaqStart Antibody, Cat. no. 5400-1; Clontech Laboratories, Palo Alto, CA, USA) that denatures and dissociates at high temperature. Because of the complex nature of multiplexed PCR, we found the FastStart *Taq* DNA Polymerase to be the best

option. Surprisingly, a similar product, AmpliTaq Gold, was found to be a poor performer for this specific multiplex formulation (*see* **Note 5**).

13. Thorough mixing of the Master Mixture is critical to the success of the assay because several components are highly viscous and will settle to the bottom by gravity if not properly dispersed.

14. Dispensing a large batch of Master Mixture is most easily accomplished with a repeating pipet such as the Eppendorf Repeater Model 4780 fitted with a 1.25-mL combi-tip set to dispense 25 µL (*see* **Note 3**).

15. Each assay should have at least three controls. Include at least two negative controls to monitor for contamination. One negative control, containing only the Master Mixture, monitors the purity of the Master Mixture batch. The second negative control, prepared last, includes 1.0–2.5 µL of the water used to dilute the unknowns. This control monitors the purity of the dilution water and monitors for the occurrence of sample contamination during the dilution and addition of template (originating from sources such as gloves, aerosolized droplets, etc.). Use a known isolate of *B. abortus* RB51 as the positive control, since RB51 will amplify all four possible products.

16. The precycle incubation of 95°C for 5 min has two purposes. First, the chemically modified polymerase is activated during the 5-min period. Second, the high heat causes a percentage of the bacteria to rupture and release the DNA that is then used as the template for amplification. If the polymerase chosen for use does not require heat activation, then the precycle incubation step can be reduced to 2 min.

17. For the analysis of large numbers of samples, a double-decked 11-cm (wide) X 14-cm (long) gel with two 20-well 1-mm thick combs (e.g., Cat. no. 11951-076; LifeTechnologies/Gibco-BRL) arranged 6-cm apart produces good resolution among fragments. Other gel sizes and comb arrangements can be used; however, the use of very small wells (< 3.5-mm wide) that are close together can make lane assignment of bands difficult and is not recommended. The best results come from wide well-spaced wells.

18. It is important that all edges of the gel wells are completely submerged in running buffer to avoid having capillary action pull the sample out of the well. When using gels and combs that are different from the sizes given, be careful that the sample volume does not exceed the well capacity since this can lead to cross-well contamination. Even minor contamination between wells can make interpretation of the data very difficult. If loading samples cleanly is difficult, skip every other well, or load each sample in duplicate adjacent wells. Cross-contamination between wells is easily detected when samples are run in duplicate.

19. Do not wait to stain and record results, as diffusion of the small fragments will occur.

20. Use gloves and safe methods when handling gels and solutions containing ethidium bromide. After staining, solutions containing ethidium bromide should not be poured down the sink, but must be deactivated and disposed of properly. For an easy method for decontaminating solutions, destaining bags (Cat. no. E732; AMRESCO), which are tea bag-like pouches containing a matrix, binds the

unused ethidium bromide. Afterwards, the solution can be poured down the sink, and only the destaining bag needs to be disposed of as hazardous chemical waste.

Acknowledgments

This work was funded by the Diagnostic Bacteriology Laboratory, [National Veterinary Services Laboratories (NVSL), Animal and Plant Health Inspection Service (APHIS), US Department of Agriculture (USDA) and CRIS 3625-32000-011-00 [National Animal Disease Center (NADC), Agricultural Research Service (ARS), USDA].

References

1. Verger, J. M., Grimont, F., Grimont, P. A. D., and Grayon, M. (1987) Taxonomy of the genus *Brucella*. *Ann. Inst. Pasteur/Microbiol.* **138,** 235–238.
2. Bricker, B. J., and Halling S. M. (1994) Differentiation of *Brucella abortus* bv. 1, 2, and 4, *Brucella melitensis*, *Brucella ovis* and *Brucella suis* bv. 1 by PCR. *J. Clin. Microbiol.* **32,** 2660–2666.
3. Bricker, B. J., and Halling S. M. (1995) Enhancement of the *Brucella* AMOS PCR assay for differentiation of *Brucella abortus* vaccine strains S19 and RB51. *J. Clin. Microbiol.* **33,** 1640–1642.
4. Ewalt, D. R., and Bricker, B. J. (2000) Validation of the abbreviated AMOS PCR as a rapid screening method for differentiation of *Brucella abortus* field strain isolates and the vaccine strains, 19 and RB51. *J. Clin. Microbiol.* **38,** 3085–3086.
5. Vemulapalli, R., McQuiston, J. R., Schurig, G. G., Sriranganathan, N., Halling, S. M., and Boyle, S. M. (1999) Identification of an IS711 element interrupting the wboA gene of *Brucella abortus* vaccine strain RB51 and a PCR assay to distinguish strain RB51 from other *Brucella* species and strains. *Clin. Diagn. Lab. Immunol.* **6,** 760–764.
6. Sangari, F. J., and Agüero, J. (1994) Identification of *Brucella abortus* B19 vaccine strain by the detection of DNA polymorphism at the *ery* locus. *Vaccine* **12,** 435–438.
7. Sangari, F. J., Garcia-Lobo, J. M., and Agüero, J. (1994) The *Brucella abortus* vaccine strain B19 carries a deletion in the erythritol catabolic genes. *FEMS Microbiol. Lett.* **121,** 337–342.
8. Prince, A. M., and Andrus, L. (1992) PCR: How to kill unwanted DNA. *BioTechniques* **12,** 358–360.
9. Alton, G. G., Jones, L. M., Pietz, D. E. (1975) *Laboratory Techniques in Brucellosis 2nd ed.,* World Health Organization Monograph Series, No. 55, World Health Organization, Geneva, pp. 11–13.
10. Alton, G. G., Jones, L. M., Angus, R. D. and Verger, J. M. (1988) Bacteriological methods, in *Techniques for the Brucellosis Laboratory.* INRA, Paris, pp. 13–62.
11. Mayfield, J. E., Bantle, J. A., Ewalt, D. R., Meador, V. P., and Tabatabai, L. B. (1990) Detection of *Brucella* cells and cell components, in *Animal Brucellosis,* (Nielsen, K. and Duncan, J. R., eds.), CRC Press, Boca Raton, FL, USA, pp. 97–120.

7

Isolation of Campylobacter and Identification by PCR

Mark D. Englen, Scott R. Ladely, and Paula J. Fedorka-Cray

1. Introduction

Campylobacter is now recognized worldwide as a leading cause of bacterial gastroenteritis in humans *(1)*. Campylobacter species are common commensals in the intestinal tracts of poultry and livestock, and food products of animal origin are frequently associated with reported cases of illness *(2)*. This chapter provides methods for the identification of *C. jejuni* and *C. coli*, which are the two species accounting for the majority of human infections. The protocols are routinely used in our laboratory and are intended to provide workers unfamiliar with Campylobacter culture and identification a useful set of methods to serve as a practical starting point.

Numerous procedures have been described for the isolation of *Campylobacter* spp. from food, feces, and environmental samples *(3–8)*. No universal method for isolating Campylobacter has yet been developed that is appropriate for all sample types. The choice depends on the expected level of Campylobacter in the sample material and any extraneous bacterial flora that may be present. In our experience, the combination of direct plating and sample enrichment often provides better recovery of Campylobacter than either technique alone. A procedure for isolating Campylobacter from livestock or poultry feces using this approach is provided in the following sections.

Several formulations of dehydrated plating media and their supplements are commercially available for the isolation of Campylobacter. Dehydrated basal media are typically supplemented with blood or charcoal and are often combined with ferrous sulfate, sodium metabisulfite, and sodium pyruvate (FBP) to diminish the toxic effects of oxygen on Campylobacter. Antimicrobials, to

From: *Methods in Molecular Biology, vol. 216: PCR Detection of Microbial Pathogens: Methods and Protocols*
Edited by: K. Sachse and J. Frey © Humana Press Inc., Totowa, NJ

which campylobacters are intrinsically resistant, are also added to inhibit competing bacterial and fungal flora found in the sample material *(9)*. The selective plating medium we prefer, Campy-Cefex *(10)*, is described below. Campy-Cefex plates are relatively translucent, and Campylobacter colonies are easier to identify and quantify compared to media formulations containing charcoal, such as the widely used modified charcoal cefoperazone desoxycholate agar (mCCDA). However, mCCDA (available from Oxoid, Ogdensburg, NY, USA), or the recently developed Campy-Line agar *(11)*, may be directly substituted for Campy-Cefex in the protocols outlined below.

Over the last decade, the polymerase chain reaction (PCR) has become a basic tool for the identification of bacterial pathogens such as Campylobacter *(12,13)*. For those unfamiliar with the use of PCR for the amplification of specific DNA sequences, a number of reviews are available (e.g., *14–16*). Beyond the basic requirement of suitable DNA sample preparations to serve as template in the PCR, a specific DNA amplification (target) sequence must be determined. Although usually a well-characterized region of the genome, this is not always a requirement for the development of a useful assay *(17)*. The range of genes reported for the identification of *Campylobacter* spp. by PCR includes 16S rRNA (18,19), 23S rRNA *(20)*, the cadF virulence gene *(21)*, and the flagellin genes, flaA and flaB *(22,23)*. The PCR we describe here is based on the hippuricase gene (hipO) for the identification of *C. jejuni* *(24)* and a siderophore transport gene (ceuE) sequence to identify *C. coli* *(25)*. A separate PCR is run for each primer pair. This PCR has also been adapted to allow the direct identification of *C. jejuni* from the sample material without prior enrichment steps *(26)*. For applications involving large numbers of samples, multiplexing *(27)* can be incorporated to reduce the total assay time and increase sample throughput.

Finally, it is important to note that some strains of Campylobacter are resistant to lysis by heating *(28)* and simply boiling cell suspensions for use in PCR may result in a significant percentage of false negatives. The DNA isolation procedure included here provides a simple method for consistently obtaining usable template DNA from Campylobacter.

2. Materials

2.1. Isolation and Culture of Campylobacter

1. Exam gloves or gloves and fecal loops (Revival Animal Health, Orange City, IA, USA).
2. Sterile tongue depressors.
3. Sterile cotton-tipped swabs.
4. Sterile spreader sticks (Simport Plastics, Beloeil, Quebec, Canada).
5. Whirl Pak bags (Nasco, Fort Atkinson, WI, USA).
6. Gallon zip-lock bags (268 × 279 mm).

7. Glass test tubes (18 × 150 mm).
8. Sterile 24-well tissue culture plates (Becton Dickinson, Franklin Lakes, NJ, USA).
9. Sterile 2-mL and 5-mL cryovials.
10. Phase-contrast microscope with100× oil immersion objective.
11. Medical gas mixture (5% O_2, 10% CO_2, and 85% N_2).
12. Horse blood agar (HBA) plates (B-D Biosciences, Sparks, MD, USA).
13. Phosphate-buffered saline (PBS), pH 7.2: 2.28 g sodium phosphate, dibasic (Na_2HPO_4), 0.46 g sodium phosphate, monobasic (NaH_2PO_4), 9.0 g sodium chloride, and distilled water (dH_2O) to 1.0 L.
14. Bolton's basal broth: 10 g meat peptone, 5 g lactalbumin hydrolyzate, 5 g yeast extract, 5 g sodium chloride, 10 mg haemin, 0.5 g sodium pyruvate, 1 g α-ketoglutaric acid, 0.5 g sodium metabisulfite, 0.6 g sodium carbonate, and 950 mL dH_2O. Supplement: 20 mg sodium cefoperazone, 50 mg sodium cycloheximide, 20 mg trimethoprim, 20 mg vancomycin, and 50 mL lysed horse blood.
Mix the basal broth ingredients in distilled water and heat with constant stirring until all components are completely dissolved. Autoclave for 15 min at 121°C. Cool to 50°C, then add horse blood and antibiotic supplement (reconstitute components as directed by the manufacturer).
Commercial Bolton's formulation: 27.6 g/L Campylobacter enrichment broth (Acumedia Manufacturing [Baltimore, MD, USA], cat. no. 7526 or Oxoid, cat. no. CM0983B). Commercial Bolton's supplement: 2 vials/L (cefoperazone, vancomycin, trimethoprim, and cycloheximide) (Malthus Diagnostics, cat. no. X131, or Oxoid, cat. no. SR183E).
15. Campy-Cefex basal agar: 43 g Brucella agar, 0.5 g ferrous sulfate, 0.2 g sodium bisulfite, 0.5 g pyruvic acid, and 950 mL dH_2O.
Supplement: 33 mg sodium cefoperazone, 0.2 g sodium cycloheximide, and 50 mL lysed horse blood.
Mix the basal agar ingredients in distilled water and heat with constant stirring until all components are completely dissolved. Autoclave the basal agar for 15 min at 121°C. Cool to 50°C, then add the horse blood and antibiotic supplement. Before adding to the cooled agar, dissolve the cefoperazone in water, and the cycloheximide in 50% methanol, then filter-sterilize (0.2 μm) both.
16. Freezing medium: 2.8 g Brucella broth, 85 mL dH_2O, and 15 mL glycerol.
Mix the Brucella broth in distilled water, and heat with constant stirring until completely dissolved. Add glycerol and mix well to combine broth and glycerol. Autoclave at 121°C for 15 min, then cool. Dispense 1-mL aliquots in 2-mL cryovials as needed.
17. Wang's transport medium: 28 g Brucella broth, 4 g agar, 950 mL dH_2O, and 50 mL lysed horse blood.
Combine the Brucella broth and agar in distilled water and heat with constant stirring until components are completely dissolved. Autoclave at 121°C for 15 min, then cool to 50°C prior to adding the horse blood. Dispense 2 to 4 mL aliquots in sterile screw-cap vials or 5-mL cryovials and allow agar to solidify in an upright position.

18. Dehydrated culture media for isolating *Campylobacter* spp. may be obtained from the following companies: Acumedia Manufacturing, Med-ox Diagnostics (Ogdensburg, NY), and Oxoid. Antimicrobials and chemicals used in media and media supplements are available from Sigma (St. Louis, MO, USA).

2.2. DNA Extraction

1. Eppendorf® tubes, 1.7-mL, nuclease-free.
2. Sterile disposable 10-µL plastic loops (Simport Plastics).
3. Nuclease-free distilled water (dH$_2$O).
4. DNAzol® reagent (Invitrogen, Carlsbad, CA, USA).
5. 100% Ethanol.
6. 70% Ethanol.
7. TE buffer: 10 mM Tris-HCl, pH 7.5, 1 mM EDTA, pH 8.0.

2.3. PCR

1. Adjustable vol pipetors (5.0–1000 µL vol range).
2. Aerosol barrier pipet tips (Molecular Bio-Products, San Diego, CA, USA).
3. Thermal cycler (Applied Biosystems; Foster City, CA, USA).
4. Thin-wall PCR tubes, 0.2-mL (Applied Biosystems).
5. Eppendorf tubes, 1.7- or 2-mL, nuclease-free.
6. Nuclease-free dH$_2$O.
7. *Taq* DNA polymerase: 5 U/µL (Invitrogen).
8. PCR buffer (10×) without MgCl$_2$: 200 mM Tris-HCl, pH 8.4, 500 mM KCl (Invitrogen).
9. MgCl$_2$: 50 mM (Invitrogen).
10. dNTP mixture: dATP, dCTP, dGTP, and dTTP, 10 mM each (Invitrogen).
11. PCR primer sets (Operon Technologies, Alameda, CA). The sequence of the *C. coli* forward primer, Col1, is 5'-ATG AAA AAA TAT TTA GTT TTT GCA-3', and that of the reverse primer, Col2, is 5'-ATT TTA TTA TTT GTA GCA GCG-3'. The C. jejuni forward primer, HIP400F, has the sequence 5'-GAA GAG GGT TTG GGT GGT-3', and the reverse primer, HIP1134R, 5'-AGC TAG CTT CGC ATA ATA ACT TG-3'. The expected amplification products are: Col1/2, 894 bp, and HIP400F/HIP1134R, 735 bp.

2.4. DNA Electrophoresis

1. 250–500 V electrophoresis power supply.
2. Horizontal gel apparatus.
3. UV transilluminator with Polaroid camera and hood (Fotodyne, Hartland, WI, USA)
4. Agarose: SeaKem® LE (FMC BioProducts, Rockland, ME, USA).
5. Molecular weight marker: 100-bp DNA ladder (Bio-Rad, Hercules, CA, USA).
6. TBE electrophoresis buffer: 89 mM Tris, 89 mM boric acid, 2 mM EDTA.
   10× stock solution: 108 g Tris base, 55 g boric acid, 40 mL 0.5 *M* EDTA, pH 8.0, and dH$_2$O to 1.0 L.
7. Ethidium bromide: 10 mg/mL. Add 100 mg ethidium bromide to 10 mL dH$_2$O.

Mix thoroughly and cover container with foil to seal out light. Store at room temperature. Caution: Ethidium bromide is a mutagen.

8. DNA loading buffer: 5% glycerol, 10 mM EDTA, pH 8.0, 0.1% sodium dodecyl sulfate (SDS), 0.01% bromophenol blue.
   10× stock solution: 5 mL glycerol, 2 mL 0.5 M EDTA, pH 8.0, 0.1 g SDS, 10 mg bromophenol blue, and dH$_2$O to 10 mL.

3. Methods

The sections below outline the steps for (i) isolating Campylobacter from feces using direct plating and sample enrichment; (ii) extraction of Campylobacter DNA for use as template in the PCR; and (iii) PCR assays specific for *C. jejuni* and *C. coli*. A brief section on agarose gel electrophoresis of the PCR amplification products using ethidium bromide staining for detection is also included.

3.1. Isolation of Campylobacter from Feces

3.1.1. Collection of Fecal Samples

1. Collect fecal samples from the cloaca or rectum using a gloved hand, a fecal loop, or cotton-tipped swab. Freshly defecated feces can be collected from the ground using a sterile tongue depressor (*see* **Note 1**).
2. Place fecal samples into Whirl Pak bags, press air from the bags, and seal.
3. Place samples on ice or cold packs for transport to laboratory or for shipment. Samples should be processed as soon as possible (*see* **Note 2**).

3.1.2. Direct Plating on Selective Media

1. Mix fecal samples by kneading bags with gloved hands.
2. Using a sterile cotton-tipped swab, transfer approx 1 g of feces into a sterile glass tube. Record sample weight, then add 3× the sample weight of PBS (*see* **Notes 3** and **4**) to the tube to obtain a 1:4 dilution of the sample. Vortex well to mix the tube contents.
3. Transfer 1.0 mL of the 1:4 dilution to a second sterile glass tube containing 9.0 mL of PBS and vortex mix well. This results in a final dilution of 1:40.
4. Dispense 0.1-mL aliquots of each dilution on duplicate Campy-Cefex plates. Spread the liquid evenly using a sterile spreader stick or sterile cotton-tipped swab.
5. Invert the plates and stack them in gallon (268 × 279 mm) zip-lock bags (approx 20 plates/bag). Flush the bags with medical gas mixture (*see* **Note 5**), then press to expel the gas mixture. Repeat 3×, then inflate and seal bags tightly to ensure a microaerobic atmosphere inside. Incubate the sealed bags for 48 h at 42°C (*see* **Note 6**).
6. Examine the plates for the presence of Campylobacter colonies (*see* **Note 7**). The number of colony forming units (cfu) of Campylobacter per g of feces is equal to 40× the number of Campylobacter colonies on the 1:4 dilution plate or 400× the number on the 1:40 dilution plate.

3.1.3. Fecal Sample Enrichment

1. An inexpensive enrichment method to enhance recovery of Campylobacter utilizes 24-well tissue culture plates *(29)*. Begin by dispensing 1.0 mL of Bolton's broth into the plate wells.
2. Dip sterile cotton-tipped swabs into the 1:4 fecal dilutions and inoculate wells with the swabs.
3. Put the plates into zip-lock bags, gas as described for direct plating (*see* **Subheading 3.1.2.**), and incubate for 48 h at 42°C.
4. Noting the direct plates that were negative for Campylobacter, streak for isolation only from those corresponding sample enrichment wells onto Campy-Cefex plates. Bag and gas the plates as described in **Subheading 3.1.2.**, and incubate for 48 h at 42°C. By limiting further isolation steps to only those samples that were negative by direct plating, unnecessary duplication of the positive samples is avoided.

3.1.4. Presumptive Identification of Campylobacter

1. Prepare wet-mounts of Campylobacter as follows (*see* **Note 8**). Place a drop of PBS or distilled water on a standard microscope slide.
2. Using a sterile loop, pick a portion of a colony to be examined. Suspend the cells in the drop of PBS by mixing.
3. Place a cover slip on the cell suspension, using care to avoid trapping air under the slip.
4. Add a drop of immersion oil on top of the cover slip, and observe the cells using phase-contrast optics at 100×.
5. Campylobacter is easily identified by its characteristic narrow, curved, or spiral rod shape and darting motility. Pairs of cells often resemble a gull's wing or the letter S. Campylobacters from cultures older than 24–48 h may become coccoid in shape and are more difficult to identify.

3.1.5. Storage and Shipment of Campylobacter Isolates

1. For frozen preservation of Campylobacter cultures, begin by streaking isolates onto HBA plates (*see* **Note 9**). Incubate the plates for 48 h at 42°C as previously described.
2. Using a sterile cotton-tipped swab, transfer a heavy inoculum from a plate to a 2-mL cryovial containing 1 mL of freezing medium.
3. Store vials at –70°C. Cultures stored in this way will remain viable for 6–12 mo. However, repeated thawing and refreezing may reduce viability significantly (*see* **Note 2**).
4. For transport or shipment of Campylobacter cultures, Wang's transport medium is recommended. Transport tubes are prepared by adding 2–4 mL of medium to sterile screw-capped glass vials or plastic 5-mL cryovials. Allow the soft agar to solidify.
5. Inoculate the vials by stabbing a loop of Campylobacter culture into the semisolid media. Place vials in a clean rack, loosen the caps and put the rack of vials in a zip-lock bag. Gas and seal the bags, and incubate for 24 h at 42°C.
6. Tighten caps, and ship according to local guidelines.

3.2. Isolation of Campylobacter DNA

As previously mentioned, some strains of Campylobacter are resistant to lysis by heating. In the method described in **Subheading 3.2.1.**, the combination of boiling the cell suspensions followed by DNA extraction, using a commercially available guanidine-based reagent, overcomes this difficulty. We have not found it necessary to quantify the isolated sample DNA to achieve reproducible PCR results.

3.2.1. DNA Extraction

1. Add 100 μL of nuclease-free dH_2O to a boil-proof 1.7 mL Eppendorf tube.
2. Using a sterile disposable 10-μL loop, scrape enough cells from a plate to make a turbid suspension in the water. Make sure cells are fully dispersed by gently vortex mixing the tube; no clumps should be visible.
3. Put the tubes in a floating tube rack and place in boiling water bath. Boil for 10 min (*see* **Note 10**).
4. Cool the tubes on ice. The boiled cell suspensions can be frozen at –20°C until needed, if necessary.
5. Add 1.0 mL DNAzol DNA extraction reagent to the boiled cell suspension. Vortex mix gently, and let the tubes stand for 10–15 min at room temperature (*see* **Note 11**).
6. Centrifuge the tubes for 10 min at 10,000g to pellet cell debris.
7. Carefully transfer supernatants to new 1.7-mL Eppendorf tubes.
8. Add 0.5 mL of 100% ethanol to the supernatant, and invert the tube several times to thoroughly mix (do not vortex the tube).
9. Let the tubes stand for 15–30 min at 4°C to allow the DNA to precipitate.
10. Pellet the precipitated DNA by centrifuging the tubes for 10 min at 12,000g.
11. Remove and discard the supernatant using a pipet or vacuum suction device. The DNA pellet often will not be visible, so use caution to prevent sample loss for this and subsequent wash steps.
12. Wash the DNA by adding 1.0 mL of 70% ethanol to the tubes, and inverting tubes several times to resuspend pellet. The DNA can be stored for long periods in the ethanol wash at –20°C.
13. Pellet the DNA by centrifuging the tubes for 2 min at 12,000g.
14. Remove and discard the ethanol wash supernatant.
15. Repeat steps 12–14 to give a total of 2 washes. Allow the tubes to dry completely after the final wash.
16. Dissolve the DNA in 40–60 μL of nuclease-free dH_2O or TE buffer, pH 7.5. Store DNA at –20°C. Use 2–5 μL of this DNA preparation in the 50 μL PCR discussed in **Subheading 3.3.2.**

3.3. Identification of Campylobacter by PCR

The primer sets for the identification of *C. coli* (Col1/2) and *C. jejuni* (HIP400F/HIP1134R) require different annealing temperatures (57° and 66°C, respectively), and a separate PCR must be run for each pair. The expected amplification products are 894 bp (Col1/2) and 735 bp (HIP400F/HIP1134R).

3.3.1. Preparation of PCR Master Mixture

To provide tube-to-tube consistency for each sample, it is best to prepare a common PCR mixture (termed a master mixture), containing all of the reaction components for each primer set. Aliquots of this mixture are added to the control and sample tubes, followed by addition of the template DNA. To calculate a PCR master mixture for your experiment, first determine the total number of PCRs (samples plus controls) needed for the experiment, allowing 10% extra for slight vol errors in pipeting. Using the formula below, multiply each component by the total number of reactions to calculate the vol required for each component.

Prepare the PCR master mixtures in nuclease-free tubes, adding the components in the order given for the 1X PCR mixture (*see* **Notes 12** and **13**). Vortex well after the addition of each component.

1X PCR mixture, 50 µL vol: 30.75 µL nuclease-free dH$_2$O, 5 µL 10X PCR buffer without MgCl$_2$, 3 µL 50 mM MgCl$_2$, 1 µL 10 mM dNTP mixture, 2.5 µL 20 µM forward primer, 2.5 µL 20 µM reverse primer, 0.25 µL Taq DNA polymerase (5 U/µL), and 45 µL plus 5 µL template DNA extract to equal 50 µL total vol.

Final concentration of components: 20 mM Tris-HCl, pH 8.4, 50 mM KCl, 3 mM MgCl$_2$, 0.2 mM dNTPs, 1 µM primers, and 1.25 U of Taq DNA polymerase.

3.3.2. Campylobacter-Specific PCR

1. Aliquot 45 µL of the master mixture into the labeled sample and control 0.2-mL reaction tubes, then add 5 µL of sample or control template DNA to tubes (*see* **Note 14**). Close the tube caps tightly (a capping tool is available from Applied Biosystems).
2. Place the tubes in the thermal cycler heat block and close the lid. Program the thermal cycling as follows. Denature at 94°C for 30 s, followed by 25–30 thermal cycles of denature at 94°C for 30 s, anneal at 57°C (Col1/2) or 66°C (HIP400F/HIP1134R) for 30 s, extend at 72°C for 1 min, final extensions at 72°C for 4 min.
3. The completed PCRs are screened for the expected amplification product using agarose gel electrophoresis (*see* **Subheading 3.3.3.**). The remainder of the reactions should be stored at –20°C.

3.3.3. Agarose Gel Electrophoresis of PCR Products

1. Prepare a 1.5% agarose gel in 1X TBE buffer. Add the dry agarose to the appropriate quantity of buffer in a flask or heat-resistant bottle containing a Teflon-coated stir bar.

2. Stir briefly, then heat the mixture in a microwave oven to melt the agarose. To prevent overheating or boiling over, check the progress at 1- to 2-min intervals, stirring occasionally.

3. Carefully remove the container of molten agarose from the microwave and place on a stir plate. While stirring the hot agarose, add ethidium bromide solution from the 10 mg/mL stock solution to yield a concentration of 0.5 µg/mL in the agarose solution (5 µL /100 mL). Allow the ethidium bromide to mix thoroughly in the molten gel.

4. Place the container in a 55°C water bath, and cool the agarose to this temperature.

5. When the agarose is sufficiently cooled, cast a gel using a comb containing the appropriate number of wells. Let stand for 45–60 min to allow the gel to solidify.

6. Assemble the horizontal gel apparatus, and fill with enough 1X TBE to cover the surface of the gel.

7. Mix PCR controls and samples (approx 10% of each reaction) with 10X DNA loading buffer. Load submerged gel wells, including one or more lanes of the 100-bp DNA ladder molecular weight marker.

8. Perform electrophoresis according to the gel apparatus manufacturer's instructions for voltage settings. Photograph gel under UV light following completion of electrophoresis. An example is shown in **Fig. 1**.

4. Notes

1. Recovery of Campylobacter from rectally collected fecal samples is significantly higher than from fecal samples collected from the ground *(30)*.

2. Campylobacter spp. are relatively fragile organisms. Cell viability is significantly reduced by environmental stresses such as prolonged exposure to room temperature, atmospheric oxygen, freezing and thawing, and desiccation.

3. It is usually desirable to obtain cell counts for the numbers of viable organisms per unit vol or weight of sample material. However, if a simple qualitative assessment is all that is necessary, proceed as directed in **Subheading 3.1.2.**, but omit weighing the samples. Also, duplicate plating is not necessary.

4. Feces are usually diluted in PBS. However, buffered peptone water (0.1%) may be substituted.

5. The medical gas mixture (5% O_2, 10% CO_2, and 85% N_2) is commercially available but may require special ordering. Gas can be conveniently dispensed into zip-lock bags using a length of plastic tubing attached to a needle valve type gas regulator.

6. Although many laboratories routinely incubate Campylobacter cultures at 37°C, we prefer to use 42°C. In our experience, the higher temperature suppresses the growth of *Arcobacter* spp. more effectively than incubation at 37°C.

7. Campylobacter colonies can be identified on Campy-Cefex agar by their translucent or cream-colored appearance. Most contaminants are not translucent.

8. Although a number of biochemical tests have been developed for the presumptive identification of Campylobacter *(31,32)*, we have found examining the cell morphology of fresh cultures to be the simplest and most reliable method for

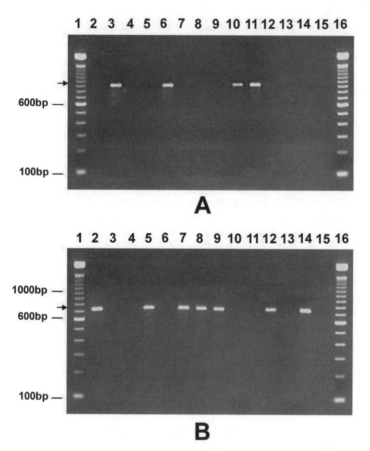

Fig. 1. Identification of *C. coli* (**A**) and *C. jejuni* (**B**) by PCR. (**A** and **B**) Lanes 1 and 16, 100-bp DNA ladder; lane 2, *C. jejuni* positive control; lane 3, *C. coli* positive control; lane 4, no DNA negative control; lanes 5, 7–9, 12, and 14, *C. jejuni*; lanes 6, 10, and 11, *C. coli*; lanes 13 and 15, blank. The reaction products were resolved on a 1.5% agarose gel containing ethidium bromide (0.5 μg/mL). Arrows indicate the positions of the expected PCR products for *C. coli* (894 bp) and *C. jejuni* (735 bp).

initial confirmation. We stress the importance of using fresh, 24–48 h cultures. In older cultures, Campylobacter becomes progressively more coccoid, making accurate identification problematic. A known reference strain of Campylobacter should also be used as an aid in identification.

9. When preparing frozen stocks or bringing cultures out of frozen storage, it is preferable to use a nonselective medium that will not suppress the growth of possible contaminating organisms. Cultures found to be contaminated should be discarded or re-isolated.

10. Once cell suspensions are prepared, the tubes should go immediately into the boiling water bath. If this is not practical, as when preparing large numbers of samples, keep the tubes on ice until ready to be boiled.

11. The boiled cell suspension/DNAzol mixture can be held overnight at 4°C if necessary. All steps in the DNA extraction procedure are carried out at room temperature unless otherwise noted.

12. All components, except the water, should be stored frozen at –20°C. The dNTP mixture and the working primer stocks should be aliquoted to avoid repeated (>5X) freeze-thaw steps. Once the Taq DNA polymerase has been added, avoid letting the master mixture sit for any length of time, as this may promote the generation of nonspecific PCR products. Ideally, the PCRs should be assembled in isolation, such as in a biocontainment hood, to minimize the possibility of contaminating the reactions. A set of pipetors dedicated to PCR work should be used in conjunction with aerosol-resistant pipet tips as a further means of reducing contamination.

13. Primer sets are usually shipped in lyophilized form from the manufacturer. Prepare a 400 μ*M* stock solution of each in TE buffer, pH 7.5. This calculation is easily made using the yield in nanomoles or micromoles for each primer provided by the manufacturer. Using the stock solutions, prepare 1:20 dilutions in water of each primer for the 20 μ*M* working solutions. Water is used as the diluent to minimize the amount of EDTA (which chelates the divalent magnesium ions) added to the PCR.

14. DNA template sample vol can range from 0.5–10 μL, depending on the concentration of DNA. The vol of water to add in the 1X PCR formula must be adjusted accordingly; 5 μL is a convenient volume to accurately pipet.

    Control tubes should always include a negative control in which water is substituted for the sample DNA template. In this way, contamination of the PCR mixture or nuclease-free water can be detected. A positive control using known template material serves as a check on the PCR.

Acknowledgments

The authors thank the technical staff of the Antimicrobial Resistance Research Unit, especially Sandra House, Leena Jain, Jennifer L. Murphy, and Jodie Plumblee.

References

1. Friedman, C.R., Neimann, J., Wegner, H. C. and Tauxe, R.V. (2000) Epidemiology of *Campylobacter jejuni* Infections in the United States and other Industrialized Nations in Campylobacter, 2nd ed., (Nachamkin, I. and Blaser, M. J., eds.), ASM Press, Washington, DC, USA, pp. 121–138.

2. Jacobs-Reitsma, W. (2000) Campylobacter in the food supply in Campylobacter, 2nd ed., (Nachamkin, I. and Blaser, M. J., eds.), ASM Press, Washington, DC, USA, pp. 467–481.

3. Goossens, H., and Butzler, J.-P. (1992) Isolation and identification of *Campylobacter* spp. in *Campylobacter jejuni*: Current Status and Future Trends, (Nachamkin, I., Blaser, M. J., and Tompkins, L. S., eds.), ASM Press, Washington, DC, USA, pp. 93–109.

4. Lastovika, A. J., Newell, D. G., and Lastovica, E. E. (1998) Campylobacter, Helicobacter & Related Organisms. Institute of Child Health, Red Cross Children's Hospital, University of Capetown: Rustica Press, Pinelands, South Africa.

5. Engberg, J., On, S. L. W., Harrington, C. S., and Gerner-Smidt, P. (2000) Prevalence of *Campylobacter, Arcobacter, Helicobacter*, and *Sutterella* spp. in human fecal samples as estimated by a reevaluation of isolation methods for Campylobacters. *J. Clin. Microbiol.* **38**, 286–291.

6. Lastovica, A. J., and le Roux, E. (2000) Efficient isolation of Campylobacteria from stools. *J. Clin. Microbiol.* **38**, 2798–2799.

7. Downes, F. P., and Ito, K. (2001) Microbiological Examination of Foods, 4th ed. American Public Health Association, Washington, DC, USA.

8. Bacteriological Analytical Manual, 8th ed. (1998) US Food and Drug Administration, Rockville, MD: AOAC International, Gaithersburg, MD, USA. This publication is also available on-line at (www.foodinfonet.com/publication/fdaBAM).

9. Corry, J. E. L., Post, D. E., Colin, P., and Laisney, M.J. (1995) Culture media for the isolation of campylobacters. *Int. J. Food Microbiol.* **26**, 43–76.

10. Stern, N. J., Wojton, B., and Kwiatek, K. (1992) A differential-selective medium and dry ice-generated atmosphere for recovery of *Campylobacter jejuni. J. Food Prot.* **55**, 514–517.

11. Line, J. E. (2001) Development of a selective differential agar for isolation and enumeration of *Campylobacter* spp. *J. Food Prot.* **64**, 1711–1715.

12. Giesendorf, B. A. J., and Quint, W. G. V. (1995) Detection and identification of *Campylobacter* spp. using the polymerase chain reaction. *Cell. Molec. Biol.* **41**, 625–638.

13. On, S. L. W. (1996) Identification methods for Campylobacters, Helicobacters, and related organisms. *Clin. Microbiol. Rev.* **9**, 405–422.

14. Gibbs, R. A. (1990) DNA amplification by the polymerase chain reaction. *Anal. Chem.* **62**, 1202–1214.

15. Bloch, W. (1991) A biochemical perspective on the polymerase chain reaction. *Biochemistry* **30**, 2735–2747.

16. Kolmodin, L. A., and Williams, J. F. (1999) Polymerase chain reaction, basic principles and routine practice in The Nucleic Acid Protocols Handbook (Rapley, R., ed.), Humana Press, Totowa, NJ, USA, pp. 569–580.

17. Vandamme, P., Van Doorn, L.-J., Al Rashid, S. T., et al. (1997) *Campylobacter hyoilei* Alderton et al. 1995 and *Campylobacter coli* Vééron and Chatelain 1973 are subjective synonyms. *Int. J. Syst. Bacteriol.* **47**, 1055–1060.

18. Van Camp, G., Fierens, H., Vandamme, P., Goosens, H., Huyghebaert, A., and De Wachter, R. (1993) Identification of enteropathogenic Campylobacter species by oligonucleotide probes and polymerase chain reaction based on 16S rRNA genes. *Sys. Appl. Microbiol.* **16**, 30–36.

19. Giesendorf, B. A. J., Quint, W. G. V., Henkens, M. H. C., Stegeman, H., Huf, F. A., and Niesters, H. G. M. (1992) Rapid and sensitive detection of *Campylobacter* spp. in chicken products by using the polymerase chain reaction. *Appl. Environ. Microbiol.* **58**, 3804–3808.

20. Fermér, C., and Engvall, E. O. (1999) Specific PCR identification and differentiation of the thermophilic Campylobacters, *Campylobacter jejuni, C. coli, C. lari,* and *C. upsaliensis. J. Clin. Microbiol.* **37**, 3370–3373.

21. Konkel, M. E., Gray, S. A., Kim, B. J., Garvis, S. G., and Yoon, J. (1999) Identification of the enteropathogens *Campylobacter jejuni* and *Campylobacter coli* based on the cadF virulence gene and its product. *J. Clin. Microbiol.* 37, 510–517.

22. Oyofo, B. H., Thornton, S. A., Burr, D. H., Trust, T. J., Pavlovskis, O. R., and Guerry, P. (1992) Specific detection of *Campylobacter jejuni* and *Campylobacter coli* by using polymerase chain reaction. *J. Clin. Microbiol.* **30**, 2613–2619.

23. Wegmüller, B., Lüthy, J., and Candrian, U. (1993) Direct polymerase chain reaction detection of *Campylobacter jejuni* and *Campylobacter coli* in raw milk and dairy products. *Appl. Environ. Microbiol.* **59**, 2161–2165.

24. Linton, D., Lawson, A. J., Owen, R. J., and Stanley, J. (1997) PCR detection, identification to species level, and fingerprinting of *Campylobacter jejuni* and *Campylobacter coli* direct from diarrheic samples. *J. Clin. Microbiol.* **35**, 2568–2572.

25. Gonzalez, I., Grant, K. A., Richardson, P. T., Park, S. F., and Collins, M. D. (1997) Specific identification of the enteropathogens *Campylobacter jejuni* and *Campylobacter coli* by using a PCR test based on the ceuE gene encoding a putative virulence determinant. *J. Clin. Microbiol.* **35**, 759–763.

26. Englen, M. D., and Kelley, L.C. (2000) A rapid DNA isolation procedure for the identification of *Campylobacter jejuni* by the polymerase chain reaction. *Lett. Appl. Microbiol.* **31**, 421–426.

27. Edwards, M. C., and Gibbs, R. A. (1994) Multiplex PCR: advantages, development, and applications. *PCR Methods Appl.* **3**, S65–75.

28. Mohran, Z. S., Arthur, R. R., Oyofo, B. A., et al. (1998) Differentiation of Campylobacter isolates on the basis of sensitivity to boiling in water as measured by PCR-detectable DNA. *Appl. Environ. Microbiol.* **64**, 363–365.

29. Musgrove, M. T., Cox, N. A., Stern, N.J., Berrang, M. E., and Harrison, M.A. (2001) Tissue culture plate enrichment method for recovery of *Campylobacter* spp. from broiler carcass rinses. *Poultry Sci. SPSS/SCAD* **Abstr. 80,** 17–18.

30. Hoar, B. R., Atwill, E. R., Elmi, C., Utterback, W. W., and Edmondson, A. J. (1999) Comparison of fecal samples collected per rectum and off the ground for estimation of environmental contamination attributable to beef cattle. *Am. J. Vet. Res.* **60,**1352–1356.

31. On, S. L. W., and Holmes, B. (1992) Assessment of enzyme detection tests useful in the identification of Campylobacteria. *J. Clin. Microbiol.* **30**, 746–749.

32. On, S. L. W., and Holmes, B. (1991) Reproducibility of tolerance tests that are useful in the identification of Campylobacteria. *J. Clin. Microbiol.* **29**, 1785–1788.

8

Detection and Differentiation of Chlamydiae by Nested PCR

Konrad Sachse and Helmut Hotzel

1. Introduction

1.1. Characteristic Features and Importance of the Agents

Chlamydiae are obligate intracellular prokaryotes with a hexalaminar cell wall that, in contrast to other gram-negative bacteria, contains no peptidoglycan. As a major antigenic constitutent, their outer membrane contains a 10-kDa lipopolysaccharide (LPS) with a trisaccharide epitope specific for the family *Chlamydiaceae (1)*. Another characteristic antigenic component is the major outer membrane protein (MOMP), a cysteine-rich protein of approx 40 kDa representing approx 60% of the weight of the outer membrane. This molecule harbors several genus- and species-specific antigenic determinants in the conserved regions and serovar-specific epitopes in variable domains *(2)*.

A unique feature of chlamydiae is the biphasic developmental cycle, in the course of which two distinct morphological forms emerge. At the extracellular stage, the smaller infectious and metabolically inactive elementary bodies are prevailing, whereas the larger metabolically active and self-replicating reticular bodies reside in vacuole-like cytoplasmic inclusions of the host cell.

Traditionally, the genus *Chlamydia (C.)* comprised four species, i.e., *C. trachomatis*, *C. psittaci*, *C. pneumoniae*, and *C. pecorum*. However, a large amount of new DNA sequence data led Everett et al. *(3)* to reassess genetic relatedness and propose taxonomic reclassification. According to this proposal, the family *Chlamydiaceae* consists of two genera, *Chlamydia* and *Chlamydophila*, with a total of nine largely host-related species. The two classification schemes are summarized in **Table 1**.

From: *Methods in Molecular Biology, vol. 216: PCR Detection of Microbial Pathogens: Methods and Protocols*
Edited by: K. Sachse and J. Frey © Humana Press Inc., Totowa, NJ

Table 1
Comparison of Old and New Taxonomic Classification Schemes
for the Family *Chlamydiaceae*

Traditional species	New species *(3)*	Host specificity[a]
Chlamydia trachomatis	*Chlamydia trachomatis*	human
	Chlamydia muridarum	mouse, hamster
	Chlamydia suis	swine
Chlamydia psittaci	*Chlamydophila psittaci*	birds, ruminants, horse
	Chlamydophila abortus	sheep, other rumi nants, swine, birds
	Chlamydophila caviae	guinea pig
	Chlamydophila felis	domestic cat
Chlamydia pecorum	*Chlamydophila pecorum*	cattle, sheep, goat, swine, koala
Chlamydia pneumoniae	*Chlamydophila pneumoniae*	human, koala, horse, amphibians

[a] Hosts are given according to present knowledge. As species-specific methods of identification will be used to a greater extent in the future, further widening of the host range can be expected in some cases.

The group of chlamydial species includes agents of important animal and human diseases. Chlamydiae are very wide spread in many host organisms, but not nearly all carriers develop symptoms of disease.

Avian strains of *C. psittaci* (new classification: *Chlamydophila psittaci*) can cause psittacosis, a systemic disease in psittacine birds of acute, protracted, chronic, or subclinical manifestation *(4,5)*. The analogous infection in domestic and wild fowl is known as ornithosis. Avian chlamydiosis is transmissible to humans, the symptoms being mainly nonspecific and influenza-like, but severe pneumonia, endocarditis, and encephalitis are also known.

Enzootic abortion in sheep and goats is caused by the ovine subtype of *C. psittaci* (new: *Chlamydophila abortus*). The disease has a major economic impact as it represents the most important cause of loss in sheep and goats in parts of Europe, North America, and Africa *(6)*. This serious and potentially life-threatening zoonosis also affects pregnant women after contact with lambing ewes and leads to severe febrile illness in pregnancy *(7,8)*.

In cattle, *C. psittaci* and *C. pecorum* were found in connection with infections of the respiratory and genital tracts, enteritis, arthritis, encephalomyelitis *(9)*, as well as endometritis and hypofertility *(10)*.

Chlamydioses in pigs are associated with three different species, i.e., *C. trachomatis* (new: *Chlamydia suis*), *C. pecorum* and *C. psittaci (11)*. A widely held view is that chlamydiae may act in concert with other agents in multifactorial infectious diseases, such as abortions in sows, polyarthritis in piglets, diarrhea in pigs, and genital disorders in boars *(12)*.

Other relevant animal diseases include conjunctivitis in cats caused by the feline serovar of *C. psittaci* (new: *Chlamydophila felis*), respiratory disorders and abortion in horses caused by *C. pneumoniae* (new: *Chlamydophila pneumoniae*) and *C. psittaci*, respectively. *C. pneumoniae* has also been isolated from diseased koalas and frogs.

Apart from zoonotic diseases mentioned above, chlamydiae are responsible for a number of diseases in humans, e.g., trachoma, respiratory infection, sexually transmitted infection of reproductive organs (*C. trachomatis*), as well as pneumonia in adults and cardiovascular diseases (*C. pneumoniae*).

1.2. Conventional Diagnostic Methods

As chlamydiae need to pass through an intracellular stage during their life cycle, they require tissue culture techniques to be isolated and propagated *(13)*. Indeed, recovery of the germs through culture in suitable cell lines, such as Buffalo Green Monkey (BGM), McCoy, and HeLa, or via inoculation of embryonated hens' eggs is still regarded as the standard method in chlamydial diagnosis. Naturally, culture is an indispensable prerequisite for demonstrating the viability of a field strain, as well as for its detailed characterization by molecular and biochemical methods. There is, however, no correlation between in vitro growth and pathogenicity of an isolate. While many avian and ovine strains of *C. psittaci* (new: *Chlamydophila psittaci/abortus*) or porcine isolates of *C. trachomatis* (new: *C. suis*) can be propagated in tissue culture relatively easily, others are more difficult to grow, e.g., strains from cattle and swine belonging to *C. psittaci*. In any case, diagnosis by cell culture requires very experienced laboratory workers and standardized protocols. Results are available within 48–72 h for well-growing isolates, but may be delayed for 2–6 wk with more difficult samples.

Other methods of antigen detection include histochemical staining, e.g., Giménez stain *(14)*, and immune fluorescence. While the methodology is relatively simple, sensitivity is lower than for culture, and as a rule, positive findings merely indicate the presence of chlamydiae in general.

Commercially available antigen enzyme linked immunosorbent assays (ELISAs) targeting chlamydial LPS only allow the detection of the genus *Chlamydia* (new: *Chlamydia* and *Chlamydophila*) without the possibility of species identification *(15)*. Likewise, tests based on MOMP as target antigen are limited in their specificity. The attainable detection limit of 10^2–10^5 inclusion-forming units (ifu) can be insufficient for certain categories of field samples.

Table 2
Primers for Detection of Chlamydiae

Denomination	Sequence[a] (5'-3')
191CHOMP	GCI YTI TGG GAR TGY GGI TGY GCI AC
CHOMP371	TTA GAA ICK GAA TTG IGC RTT IAY GTG IGC IGC
201CHOMP	GGI GCW GMI TTC CAA TAY GCI CAR TC
CHOMP336s[b]	CCR CAA GMT TTT CTR GAY TTC AWY TTG TTR AT
218PSITT	GTA ATT TCI AGC CCA GCA CAA TTY GTG
TRACH269	ACC ATT TAA CTC CAA TGT ARG GAG TG
PNEUM268	GTA CTC CAA TGT ATG GCA CTA AAG A
204PECOR	CCA ATA YGC ACA ATC KAA ACC TCG C

[a] Degenerate nucleotides: K = G, T; M = A, C; R =A, G; W = A, T; Y = C , T; I = inosine.
[b] Modified from **ref. (21)**.

Among serodiagnostic tests, the antibody ELISA is easy to handle, suitable for high sample throughput and more sensitive and faster than the complement fixation test *(16)*. The latter is still widely used despite being rather laborious, poorly reproducible between different laboratories, and having low specificity. Generally, serological tests are based on the two main cross-reactive antigens present in all chlamydial species, LPS and MOMP, and so are not species-specific. New tests using highly specific capture antigens have been developed *(17)*, but still need to be validated.

1.3. PCR-Based Detection and Differentiation

The possibilities of diagnostic detection of chlamydiae have considerably improved with the introduction of molecular methods, particularly the polymerase chain reaction (PCR), which permits direct identification from clinical specimens and genetically based differentiation of species.

Among the large number of tests published in the literature, only a few were designed to cover all chlamydial species. They utilize two different genomic target regions for amplification, i.e., the ribosomal RNA gene region *(18–20)* and the gene encoding the MOMP antigen designated *omp*1 or *omp*A *(21,22)*. The latter harbors four variable domains known as VD I–IV and five conserved regions. Genus- and species-specific antigenic determinants are encoded by the conserved regions, and serovar-specific segments are located in VD I and VD II. This heterogeneous primary structure makes the *omp1* gene an ideal target for diagnostic PCR.

In the present chapter, a nested-PCR assay targeting the *omp*1 gene is described. The methodology is based on the paper of Kaltenböck et al. *(21)*, only one primer was modified (*see* **Table 2**), but amplification profiles and

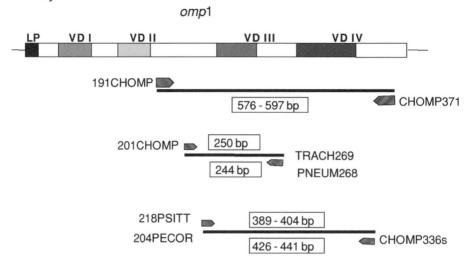

Fig. 1. Principle of detection and differentiation of chlamydial species using a nested PCR system targeting the *omp1* gene. The upper bar represents the *omp1* gene with the leader peptide (LP) and VD I–IV. In the first amplification (outer primer pair 191CHOMP/CHOMP371), a genus-specific product is generated. The second round of PCR involves one genus-specific inner primer (201CHOMP or CHOMP336s) and one species-specific primer (TRACH269/PNEUM268 or 218PSITT/204PECOR, respectively). Boxed numbers denote amplicon sizes.

pre-PCR processing of samples were optimized in our laboratory. In the first amplification, primer binding sites are located in conserved regions between VD II/VD III and downstream VD IV, whereas in the second round, primers flanking VD III (*C. trachomatis*, *C. pneumoniae*) and VD III or VD IV (*C. psittaci*, *C. pecorum*), respectively, are used. The principle of the nested amplifications and species differentiation is depicted in **Fig. 1**. Although the procedure identifies chlamydial species according to the traditional classification, we found it very robust for routine use and the most sensitive among several protocols (*see* **Note 1**).

2. Materials

2.1. DNA Extraction

1. Water. Deionized water must be used for all buffers and dilutions.
2. Lysis buffer: 100 mM Tris-HCl, pH 8.5, 0.05% (v/v) Tween® 20.
3. Proteinase K: 10 mg/mL in water
4. Phosphate-buffered saline (PBS): 10 mM Na$_2$HPO$_4$, 10 mM NaH$_2$PO$_4$, 145 mM NaCl, pH 7.0. Adjust pH by adding NaH$_2$PO$_4$.
5. Tris-EDTA (TE) buffer: 10 mM Tris-HCl, pH 8.0, 1 mM EDTA (ethylene diamine tetraacetic acid).
6. Sodium dodecyl sulfate (SDS) solution: 10 mg/mL in TE.

7. Phenol: saturated solution in TE buffer. If two separate phases are visible, use the lower phase only.
8. Chloroform-isoamyl alcohol: 24:1 (v/v).
9. Isopropanol, analytical, or molecular biology grade.
10. Commercially available DNA extraction kit for PCR template preparation (*see* **Note 2**). In our hands, the following products worked well: High Pure PCR Template Preparation Kit (Roche Diagnostics, Mannheim, Germany), QIAamp® DNA Mini Kit (Qiagen, Hilden, Germany), E.Z.N.A. Tissue DNA Kit II (Peqlab, Erlangen, Germany).

2.2. PCR

1. *Taq* DNA polymerase. We use MasterTaq (5 U/µL) from Eppendorf (Hamburg, Germany).
2. 10X Reaction buffer for *Taq* DNA polymerase: provided by the manufacturer of the enzyme.
3. dNTP mixture: dATP + dGTP + dCTP + dTTP, 2 mM each. Store in aliquots at −20°C.
4. Primer oligonucleotides according to **Table 2**.

2.3. Electrophoresis and Visualization

1. Agarose, molecular biology-grade: 1% gels for PCR products of 300–1000 bp, 2% gels for products below 300 bp.
2. Tris-borate EDTA electrophoresis buffer (TBE): 0.09 M Tris-borate, 0.002 M EDTA, pH 8.0. For 1 L of 10X TBE, mix 108 g Tris-base, 55 g boric acid, and 80 mL of 0.25 M EDTA, make up with water. Dilute 1:10 before use.
3. Gel loading buffer (GLB): 20% (v/v) glycerol, 0.2 M EDTA, 0.01% (w/v) bromophenol blue, 0.2% (w/v) Ficoll® 400.
4. Ethidium bromide stock solution: 1% (10 mg/mL) solution in water.
   CAUTION: The substance is presumed to be mutagenic. Avoid direct contact with skin. Wear gloves when preparing solutions and handling gels.
5. DNA size marker. We mostly use the 100-bp DNA ladder (Gibco/Life Technologies, Eggenstein, Germany). For large fragments, *Hin*dIII-digested λ DNA (Roche Diagnostics) may be used.

2.4. General Equipment and Consumables

1. Thermal cycler. We use the T3 Thermal cycler (Biometra, Goettingen, Germany) and the Mastercycler (Eppendorf).
2. Vortex shaker, e.g., MS1 Minishaker (IKA Works, Wilmington, DE, USA).
3. Benchtop centrifuge with Eppendorf rotor, e.g., Model 5402 (Eppendorf) and/or a mini centrifuge, e.g., Capsule HF-120 (Tomy Seiko, Tokyo, Japan).
4. Heating block, for incubation of Eppendorf tubes, adjustable temperature range 30–100°C.
5. Apparatus for horizontal gel electrophoresis.
6. UV transilluminator, 254 nm and/or 312 nm.

7. Video documentation or photographic equipment.
8. Set of pipets covering the whole vol range from 0.1–1000 µL. We use the Eppendorf Research series (Eppendorf).
9. Aerosol-resistant pipet tips (filter tips),
10. Plastic tubes 0.2 or 0.5 mL, sterile, thin-walled, DNase-, and RNase-free (Eppendorf) for PCR.
11. Plastic tubes 1.5 and 2.0 mL for pre-PCR operations.

3. Methods

3.1. DNA Extraction from Different Sample Matrixes

3.1.1. Broth Culture

The simplest method to release DNA suitable for PCR from chlamydial cell cultures is 5-min boiling. After removal of cellular debris by centrifugation at 12,000*g* for 1 min, the supernatant can be used as template. Failure to amplify a specific target could be due to the presence of PCR inhibitors. In these instances, a commercial DNA extraction kit should be tried (*see* **Note 2**).

3.1.2. Swabs (e.g., Nasal, Vaginal, Conjunctival) and Mucus, Bronchoalveolar Lavage or Sputum

1. Pipet 500 µL of lysis buffer into a 2-mL Safe-Lock tube containing the cotton swab.
2. Vortex mix thoroughly for 1 min
3. Centrifuge at 12,000*g* for 30 s.
4. Put the swab into a 1-mL pipet tip whose lower half was cut off and place it all into a fresh tube.
5. Centrifuge at 12,000*g* for 1 min to press the remaining liquid out of the cotton.
6. Add the liquid to that in the first tube from **step 3**. If you have samples of mucus, bronchoalveolar lavage or sputum, start with **step 7**.
7. Centrifuge the liquid or mucus at 12,000*g* for 15 min.
8. Discard the supernatant and resuspend the pellet in 50 µL of lysis buffer.
9. Add 20 µL of proteinase K and incubate at 60°C for 2 h.
10. Inactivate the proteinase K by heating at 97°C for 15 min.
11. Centrifuge at 12,000*g* for 5 min to remove debris.
12. Use 5 µL of the supernatant for PCR.

3.1.3. Tissue from Lung, Tonsils, Lymphnodes, Spleen, Liver, and Other Organs

1. Boil 100 mg of homogenized tissue in 200 µL of water in a plastic tube for 10 min. Subsequently, allow the tube to cool to room temperature.
2. Optionally, proteinase digestion can be carried out to increase the final yield of DNA: add 200 µL SDS solution and 20 µL of proteinase K to the tube and incubate at 55°C for 1 h.
3. Add 200 µL of phenol.

4. Vortex mix vigorously for 1 min.
5. Centrifuge at 12,000g for 5 min.
6. Transfer the (upper) aqueous phase into a fresh tube.
7. Add 200 μL of chloroform-isoamyl alcohol.
8. Vortex mix at highest intensity for 1 min.
9. Pipet the (upper) aqueous phase into a fresh tube.
10. Precipitate DNA by adding 120 μL of isopropanol. Thoroughly mix the reagents and incubate at room temperature for 10 min.
11. Collect DNA by centrifugation at 12,000g for 10 min. Discard supernatant.
12. Allow DNA pellet to air-dry for 30 min.
13. Redissolve pellet in 20 μL of water. Use 1 μL for an amplification reaction.

The described procedure is the simplest method for DNA extraction from tissue specimens. Alternatively, commercial DNA extraction kits can be used (*see* **Note 2**).

3.1.4. Feces (see **Notes 2** and **3**)

1. Add 200 μL of water to 100 mg of feces and vortex mix vigorously for 1 min.
2. Boil the suspension for 10 min.
3. Add 300 μL of phenol to each tube for DNA extraction and vigorously vortex mix the mixture for 1 min.
4. Centrifuge at 14,000g for 5 min.
5. Transfer the (upper) aqueous phase into fresh tubes.
6. Add 300 μL of chloroform-isoamyl alcohol to each tube.
7. Vortex mix at highest intensity for 1 min.
8. Centrifuge at 14,000g for 5 min.
9. Transfer the (upper) aqueous phase into a fresh tubes.
10. Add 200 μL of isopropanol for DNA precipitation.
11. Mix reagents and incubate at room temperature for 10 min.
12. Centrifuge at 14,000g for 10 min. Discard supernatant.
13. Allow pellet (with DNA) to air-dry for 30 min.
14. Dissolve pellet in 20 μL water. Use 1 μL in an amplification reaction.

3.1.5. Semen (see **Note 2**)

1. Dilute 50 μL of semen in a plastic tube with 150 μL of SDS and homogenize by intensive vortex mixing.
2. Digest proteins by adding 20 μL of proteinase K solution and vortex mix for 1 min.
3. Incubate at 55°C for 1 h.
4. Continue extraction procedure as described for tissue (*see* **Subheading 3.1.3.**) beginning with **step 3**.

3.1.6. Milk Samples

We use the QIAamp DNA Stool Kit (Qiagen) according to the instructions of the manufacturer.

3.2. DNA Amplification

3.2.1. Genus-Specific Detection of Chlamydiae

1. Prepare a master mixture of reagents for all amplification reactions of the series. It should contain the following ingredients per 50-µL reaction: 1µL dNTP mixture (2 m*M* each), 1 µL primer 191CHOMP (20 pmol/µL), 1 µL primer CHOMP371 (20 pmol/µL), 5 µL reaction buffer (10X), 0.2 µL *Taq* DNA polymerase (5 U/µL), and 40.8 µL H$_2$O (36.8 µL in case of swab specimens).
2. Add template to each reaction vessel: 1 µL of DNA extract from infected tissue or 5 µL of extract from swab samples.
3. Include amplification controls: DNA of a chlamydial reference strain (positive control) and water (negative control 1) instead of sample extract.
4. Run PCR according to the following temperature–time profile: Initial denaturation at 95°C for 30 s, 35 cycles of denaturation (95°C for 30 s), primer annealing (50°C for 30 s), and primer extension (72°C for 30 s).
5. Correct amplification leads to the formation of a 576–597-bp product specific for the genus *Chlamydia* (according to the new taxonomy: *Chlamydia* and *Chlamydophila*).

3.2.2. Species-Specific Detection of Chlamydiae

1. Choose primer pairs according to the scheme in **Fig. 1** and **Table 2**. Prepare a master mixture of reagents for all amplification reactions of the series. It should contain the following ingredients per 50-µL reaction: 1 µL dNTP mixture (2 m*M* each), [1 µL forward primer 201CHOMP (20 pmol/µL) plus 1 µL reverse primer TRACH269 or PNEUM268 (20 pmol/µL)] or [1 µL forward primer 204PECOR or 218PSITT (20 pmol/µL) plus 1 µL reverse primer CHOMP336s (20 pmol/µL)], 5 µL reaction buffer (10X), 0.2 µL *Taq* DNA polymerase (5U/µL), and 40.8 µL H$_2$O.
2. Add 1 µL of the product from genus-specific PCR (*see* **Subheading 3.2.1.**) as template to each reaction vessel.
3. Subject the products of positive control and negative control 1 (1 µL of each) from the previous amplification to the second round of nested PCR. Additionally include a fresh reagent control (negative control 2).
4. Run PCR according to the following temperature–time profile: initial denaturation at 95°C for 30 s, 20 cycles of denaturation (95°C for 30 s), primer annealing (60°C for 30 s) and primer extension (72°C for 30 s), (*see* **Note 4**).
5. The correct sizes of species-specific amplicons are 250 bp for *C. trachomatis*, 244 bp for *C. pneumoniae*, 389–404 bp for *C. psittaci*, and 426–441 bp for *C. pecorum* (*see* **Fig. 1**).

The specificity of this procedure is illustrated in **Fig. 2** and discussed (*see* **Note 5**). The detection limit of the first-round PCR is in the order of 10^2 ifu, and nested amplification allowed the detection of 10^{-1} ifu (*see* **Fig. 3** and **Note 6**).

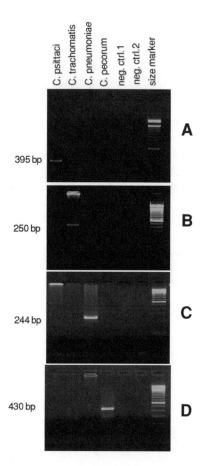

Fig. 2. Specificity of the nested amplification assay. Chromosomal DNA of field isolates of *C. psittaci*, *C. trachomatis*, *C. pneumoniae*, and *C. pecorum* was amplified according to the protocols given in **Subheadings 3.2.1.** and **3.2.2.** In the second round, the following primer pairs were used: 218PSITT/CHOMP336s (**A**), 201CHOMP/TRACH269 (**B**), 201CHOMP/PNEUM268 (**C**), and 204PECOR/CHOMP336s (**D**). Amplicon sizes are given at the left-hand margin.

3.3. Electrophoresis and Visualization

1. Prepare 1 or 2% (w/v) solution of agarose in TBE. Store gel in Erlenmeyer flasks at room temperature.
2. Liquefy gel by microwave heating (approx 30 s at 600 W) prior to use.
3. Pour gel on a horizontal surface using an appropriately sized frame.
4. Fill electrophoresis tank with TBE buffer.
5. Run the gel at a voltage corresponding to 5 V/cm of electrode distance for approx 30 min.

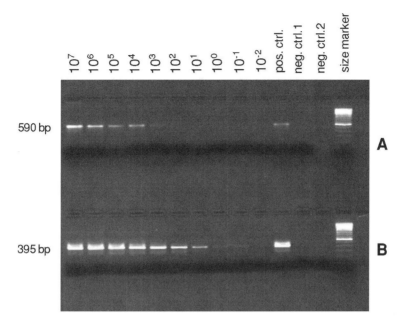

Fig. 3. Sensitivity of detection of *C. psittaci*. Chromosomal DNA was prepared from a culture of strain C5/98 of *C. psittaci* (new classification: *Chlamydophila psittaci*) in BGM cells containing 10^8 ifu/mL. Aliquots from a dilution series were amplified according to the protocols given in **Subheadings 3.2.1.** and **3.2.2.** **(A)** Genus-specific amplification using primer pair 191CHOMP/CHOMP371. **(B)** Nested PCR using primers 218PSITT/CHOMP336s in the second round.

6. Load each well with 10 µL of PCR product mixed with 5 µL GLB.
7. Stain DNA bands by immersing the gel in ethidium bromide solution containing 5 µL of stock solution in 200 mL of water. (Alternatively, ethidium bromide-containing agarose gels can be used. Add 5 µL of ethidium bromide solution to 100 mL of melted agarose in TBE buffer.)
8. Visualize bands under UV light using a transilluminator.

4. Notes

1. For differentiation of species according to the new taxonomic classification, the method of Everett and Andersen *(19)* is recommended, which involves amplification of a approx 600-bp fragment of the 16–23S intergenic spacer region of the rRNA operon with subsequent restriction enzyme analysis. However, the method is less sensitive than the present *omp*1-PCR by two orders of magnitude *(23)*, which makes it unsuitable for most field samples. In our hands, its main area of application is the identification and differentiation of cultured strains.
2. The use of commercial DNA preparation kits can be recommended for samples of organ tissue, broth culture, and, with some qualification, also for semen and feces.

In the latter instances, the kit should be tested with a series of spiked samples containing defined numbers of chlamydia cells in order to examine its suitability.

Most commercial kits are easy to work with. They contain a special buffer reagent for lysis of the bacterial and tissue cells, the effectiveness of which is decisive for the kit's performance. An optional RNase digestion is intended to remove cellular RNA. The lysate is then centrifuged through a mini-column, where the released DNA is selectively bound to a solid phase (modified silica, hydroxyl apatite, or special filter membrane). After washing, the DNA can be eluted with an elution buffer or water. DNA prepared in this manner is usually of high purity and free of PCR inhibitors.

It should be noted, however, that the yield of extracted DNA is limited by the binding capacity of the mini-column. If maximum recovery of chlamydial DNA from high-titer samples is important, e.g., in quantitative assays or for preparation of reference DNA, one of the multi-step extraction protocols should be followed.

3. Commercial DNA extraction kits for human stool specimens can be used in principle. In our hands, the QIAamp DNA Stool Kit and the Invitek Spin Stool DNA Kit (InVitek, Berlin, Germany) worked reasonably well. However, chlamydial DNA recovery from specimens containing less than 10^3 ifu proved difficult. For these cases, we recommend phenol-chloroform extraction as described in **Subheading 3.1.4.**

4. Practical experience from routine use of the present methodology led us to set the number of cycles to 20 in the second round of amplification. Running 30 cycles or more would yield an increase in sensitivity, but the nested assay would be more vulnerable to carryover contamination.

5. To assess the performance of the present PCR assay, a validation study was conducted in our laboratory *(23)*. In a series of 99 samples, PCR proved more sensitive than cell culture, whereas both methods were nearly equivalent in terms of specificity. The figures for cell culture were: sensitivity 74%, specificity 96%.

6. For correct interpretation of the sensitivity figures, it has to be noted that the count of ifu is representative of the number of elementary bodies, but does not comprise all chlamydial cells present in the culture or sample. Reticular bodies residing in cellular inclusions also provide target DNA. Consequently, a content of 10^{-1} ifu in infected cell culture or tissue does not necessarily mean that there is less than one chlamydial cell.

References

1. Brade H., Brade L., and Nano, F. E. (1987) Chemical and serological investigations on the genus-specific lipopolysaccharide epitope of *Chlamydia. Proc. Natl. Acad. Sci. USA* **84,** 2508–2512.

2. Conlan, J. W., Clarke, I. N., and Ward, M. E. (1988) Epitope mapping with solid-phase peptides: identification of type-, subspecies-, species-, and genus-reactive antibody binding domains on the major outer membrane of *Chlamydia trachomatis. Mol. Microbiol.* **2,** 673–679.

3. Everett, K. D. E., Bush, R. M., and Andersen, A. A. (1999) Emended description of the order *Chlamydiales*, proposal of *Parachlamydiaceae* fam. nov. and *Simkaniaceae* fam. nov., each containing one monotypic genus, revised taxonomy of the family *Chlamydiaceae*, including a new genus and five new species, and standards for the identification of organisms. *Int. J. Syst. Bacteriol.* **49,** 415–440.

4. Kaleta, E. F. (1997) Aktuelle Fragen der Diagnose und Bekämpfung der Psittakose. *Tierärztl. Umschau* **52,** 36–44.

5. Vanrompay, D., Ducatelle, R., and Haesebrouck, F. (1995) *Chlamydia psittaci* infections: a review with emphasis on avian chlamydiosis. *Vet. Microbiol.* **45,** 93–119.

6. Aitken, I. D. (2000) Chlamydial abortion, in *Diseases of sheep* (Martin, W. B., and Aitken, I. D., eds.), Blackwell Science, Oxford, UK, pp. 81–86.

7. Buxton, D. (1986) Potential danger to pregnant women of *Chlamydia psittaci* from sheep. *Vet. Rec.* **118,** 510–511.

8. Kampigna, G. A., Schroder, F. P., Visser, I. J., Anderson, J. M., Buxton, D., and Moller, A. V. (2000) Lambing ewes as a source of severe psittacosis in a pregnant woman. *Ned. Tijdschr. Geneeskd.* **144,** 2500–2504.

9. Storz, J., and Kaltenböck, B. (1993) Diversity of chlamydia-induced diseases, in *Rickettsial and Chlamydial Diseases of Domestic Animals* (Woldehiwet, Z. and Ristic, M., eds.), Pergamon Press, Oxford, UK, pp. 363–393.

10. Wittenbrink, M. M., Schoon, H. A., Bisping, W., and Binder, A. (1993) Infection of the bovine female genital tract with *Chlamydia psittaci* as a possible cause of infertility. *Reprod. Dom. Anim.* **28,** 129–136.

11. Wittenbrink, M. M., Wen, X., Böhmer, N., Amtsberg, G., and Binder, A. (1991) Bakteriologische Untersuchungen zum Vorkommen von *Chlamydia psittaci* in Organen von Schweinen und abortierten Schweinefeten. *J. Vet. Med.* **B38,** 411–420.

12. Szeredi, L., Schiller, I., Sydler, T., et al. (1996) Intestinal *Chlamydia* in finishing pigs. *Vet. Pathol.* **33,** 369–374.

13. Thejls, H., Gnarpe, J., Gnarpe, H., Larsson, P. G., Platz-Christensen, J. J., Östergaard, L., and Victor, A. (1994) Expanded gold standard in the diagnosis of *Chlamydia trachomatis* in a low-prevalence population: diagnostic efficacy of tissue culture, direct immunofluorescence, enzyme immunoassay, PCR and serology. *Genitourin. Med.* **70,** 300–303.

14. Vanrompay, D., Ducatelle, R., and Haesebrouck, F. (1992) Diagnosis of avian chlamydiosis: specificity of the modified Gimenez staining on smears and comparison of the sensitivity of isolation in eggs and three different cell cultures. *J. Vet. Med.* **B39,** 105–112.

15. Vanrompay, D., van Nerom, A., Ducatelle, R., and Haesebrouck, F. (1994) Evaluation of five immunoassays for detection of *Chlamydia psittaci* in cloacal and conjunctival specimens from turkeys. *J. Clin. Microbiol.* **32,** 1470–1474.

16. Anonymous (1996) Avian Chlamydiosis, in *O.I.E. Manual of Standards for Diagnostic Tests and Vaccines, 3rd edition Ch. 3.6.4.,* Office International des Epizooties, Paris, France, pp. 526–530.

17. Longbottom, D., Psarrou, E., Livingstone, M., and Vretou, E. (2001) Diagnosis of ovine enzootic abortion using an indirect ELISA (rOMP91B iELISA) based on a recombinant protein fragment of the polymorphic outer membrane protein POMP91B of *Chlamydophila abortus. FEMS Microbiol. Lett.* **195,** 157–161.

18. Messmer, T. O., Skelton, S. K., Moroney, J. F., Daugharty, H., and Fields, B. S. (1997) Application of a nested, multiplex PCR to psittacosis outbreaks. *J. Clin. Microbiol.* **35,** 2043–2046.

19. Everett, K. D. E., and Andersen, A. A. (1999). Identification of nine species of the *Chlamydiaceae* using RFLP-PCR. *Int. J. Syst. Bacteriol.* **49,** 803–813.

20. Madico, G., Quinn, T. C., Bomann, J., and Gaydos, C. A. (2000) Touchdown enzyme release-PCR for detection and identification of *Chlamydia trachomatis, C. pneumoniae,* and *C. psittaci* using the 16S and 16S–23S spacer rRNA genes. *J. Clin. Microbiol.* **38,** 1085–1093.

21. Kaltenböck, B., Schmeer, N., and Schneider, R. (1997) Evidence for numerous *omp*1 alleles of porcine *Chlamydia trachomatis* and novel chlamydial species obtained by PCR. *J. Clin. Microbiol.* **35,** 1835–1841.

22. Yoshida, H., Kishi, Y., Shiga, S., and Hagiwara, T. (1998) Differentiation of *Chlamydia* species by combined use of polymerase chain reaction and restriction endonuclease analysis. *Microbiol. Immunol.* **42,** 411–414.

23. Sachse, K. and Hotzel, H. (2001) Unpublished results.

9

Detection of Toxigenic Clostridia

Michel R. Popoff

1. Introduction

Clostridia are anaerobic spore-forming bacteria that are widespread in the environment. They produce many extracellular hydrolytic enzymes and are especially involved in the decomposition of carcasses and plants in natural conditions. Some species produce potent toxins and are pathogenic for man and animals. Clostridia do not invade healthy cells nor multiply within them. They are able to enter host organisms by two ways, the oral route and wounds, but their proliferation in the intestinal content or in wounds requires the presence of risk factors. Thus, incomplete or nonfunctional digestive microflora in newborns, perturbation of the digestive microflora by antibiotics, overfeeding, intestinal stasis, or malignancy of the intestinal wall represent common factors permitting clostridial growth. Deep wounds forming a small hole on the outside and harboring necrotic tissues enable their implantation in connective and muscular tissues. Toxins as the main virulence factors are responsible for all symptoms and lesions observed in clostridial diseases. Consequently, toxins are the main target for diagnosis of clostridial diseases, as well as the basis for efficient vaccines.

Two main classes of clostridial diseases can be distinguished according to the mode of acquisition of the pathogen: (*i*) digestive and food-borne diseases which are a consequence of the clostridia's entry by the oral route; and (*ii*) the affections due to the penetration of the agent through the teguments. The former group of diseases includes enteritis, necrotic and/or hemorrhagic enteritis, and enterotoxemia. Necrotic hepatitis, bacillary hemoglobinuria, blackleg and a nervous affection characterized by flaccid paralysis (botulism) are also caused by orally acquired clostridia. Entry of clostridia through the skin or mucosa may result in gangrene, malignant edema and also tetanus, a nervous disease with

From: *Methods in Molecular Biology, vol. 216: PCR Detection of Microbial Pathogens: Methods and Protocols*
Edited by: K. Sachse and J. Frey © Humana Press Inc., Totowa, NJ

spastic paralysis. The general paradigm of pathogenicity consists in overgrowth of *Clostridium* spp. on the site of infection, mainly intestine or wound, production and diffusion of toxin(s), which specifically interact with cells of the host organism. One exception is botulism, in which the toxin can be produced during growth of *Clostridium botulinum* in food. Ingestion of food containing the mere botulinum toxin can be sufficient for the outbreak of the disease.

Diagnosis of clostridial infections is based on the toxins and/or phenotypic characters. *Clostridium* spp. can be identified by classical methods of bacteriology including colony isolation, characterization of the isolated clones, biochemical properties, gas chromatography, and toxin production. The main difficulty is that most of the pathogenic *Clostridium* spp. are strictly anaerobic bacteria, which hampers their isolation from biological samples and identification by culture. Most of the clostridial toxin genes are well characterized. This permits the use of molecular methods for detection and identification of toxigenic *Clostridium* species from biological and food samples. The present chapter is focused on the identification of toxigenic *Clostridium* spp. by polymerase chain reaction (PCR) methods targeting toxin genes.

2. Materials

1. Culture (10 mL) of *Clostridium* strain in tryptone glucose yeast extract (TGY) broth medium (trypticase 30 g/L, glucose 5 g/L, yeast extract 20 g/L, cysteine-HCl 0.5 g/L, pH 7.4).
2. Culture of *Clostridium* strain in agar medium, TGY containing 15g/L bacto-agar or other regular agar medium for anaerobic bacteria.
3. Instagen (Bio-Rad, Paris, France).
4. QIAamp® DNA Stool Mini Kit (Qiagen, Courtaboeuf, France).
5. Phenol-chloroform-isoamyl alcohol: 49.5:49.5:1 (v/v/v).
6. PCR equipment.
7. Oligonucleotide primers (*see* **Tables 1–4**): final working solution of 50 pmol/μL in water.
8. dATP: 10 mM.
9. dCTP: 10 mM.
10. dGTP: 10 mM.
11. dTTP: 10 mM.
12. *Taq* DNA polymerase: 5 U/μL.
13. *Taq* reaction buffer (10X): 100 mM Tris-HCl, pH 8.3, 500 mM KCl, 15 mM MgCl$_2$, 1% gelatin (10X *Taq* reaction buffer provided by the supplier of *Taq* DNA polymerase can be used instead).
14. Agarose and DNA electrophoresis equipment.
15. TAE buffer 50X: 2 M Tris-acetate, 50 mM EDTA, 242 g Tris-base, 57.1 mL glacial acetic acid, 100 mL 0.5 M EDTA, pH 8.0, for 1 L.
16. Ethidium bromide: 10 mg/mL.
17. DNA ladder (size marker).

Table 1
Primers for Toxin Gene Detection in *Clostridium botulinum* A, B, E, F and G

Toxin	Primer	Sequence (5' → 3')	Positions[a]	Size (bp)	T°C of annealing	Accession number
BoNT/A	P478	CAGGTATTTTAAAAGACTTTTGGGGG	3638-3662			M30196
BoNT/A	P479	TATATACACGATCATTATTTCTAAC	3877-3901 R	263	50	
BoNT/B	P482	GCGAATATTTAAAAGATTTTTGGGG	3298-3322			M81186
BoNT/B	P483	CTAGATATATATAAATCTTCTTTTCTAA	3538-3564 R	266	50	
BoNT/E	P480	CAAATATTTTGAAGGATTTTTGGGG	3434-3458			X62089
BoNT/E	P481	TATATACCTGATCATTCTTTCTAAC	3652-3676 R	242	50	

[a] R indicates reverse primer.

Table 2
Primers for Toxin Gene Detection in *C. botulinum* C and D

Toxin	Primer	Sequence (5' → 3')	Positions[a]	Size (bp)	T°C of annealing	Accession number
BoNT/C1-D	P293	TWA TTC CMT ATA TAG GAC C	2097–2115 from *bont/C1* 1917–1936 from *bont/D*	340		X53751X54254
BoNT/C1-D	P294	TTT AGC TTT GAT TGC ACC TGC CTG	2413–2437 R from *bont/C1* 2258–2234 R from *bont/D*		50	X53751X54254
BoNT/C1	P295	GAA GCA TTT GCA GTT A	2158–2174	199		X53781
BoNT/C1	P296T	GG ATA ACC ACG TTC CCA T	2338–2357 R		50	X53781
BoNT/D	P297	CAA GCA TTT GCA ACA G	1979–1995	199		X54254
BoNT/D	P298	TTG ACA ACC AAT TTG ATA C	2159–2178 R			X54254

[a] R indicates reverse primer.

Table 3
Primers for Toxin Gene Detection in *C. perfringens*

Toxin	Primer	Sequence (5' → 3')	Positions[a]	Size (bp)	T°C of annealing	Reference or accession number
α	PL3	AAG TTA CCT TTG CTG CAT AAT CCC	1676–1699 R	236	50	Infect. Immun. 1989, 57, 367
α	PL7	ATA GAT ACT CCA TAT CAT CCT GCT	1418–1440			
Enterotoxine	P145	GAA AGA TCT GTA TCT ACA ACT GCT GGT CC	472–500	425	50	Ant. Leeuw. 1989 56, 181–190
Enterotoxine	P146	GCT GGC TAA GAT TCT ATA TTT TTG TCC AGT	868–897 R			
β1	P463	CTA ATA TGT CTG TAG TTC TAA CTG CTC CTA	1127–1156	102	50	L13198
β1	P464	TAT CTA CAT TTG GGG TAT CAA AAG CTA GCC	1287–1258 R			
β2	P465	TTT TCT ATA TAT AAT CTT ATT TGT CTA GCA	978–948 R	277	50	L77965
β2	P466	AGT TTG TAC ATG GGA TGA TGA ACT AGC ACA	723–752			
ε	P497	GTC CCT TCA CAA GAT ATA CTA GTA CC	1051–1101	172	50	M80837
ε	P498	CCT AGG AAA AGC TAA ATA ACT AGG	1183–1223 R			
ι Ia	P245	GCT TTT ATT GAA AGA CCA GAA	1588–1609	642	50	X73562
ι Ia	P240	CAT CTT TAA AAT CAA GAC TG	2233–2214 R			
ι Ib	P181	GTG TGA GAA TAG CTT GTA GTT	3945–3925 R	480	50	X73562
ι Ib	P187	TGA AAA TGA TCC GTT TAT ACC	3445–3465			

[a] R indicates reverse primer.

Table 4
Primers for Toxin Gene Detection in other *Clostridium* spp.

Toxin	Primer	Sequence (5' —> 3')	Positions[a]	T°C of annealing	Size(bp)	Accession number
ToxA *C. difficile*	P533	CCT AAT ACA GCT ATG GGT GCG AAT GG	7919–7944			M30307
ToxA *C. difficile*	P534	GGG GCT TTT ACT CCA TCA ACA CCA AAG	8251–8277 R	55	358	X53138
ToxB *C. difficile*	P537	AAA AAT GGA GAG TCA TTC AAC	1137–1157			
ToxB *C. difficile*	P538	GCC CTT GAT TTA TAA TAC CC	1538–1557 R	55	420	L76081
CDTa *C. difficile*	P368	GAA GCA GAA AGA ATA GAG C	312–331			
CDTa *C. difficile*	P589	GGT TTT TCA TCA CCT TTT CCA GG	675–697 R	55	385	L76081
CDTb *C. difficile*	P304	TAA ACA AAG GAG AAT CTG C	2680–2699			
CDTb *C. difficile*	P590	TTT CTA ATT TAA TTT GCT TTC CAG C	2953–2977 R	55	297	Z48636
Tox α *novyi*	P531	GAA GGA GAT AAA AGT GCT ATA AAT TAT AAA GG	6146–6177			
Tox α *novyi*	P532	CAC CTA ATA CTC GCC AAC CCG TTA CTG CAC	6555–6585 R	55	439	D17668
Tox α *septicum*	P512	CTT ACA AAT CTT GAA GAG GGG GG	654–676			
Tox α *septicum*	P513	CAT TTG GAT TGT ATC TAG CAG	924–904 R	55	239	X82638
LT *C. sordellii*	P535	CTA GAA AAA TTT GCT GAT GAG GAT TTG GTA AG	1202–1233			
LT *C. sordellii*	P536A	GG CTT GAT TTA TAA CAG AGT TAT TGG C	1607–1634 R	55	432	U77593
Neuraminidase *C. sordellii*	P708	GACTTTGGCAGATGGTACTATGCTAGC	180–206			
Neuraminidase *C. sordellii*	P709	CATCAGAATAAACCATTTGAACAGACC	458–481R	55	301	M59091
rRNA gene *C. chauvoei*	CC16S-L	GTCGAGCGAGGAGAGTTC	54–71	58	960	
rRNA gene *C. chauvoei*	CC16S-R	TCATCCTGTCTCCGAAGA	996–1013R			X04436
Tetanus toxin *C. tetani*	P476	ATGCCAATAACCATAAATAATTTTAGAGATATAG	397–P292	55	1475	
Tetanus toxin *C. tetani*	P477	TTCATCTTGAAATGGTTCTTCTG	1734–1756R			

[a] R indicates reverse primer.

18. Agarose gel loading dye: 100 m*M* EDTA, 30% glycerol, 0.25 % bromophenol blue.
19. Denaturing solution: 200 m*M* NaOH, 500 m*M* EDTA.
20. Hybridization buffer: 5X saline sodium citrate (22 g/L sodium citrate, 43.8 g/L NaCl), 0.1% laurylsarcosine sodium, 0.02% sodium dodecyl sulfate, 1% blocking reagent (Roche Diagnostics, Penzberg, Germany).
21. Washing buffer: 100 m*M* Tri-HCl, pH 7.5, 150 m*M* NaCl, 0.5 g/L Tween 20®, 1% blocking reagent, 100 µg/mL fish sperm DNA.
22. Revelation buffer: 100 m*M* maleic acid, pH 7.5, 150 m*M* NaCl, 1% blocking reagent, 1% peroxidase-labeled anti-digoxigenin antibody (Roche Diagnostics).
23. 3,3',5,5'-tetramethylbenzidine (Roche Diagnostics).
24. Stop solution: 1.5 *M* H$_2$SO$_4$.

3. Methods

3.1. Cultures

3.1.1. Broth Medium

Inoculate one colony from an agar medium culture in 10 mL of broth medium.

3.1.2. Agar Medium

Use an isolated colony from TGY agar medium or other solid medium for anaerobic bacteria for PCR detection.

3.1.3. Enrichment Culture from Food or Biological Samples

Food: Inoculate 25 g in 225 mL of TGY broth.

Biological samples: The content from small intestine is the preferred sample for diagnosis of clostridial gastro intestinal diseases. Use muscle or exudation from gangrenous lesions. Inoculate 1 mL or 1 g in 9 mL of TGY broth.

Culture in anaerobic conditions at 37°C for 18 h. Anaerobic jars containing gaspack or H$_2$-CO$_2$ (95:5), or an anaerobic chamber can be used.

3.2. DNA Extraction

3.2.1. DNA Extraction from Broth Culture

1. Centrifuge 1 mL culture for 10 min at 13,000*g*.
2. Wash the pellet in 1 mL of distilled water.
3. Suspend the pellet in 200 µL of Instagen.
4. Incubate at 56°C for 30 min and vortex mix vigorously for 10 s.
5. Incubate for 10 min at 95°C and vortex mix vigorously for 10 s.
6. Centrifuge for 10 min at 13,000*g*. The supernatant is immediately used for PCR.

3.2.2. DNA Extraction from Colonies Grown on Agar Medium

Resuspend one colony in 200 µL of Instagen and proceed with the Instagen protocol as above (*see* **Subheading 3.2.1., steps 4–6**).

3.2.3. DNA Extraction from Biological Samples (see **Note 1**)

3.2.3.1. INSTAGEN METHOD

Use the same protocol as for DNA extraction from broth culture. Liquid samples can be processed as broth culture, and solid samples are homogenized in phosphate buffered saline (PBS) 1:10 (w/v).

3.2.3.2. QIAGEN METHOD

For those samples containing PCR inhibitors, a more reliable method of DNA extraction is required. The QIAamp DNA Stool Mini Kit has proved suitable for a variety of sample matrixes.

1. Pipet 200 mL of liquid sample or weigh 180–220 mg of solid sample in a 2-mL microcentrifuge tube.
2. Add 1.2 mL of ASL buffer (provided in the kit), vortex mix for 1 min or until the sample is thoroughly homogenized.
3. Incubate at 70°C for 5 min.
4. Vortex mix for 15 s and centrifuge (18000g) for 1 min.
5. Pipet 1.2 mL of the supernatant in another 2-mL microtube and discard the pellet.
6. Add 1 inhibiEx tablet (provided in the kit) and vortex mix until the tablet is completely suspended. Incubate 1 min at room temperature.
7. Centrifuge at high speed for 3 min.
8. Pipet the supernatant in a 1.5 mL microtube and discard the pellet. Centrifuge at high speed for 3 min.
9. Pipet 15 µL of proteinase K solution (provided in the kit) into a fresh 1.5-mL microtube.
10. Pipet 200 µL of supernatant from step 8 into the microtube containing proteinase K.
11. Add 200 µL of buffer AL and vortex mix for 15 s.
12. Incubate at 70°C for 10 min.
13. Add 100 µL of ethanol and vortex mix to homogenize.
14. Apply the complete lysate from **step 13** in a QIAamp spin column sitting in a 2-mL microtube, and centrifuge for 1 min. Place the QIAamp spin column in a new 2-mL microtube and discard the filtrate.
15. Add 500 µL of buffer AW1 into the column, centrifuge for 1 min. Place the QIAamp spin column in a new 2-mL microtube and discard the filtrate.
16. Add 500 µL of buffer AW2, centrifuge for 2 min and discard the filtrate.
17. Transfer the QIAamp column into a new 1.5-mL microtube, and add 200 µL of buffer AE ŏnto the QIAamp membrane. Incubate for 1 min and centrifuge for 1 min. Use 5 µL for PCR. Add bovine serum albumin (BSA) to a final concentration of 0.1 µg/µL in the PCR mixture. The DNA extract can be stored at −20°C for later use.

3.3. Amplification by PCR (General Protocol)

3.3.1. Preparation of the Master Mixture

Prepare a master mix for all amplification reaction of the series. It should contain the following concentrations of reagents in each 50-µL reaction:

Taq reaction buffer 1X, deoxynucleotide triphosphates 200 µ*M* each, primers forward and reverse 0.5 µ*M* each, *Taq* polymerase 2.5 U, and template as described previously (3 µL).

For instance, a 1-mL master mixture will contain: 100 µL *Taq* reaction buffer (10X), 20 µL dATP (10 m*M*), 20 µL dCTP (10 m*M*), 20 µL dGTP (10 m*M*), 20 µL dTTP (10 m*M*), 10 µL Forward primer (50 pmol/µL), 10 µL Reverse primer (50 pmol/µL), 800 µL Distilled water.

The master mixture can be aliquoted and stored at –20°C until use.

For PCR, add 46 mL of master mixture, 1 mL of *Taq* polymerase, and 3 µL of template (DNA extract) to each tube.

3.3.2. Amplification Cycles

Run the PCR according to the following temperature–time profile: Initial denaturation 95°C for 5 min, 30 cycles of denaturation (95°C for 20 s), primer annealing (for annealing temperature *see* **Tables 1–4**, for a duration of 20 s), elongation (72°C for 20 s), and final extension at 72°C for 5 min.

3.4. Strategy for Identification of Clostridium botulinum *Toxin Gene Types A, B, E, F, and G*

The identification of *C. botulinum* A, B, E, F, and G using specific primers for each toxin gene has been described by several authors *(1–9)*. For certain applications, mainly food control, it is more useful to detect neurotoxigenic *C. botulinum*, *C. butyricum*, and *C. baratii* in a multiplex PCR. For this purpose, a degenerate set of primers flanking a specific neurotoxin gene fragment from *C. botulinum* A, B, E, F, G, *C. butyricum* E, and *C. baratii* F has been designed *(10–14)*. We have also proposed a method consisting of a PCR amplification using one degenerate primer pair and identification of the amplicons by hybridization with 5 individual probes specific to each toxin gene A, B, E, F, and G, respectively *(14)*.

We now prefer to use a mixture of primers specific for neurotoxin genes A, B, and E instead of the degenerate primer pair (first amplification). When positive results are obtained, identification of the toxin gene type is achieved with a second PCR, using specific primers for each toxin gene, preferentially A, B, and E, which are the most common types. The recommended protocol is given in the following paragraphs.

3.4.1. Preparation of Primers BotU and BotR (see *Table 2*)

For BotU, mix: 50 pmol/µL P478, 50 pmol/µL P480, and 100 pmol/µL P482, (final concentrations).

For BotR, mix: P479 50 pmol/mL, P481 50 pmol/µL, and P483 100 pmol/µL (final concentrations).

3.4.2. First Amplification

Proceed as described in **Subheading 3.3.**

3.4.3. Second Amplification

In case of a positive result, three more PCRs are performed with the following primers:

P478/P479 (type A), P482/P483 (type B), and P480/P481 (type E).

3.5. Identification of Clostridium botulinum
Toxin Gene Types C and D

C. botulinum C and D are responsible for animal botulism. A nested PCR procedure is recommended for the detection and identification of each toxin type *(15)*.

In the first round of PCR, primers P293 and P294 (**Table 2**) are used to amplify a 340-bp fragment that is common to both *C. botulinum* C and D neurotoxin genes.

In the second round, 3 µL of the first PCR are diluted 1:10 in distilled water and used for two parallel amplifications with: P295/P296 specific for *C. botulinum* C and P297/P298 specific for *C. botulinum* D.

3.6. Strategy for Identification of Clostridium perfringens

Strains of *C. perfringens* produce numerous toxins, and various toxin types have been described *(16)*. While the accepted standard method for *C. perfringens* toxin typing includes a mouse bioassay, molecular biological methods based on amplification by PCR are more reliable, accurate, and rapid. In particular, detection of the *C. perfringens* enterotoxin by biological or immunological methods requires sporulating cultures, since this toxin is only produced during the sporulation phase. However, *C. perfringens* isolates very often do not sporulate in regular or even sporulation culture medium. In contrast, detection of the *C. perfringens* enterotoxin gene by PCR can be performed on both sporulating or nonsporulating cultures. Specific *C. perfringens* toxin gene probes are listed in **Table 3**. Note that iota toxin, and also *C. difficile* CDT and *C. spiroforme* toxin, represent binary toxins consisting in two independent proteins encoded by two separate genes. Probes for enzymatic component genes (Ia, CDTa, and Sa) and for binding component genes (Ib, CDTb, and Sb) are included.

Numerous methods for *C. perfringens* toxin gene detection by PCR have been described *(17–28)*. The classical method consists of a single reaction for each toxin gene. Combinations of two or three primer pairs were successfully performed for simultaneous detection of two or three toxin genes. For example, the identification of α toxin and enterotoxin genes by duplex amplification was shown to be useful in food microbiology. Since the α toxin is present in almost all *C. perfringens* strains, it represents a marker of *C. perfringens* species, and the enterotoxin gene, which is present only in some strains responsible for food intoxication, is a suitable target for the identification of enterotoxigenic strains *(20)*. In some circumstances, a quantitative result is required (*see* **Note 2**).

The standard protocol of *C. perfringens* typing includes: DNA extraction from broth culture (*see* **Subheading 3.2.1.**) or colonies (*see* **Subheading 3.2.2.**). PCR amplification (*see* **Subheading 3.3.**) using primers specific for each toxin gene (**Table 3**).

3.7. Differentiation of C. chauvoei and C. septicum

C. chauvoei and *C. septicum* are two closely related pathogenic bacteria. The gene of the α-toxin, which is the main toxin produced by *C. septicum*, has been characterized and can be used as target for PCR detection (*see* **Table 4**). However, no toxin gene of *C. chauvoei* has been cloned or sequenced yet.

The method proposed here targets the ribosomal RNA gene region *(29,30)*. Primers CC16S-L/CC16S-R (**Table 4**) give rise to an amplification product of 960 bp for *C. chauvoei* and, to a lesser extent, for *C. septicum*. The distinction between *C. chauvoei* and *C. septicum* can be made by analysis of the restriction digestion pattern of PCR products using *HincII/SspI* enzymes and 2%-agarose gel electrophoresis. Digestion by *HincII/SspI* will produce 118, 182, 186, and 474-bp fragments for *C. chauvoei*, and 182, 304, and 474-bp fragments for *C. septicum*.

3.8. Identification of C. tetani

Samples (suppuration, necrotic tissue) from the inoculation wound are processed for enrichment culture as described in **Subheading 3.1.3.**, and subsequent PCR is conducted with specific primers (P476–P477) for the tetanus toxin gene (**Table 4**).

3.9. Strategy for Identification of Other Toxigenic Clostridium Species

Detection of other toxigenic *Clostridium* species from intestinal content or organ samples is based on probes specific for toxin genes using the general protocols of enrichment culture (*see* **Subheading 3.1.**), PCR (*see* **Subheading 3.3.**), and hybridization (*see* **Subheading 3.10.2.**). The probes for *Clostridium* toxin genes are listed in **Table 4**.

Some *Clostridium* strains can easily lose their toxin gene after subculture. This is the case with *C. sordellii*, which often becomes nontoxic after subculture. Specific probes derived from the sialidase gene, a stable gene in all *C. sordellii* strains, are used for detection of this bacterium, and toxin gene probes permit the identification of toxigenic strains. For the detection of *C. sordellii* from a sample, we recommend a PCR with primers P708–P709 (sialidase gene) to identify the species *C. sordellii*, and another amplification using primers P535–P536 (lethal toxin gene) (**Table 4**) to identify the toxigenic strains.

C. difficile is responsible for pseudomembranous colitis and postanti-biotherapy diarrhea, which are the most prominent nosocomial affections in hospitalized patients. This pathogen is also involved in some digestive diseases in animals. *C. difficile* produces two main toxins called ToxA and ToxB, but some strains only synthesize ToxA or ToxB, and others produce another toxin (*C. difficile* transferase or CDT) which probably represents an additional virulence factor. Specific ToxA, ToxB, and CDT gene probes are listed in **Table 4**.

3.10. Detection of PCR Products

3.10.1. Agarose Gel Electrophoresis

Agarose gel electrophoresis is the method of choice for checking the size and purity of a PCR product.

1. To prepare a 2% agarose minigel, place 1 g of DNA-grade agarose in a 100-mL flask or bottle.
2. Add 50 mL of 1X TAE buffer (49 mL distilled water and 1 mL of 50X TAE buffer).
3. Heat the mixture in boiling water or in a microwave for 2–5 min for the agarose to dissolve completely.
4. Cool the solution on the bench top for few min.
5. Add 10 mL of 10 mg/mL ethidium bromide stock solution.
6. Pour onto a gel casting stand with well comb(s) in place, and allow about 20 min for the gel to set. Remove the comb(s) and transfer the solid gel to a gel tank and add enough 1X TAE buffer (usually 500 mL) to cover the gel by at least several millimeters.
7. Mix 10 μL of PCR product and 2 μL of gel-loading dye, and load the samples into the wells. A sample DNA ladder is loaded in each agarose gel.
8. Run the gel at constant voltage of 100–120 V for approx 1 h until the dye front is one-half to two-thirds of the way down the gel.
9. View the DNA by placing the stained gel on a UV lightbox.
10. (Optional) Photograph the gel on a UV illuminator using a Polaroid camera.

3.10.2. Hybridization

When many samples have to be routinely analyzed, agarose gel electrophoresis is not an appropriate method. This is the case for *C. botulinum* A, B, and E examination of food samples. We have proposed a method that includes transfer of PCR products onto nitrocellulose membrane and hybridization with internal probes specific of each toxin type A, B, and E, respectively (*31*). A preferred method is sandwich hybridization which can be automated (*32*).

The following protocol is from Fach et al. (manuscript submitted):

1. PCR amplification with BotU and BotR (*see* **Subheading 3.3.**).

2. Add to each PCR tube 100 μL of denaturing solution and incubate at room temperature for 10 min.
3. Transfer 50 μL of each reaction into a well of a 96-well microtiter plate coated with streptavidin (Roche Diagnostics).
4. Add 200 μL of hybridization buffer and 12 pmol/mL of the capture probe (with the 5' end labeled with biotin) and 12 pmol/mL of the detection probe (with the 3' end-labeled with digoxygenin). Capture and detection probes are internal primers to the DNA fragment flanked by BotU and BotR.
5. Wash 6× with washing buffer.
6. Add 100 μL of revelation buffer.
7. Incubate at 37°C for 30 min.
8. Washings (*see* **step 5**).
9. Add 200 μL of 3,3',5,5'-tetramethylbenzidine.
10. Incubate at 37°C for 30 min.
11. Add 100 μL of stop solution.
12. Read the absorbance at 450 nm in a microplate reader. Twice the highest signal obtained from negative samples or PCR water blanks is taken as background.

3.11. Interpretation of the Results

Evidence of a PCR product migrating at the expected size indicates the presence of the corresponding toxin gene in the examined bacterial strain. Further confirmation can be achieved by hybridization with an internal probe of those used for PCR. However, the presence of a toxin gene does not necessarily mean that the toxin is effectively produced by this strain. Some toxin genes can be silent (*see* **Note 3**).

4. Notes

1. The quality of DNA extraction is essential to avoid false negative results and insure reproducibility. A positive PCR standard should be included for each DNA extract. For this purpose, a known DNA (recombinant or bacterial wild-type DNA) and a suitable primer pair are diluted in the DNA extract sample as for PCR with the diagnostic primers. An internal standard is typically constituted of a recombinant DNA, the ends of which are complementary to the primers used for the diagnostic PCR (P1 and P2, for example). The size of the recombinant DNA standard has to be different from that of the expected PCR product of a positive sample. The standard DNA is added to the DNA extract sample, and PCR is performed with primers P1 and P2. In the absence of PCR inhibitors, positive samples show two DNA bands on agarose gel electrophoresis one corresponding to the internal standard and the other to the target DNA, and negative samples only show one band corresponding to the internal standard. No PCR amplification product with standard DNA and primers indicate that PCR inhibitors are probably present in the sample. In that case, a more efficient method of DNA extraction (e.g., QIAamp DNA Stool Mini Kit) has to be used with this sample.

2. Quantification of toxigenic *Clostridium* spp. is required for certain biological or food samples. For example, monitoring levels of *C. perfringens* is important in the food industry, since low titers (50–200 colony forming units (cfu/g) of this bacterium are tolerable in some food products. The standard detection by PCR as described in the present chapter is not quantitative. The present quantitative PCR technology is based on the detection of amplicons using hybridization probes labeled with different fluorescent dyes (e.g., TaqMan® and LightCycler® technology). A quantitative procedure based on the most probable number method, consisting of inoculating serial dilutions of food samples into enrichment medium and performing PCR with each dilution culture, which has been proposed for evaluation of *C. botulinum* contents *(7)*, could also be used for other toxigenic *Clostridium* species. Another method developed for quantitative detection of *C. botulinum* E *(33)* allows evaluation of *C. botulinum* in the range from 10^2 to 10^8 cfu/g within 1 or 2 h.

3. Evidence of a toxin gene does not mean that the toxin is actually produced by the strain. A toxin gene can be silent. This is the case for the *C. botulinum* neurotoxin B gene in some *C. botulinum* type A strains *(34)*, as well as for the enterotoxin gene in *C. perfringens* type E strains *(35)*. Actual expression of toxin genes can be shown by reverse transcription PCR (RT-PCR) targeting clostridial mRNA. This methodology has been used to monitor neurotoxin E production of *C. botulinum (36)*.

References

1. Szabo, E. A., Pemberton, J. M., and Desmarchelier P. M. (1992) Specific detection of *Clostridium botulinum* type B by using the polymerase chain reaction. *Appl. Environ. Microbiol.* **58,** 418–420.

2. Szabo, E. A., Pemberton, J. M., and Desmarchellier, P. M. (1993) Detection of the genes encoding botulinum neurotoxin types A to E by the polymerase chain reaction. *Appl. Environ. Microbiol.* **59,** 3011–3020.

3. Szabo, E. A., Pemberton, J. M., Gibson, A. M., Eyles, M. J., and Desmarchellier P. M. (1994) Polymerase chain reaction for detection of *Clostridium botulinum* types A, B and E in food, soil and infant faeces. *J. Appl. Bacteriol.* **76,** 539–545.

4. Franciosa, G., Ferreira, J. L., and Hatheway C. L. (1994) Detection of type A, B, and E botulism neurotoxin genes in *Clostridium botulinum* and other *Clostridium* species by PCR: evidence of unexpressed type B toxin genes in type A toxigenic organisms. *J. Clin. Microbiol.* **32,** 1911–1917.

5. Takeshi, K., Fujinaga, Y., Inoue, K., et al. (1996) Simple method for detection of *Clostridium botulinum* type A to F neurotoxin genes by polymerase chain reaction. *Microbiol. Immunol.* **40,** 5–11.

6. Ferreira, J. L., Baumstark, B. R., Hamdy, M. K., and McCay, S. G. (1993) Polymerase chain reaction for detection of type A *Clostridium botulinum* in food samples. *J. Food Prot.* **56,** 18–20.

7. Hielm, S., Hyyttia, E., Ridell, J., and Korkeala, H. (1996) Detection of *Clostridium botulinum* in fish and envionmental samples using polymerase chain reaction. *Int. J. Food Microbiol.* **31,** 357–365.
8. Ferreira, J. L., Hamdy, M. K., McGay, S. G., Hemphill, M., Kirma, N., and Baumstark, B. R. (1994) Detection of *Clostridium botulinum* type F using the polymerase chain reaction. *Mol. Cel. Probes* **8,** 365–373.
9. Kakinuma, H., Maruyama, H., Yamakawa, K., Nakamura, S., and Takahashi, H. (1997) Application of nested polymerase chain reaction for the rapid diagnosis of infant botulism type B. *Acta Paed. Jap.* **39,** 346–348.
10. Cordoba, J. J., Collins, M. D., and East, A. K. (1995) Studies on the genes encoding botulinum neurotoxin type A of *Clostridium botulinum* from a variety of sources. *System. Appl. Microbiol.* **18,** 13–22.
11. Campbell, K. D., Collins, M. D., and East, A. K. (1993) Gene probes for identification of the botulinal neurotoxin gene and specific identification of neurotoxin types B, E and F. *J. Clin. Microbiol.* **31,** 2255–2262.
12. Aranda, E., Rodriguez, M. M., Asensio, M. A., and Cordoba, J. J. (1997) Detection of *Clostridium botulinum* types A, B, E, and F in foods by PCR and DNA probe. *Lett. Appl. Microbiol.* **25,** 186–190.
13. Broda, D. M., Boerema, J. A., and Bell, R. G. (1998) A PCR survey of psychrotophic *Clostridium botulinum*-like isolates for the presence of BoNT genes. *Let. Appl. Microbiol.* **27,** 219–223.
14. Fach, P., Gibert, M., Grifais, R., Guillou, J. P., and Popoff, M. R. (1995) PCR and gene probe identification of botulinum neurotoxin A-, B-, E-, F-, and G-producing *Clostridium* spp. and evaluation in food samples. *Appl. Environ. Microbiol.* **61,** 389–392.
15. Fach, P., Gibert, M., Griffais, R., and Popoff, M. R. (1996) Investigation of animal botulism outbreaks by PCR and standard methods. *FEMS Immunol. Med. Microbiol.* **13,** 279–285.
16. Petit, L., Gibert, M., and Popoff, M. R. (1999) *Clostridium perfringens*: toxinotype and genotype. *Trends Microbiol.* **7,** 104–110.
17. Daube, G., China, B., Simon, P., Hvala, K., and Mainil, J. (1994) Typing of *Clostridium perfringens* by in vitro amplification of toxin genes. *J. Appl. Bacteriol.* **77,** 650–655.
18. Moller, K., and Ahrens, P. (1996) Comparison of toxicity neutralization-, ELISA, and PCR tests for typing of *Clostridium perfringens* and detection of the enterotoxin gene by PCR. *Anaerobe* **2,** 103–110.
19. Yamagishi, T., Sugitani, K., Tanishima, K., and Nakamura, S. (1997) Polymerase chain reaction test for differentiation of five toxin types of *Clostridium perfringens*. *Microbiol. Immunol.* **41,** 295–299.
20. Fach, P., and Popoff, M. R. (1997) Detection of enterotoxigenic *Clostridium perfringens* in food and fecal samples with a duplex PCR and the slide agglutination test. *Appl. Environ. Microbiol.* **63,** 4232–4236.
21. Songer, J. G., and Meer, R. R. (1996) Genotyping of *Clostridium perfringens* by polymerase chain reaction is a useful adjunct to diagnosis of clostridial enteric disease in animals. *Anaerobe* **2,** 197–203.

22. Schoepe H., Potschka H., Schlapp T., Fiedler J., Schau H., and Baljer G. (1998) Controlled multiplex PCR of enterotoxigenic Clostridium perfringens strains in food samples. *Mol. Cell. Probes* **12**, 359–365.

23. Yoo, H. S., Lee, S. U., Park, K. Y., and Park, Y. H. (1997) Molecular typing and epidemiological survey of prevalence of *Clostridium perfringens* types by multiplex PCR. *J. Clin. Microbiol.* **35**, 228–232.

24. Kanakaraj, R., Harris, D. L., Songer, J. G., and Bosworth, B. (1998) Multiplex PCR assay for detection of Clostridium perfringens in feces and intestinal contents of pigs and in swine feed. *Vet. Microbiol.* **63**, 29–38.

25. Miserez, R., Frey, J., Buogo, C., Capaul, S., Tontis, A., Burnens, A., and Nicolet, J. (1998) Detection of α- and ε-toxigenic *Clostridium perfringens* type D in sheep and goats using a DNA amplification technique (PCR). *Lett. Appl. Microbiol.* **26**, 382–386.

26. Kadra, B., Guillou, J. P., Popoff, M. R., and Bourlioux, P. (1999) Typing of sheep clinical isolates and identification of enterotoxigenic Clostridium perfringens strains by classical methods and polymerase chain reaction (PCR). *FEMS Immunol. Med. Microbiol.* **24**, 259–266.

27. Garmory, H. S., Chanter, N., French, N. P., Bueschel, D., Songer, J. G., and Titball, R. W. (2000) Occurence of *Clostridium perfringens* β2-toxin amongst animals, determined using genotyping and subtyping PCR assays. *Epidemiol. Infect.* **124**, 61–67.

28. Gkiourtzidis, K., Frey, J., Bourtzi-Hatzopoulou, E., Iliadis, N., and Sarris, K. (2001) PDR detection and prevalence of α-, β-,β2-,ε-, ι-, and enterotoxin genes in *Clostridium perfringens* isolated from lambs with clostridial dysentery. *Vet. Microbiol.* **82**, 39–43.

29. Kunert, P., Krampe, M., Capaul, S. E., Frey, J., and Nicolet, J. (1997) Identification of *Clostridium chauvoei* in cultures and clinical material from blacleg using PCR. *Vet. Microbiol.* **51**, 291–298.

30. Kunert, P., Capaul, S. E., Nicolet, J., and Frey, J. (1996) Phylogenetic positions of *Clostridium chauvoei* and *Clostridium septicum* based on 16S rRNA gene sequences. *Int. J. Syst. Bacteriol.* **46**, 1174–1176.

31. Fach, P., Gibert, M., Grifais, R., Guillou, J. P., and Popoff, M. R. (1995) PCR and gene probe identification of botulinum neurotoxin A-, B-, E-, F-, and G-producing *Clostridium* spp. and evaluation in food samples. *Appl. Environ. Microbiol.* **61**, 389–392.

32. Kemp, D. J., Smith, D. B., Foote, S. J., Samaras, N., and Peterson, M. G. (1989) Colorimetric detection of specific DNA segments amplified by polymerase chain reaction. *Proc. Natl. Acad. Sci. USA* **86**, 2423–2427.

33. Kimura, B., Kawasaki, S., Nakano, H., and Hujii, T. (2001) Rapid, quantitative PCR monitoring of growth of *Clostridium botulinum* type E in modified-atmosphere-packaged fish. *Appl. Environ. Microbiol.* **67**, 206–218.

34. Hutson, R. A., Zhou, Y., Collins, M. D., Johnson, E. A., Hatheway, C. L., and Sugiyama, H. (1996) Genetic characterization of Clostridium botulinum type A containing silent type B neurotoxin gene sequences. *J. Biol. Chem.* **271**, 10,786–10,792.

35. Wieckowski, E., Billington, S., Songer, G., and McClane, B. (1998) Clostridium perfringens type E isolates associated with veterinary enteric infections carry silent enterotoxin gene sequences. *Zbl. Bakteriol.* **S29,** 407–408.
36. McGrath, S., Dooley, J. S. G., and Haylor, R. W. (2000) Quantification of Clostridium botulinum toxin gene expression by competitive reverse trasncription PCR. *Appl. Environ. Microbiol.* **66,** 1423–1428.

10

PCR-Based Detection of *Coxiella burnetii* from Clinical Samples

Mustapha Berri, Nathalie Arricau-Bouvery, and Annie Rodolakis

1. Introduction

Q fever is caused by *Coxiella burnetii*, an organism widely found in nature and responsible for infections in arthropods, pets, domestic and wild animals, as well as humans *(1,2)*. Conventional diagnosis of Q fever is mainly based on serological tests, such as immunofluorescence, enzyme-linked immunosorbent assay, and complement fixation *(3)*. Isolation of *C. burnetii* is performed in cell culture, animals or embryonated chicken eggs, however, the procedure is time-consuming and hazardous and therefore restricted to specialized laboratories. The highly sensitive polymerase chain reaction (PCR) was shown to be a useful tool for the detection of *C. burnetii*-specific genomic DNA sequences in biological samples *(4–9)*. A PCR assay designated Trans-PCR, which targets a repetitive, transposon-like element *(5–8)* proved to be specific and sensitive. However, the numerous DNA extraction steps required are time-consuming and carry a high risk of carryover contamination, thus reducing the assay's sensitivity.

In the present chapter, we describe a simple DNA preparation method combined with a Trans-PCR assay for sensitive and specific detection of *C. burnetii* from ovine genital swabs, placenta, milk and fecal samples *(9)*. Examinations performed on samples taken from infected animals demonstrated that this method could be suitable for routine diagnosis, particularly for the collection of data about intermittent shedding of *C. burnetii* by animals and the elucidation of transmission routes *(9, 10)*.

From: *Methods in Molecular Biology, vol. 216: PCR Detection of Microbial Pathogens: Methods and Protocols*
Edited by: K. Sachse and J. Frey © Humana Press Inc., Totowa, NJ

2. Materials

1. *C. burnetii* Nine Mile phase I and II (Cote Ouest, Arcachon, France).
2. QIAamp® Tissue Kit 250 (Qiagen S.A., Courtaboeuf, France).
3. Perfect gDNA Blood Mini Kit (Eppendorf, Sartroville, France).
4. Proteinase K (Qiagen): working solution 20 mg/mL.
5. Oligonucletide primers (Eurogentec, Seraing, Belgium): stock solution 100 µ*M*, working solution 20 µ*M*.
6. dNTP mixture: stock solution 10 m*M*, working solution 2 m*M*.
7. Magnesium chloride: 25 m*M*.
8. PCR 10X reaction buffer: 10 m*M* Tris-HCl, 50 m*M* KCl, 0.1% Triton® X-100.
9. *Taq* DNA polymerase (Promega, Charbonnières, France).
10. Agarose, molecular biology grade: for 1.5% (w/v) gels.
11. Ethidium bromide: 10 mg/mL.
12. DNA size marker: 100-bp ladder (Promega).
13. UV transilluminator.
14. Thermal cycler: UNO Thermobloc (Biometra, Göttingen, Germany).
15. *Taq*I and *Alu* restriction endonucleases (Appligen Ancor, France): 10 U/µL.
16. DNA sequencing equipment.
17. TA cloning kit (Invitrogen, Groningen, Netherlands).
18. Physiological saline solution: 145.4 m*M* NaCl, 7.4 m*M* KH$_2$PO$_4$, 11.5 m*M* K$_2$HPO$_4$, pH 6.85.
19. Phosphate-buffered saline (PBS): 136.8 m*M* NaCl, 2.7 m*M* KCl, 1.5 m*M* KH$_2$PO$_4$, 8.09 m*M* Na$_2$HPO$_4$/2H2O, pH 7.2–7.4.
20. Confined level L3 laboratory.

3. Methods
3.1. Coxiella Cultures

C. burnetii Nine Mile phase I and II isolate as standard was prepared. The concentration of the bacterial suspension was determined by the manufacturer and estimated to be 10^{10} *Coxiella* cells/mL. Serial dilutions were prepared from the same bacterial suspension with determined concentration. *C. burnetii* suspensions were stored at –20°C until use.

3.2. Collection of Samples

To make a valid diagnosis, samples must be collected correctly at the right time. They must be clearly labeled and transported to the laboratory as quickly as possible, having been cooled and packaged in a waterproof container holding sufficient absorbent material to avoid any loss of liquid. It must be kept in mind that these biological samples are potentially hazardous to humans and, therefore, any possible leakage must be prevented during transportation *(11)*. If the samples are not examined immediately on arrival to the laboratory, they must be stored at −20°C (*see* **Note 1**).

3.2.1. Vaginal Swabs

Vaginal secretions, sampled immediately after abortion by swabbing provide a suitable material for isolation and PCR analysis of abortive organisms. They are not usually as heavily infected as the cotyledons, but they reflect moderate infection of the placenta and are less hazardous to the handler. Samples should be collected as soon as possible after abortion. Vaginal excretion, often abundant during the first few days, can decrease rapidly or become intermittent making testing inaccurate. Vaginal swabs are taken by the insertion of a dry, sterile cotton wool swab (10 cm) into the vagina of each animal and should be sent to the laboratory as they are without any transport medium.

3.2.2. Milk

Colostrum and milk from the quarters should be collected aseptically in a sterile container. Before sampling, the teats have to be cleaned, and the first two jets of milk should be discarded. The milk sample may stored at −20°C before use.

3.2.3. Feces

As for milk samples, fecal material should also be taken directly from the animal and collected in a sterile container. If the samples are not examined immediately, they must be stored at −20°C.

3.2.4. Placental Tissue

The placenta, when available and not soiled, is the best sample for detection of the abortive agents (*see* **Note 1**). Since the entire placenta is difficult and hazardous to handle, it is better to sample the cotyledons. Thus, 5 or 6 cotyledons and their intercotyledonary membranes should be collected and placed in a sterile watertight container. Rinse them with physiological saline and dry with filter paper. The samples may be stored at 4°C for several days or at −20°C for several months.

3.3. DNA Extraction from Different Types of Samples

The sample preparation steps are performed in a confined level L3 laboratory. The DNA extraction steps can be conducted outside the safety laboratory after proteinase K treatment.

3.3.1. DNA Extraction from C. burnetii Nine-Mile Reference Strain

1. Boil 100 µL of the bacterial suspension for 10 min.
2. Centrifuge the solution at 13,000g for 5 min.
3. After centrifugation, the supernatant can be used either as a positive amplification control or for the evaluation of the sensitivity of the PCR assay.

3.3.2. Vaginal Swabs

1. Wash the genital swab in 1 mL of PBS.
2. Digest a total of 200 µL of vaginal swab extract by proteinase K (final concentration 200 µg/mL) at 56°C during 3 h.
3. Heat the reaction mixture at 100°C for 10 min.
4. Instead of **steps 1–3**, 200 µL of the vaginal extract may be boiled for 10 min (simple alternative).
5. Centrifuge the solution at 13,000*g* for 5 min.
6. Use 2.5 µL of the supernatant directly in the PCR assay or keep the DNA extract at –20°C until use.

3.3.3. Milk Samples

1. Centrifuge 1 mL of milk sample at 13,000*g* for 60 min.
2. Remove cream and milk layers and wash the pellet 2× with sterile water.
3. Extract the DNA from the pellet with the QIAamp Tissue Kit 250 according to the instructions of the manufacturer (for slight modifications, *see* **Note 2** and **Ref. 9**).
4. Use 2.5 µL of DNA extract in the PCR assay.

3.3.4. Fecal Samples

1. Treat 20 mg of fecal sample directly with proteinase K (final concentration 200 µg/mL) in 1 vol of ATL lysis buffer (provided with the QIAamp Tissue Kit) overnight at 56°C.
2. Add AL lysis buffer (QIAamp Tissue Kit) and heat the solution at 100°C for 10 min.
3. Centrifuge the reaction mixture at 13,000*g* for 5 min.
4. Collect the supernatant and mix it with 0.525 vol of ethanol.
5. Follow the QIAamp Tissue Kit protocol and use 2.5 µL of extracted DNA solution in Trans-PCR assay.
6. Dilute the eluted DNA before use in the PCR assay as described in **Subheading 3.4.4.**

3.3.5. Placental Tissue Samples

1. Grind 1 to 2 g of sampled cotyledon with 10 vol of a sterile physiological saline solution.
2. Centrifuge the mixture for 30 min at 3000*g* at 4°C.
3. Treat 50 µL of the supernatant with proteinase K (20 mg/mL) at 70°C for 30 min.
4. Follow the instructions of the Perfect gDNA Blood Mini Kit (Eppendorf) starting from the proteinase K treatment step (*see* **Note 3**).
5. Elute the DNA as recommended and use 2.5 µL in PCR assay.

3.4. PCR Amplification

In order to amplify DNA of *C. burnetii* from the samples, a pair of 21-mer oligonucleotide primers were designed based on the published DNA sequence of the gene encoding a transposon-like repetitive region of the *C. burnetii*

genome *(12)*: Trans-1 (5'-TAT GTA TCC ACC GTA GCC AGT C-3') and
Trans-2 (5'-CCC AAC AAC ACC TCC TTA TTC-3'). The expected product
of amplification of the target sequence will be 687 bp in length.

3.4.1. PCR Procedure

1. Perform the assay in 25-μL reactions, each of which should contain: 2 μM of
   each primer, 200 μ*M* of each deoxynucleoside triphosphate, 3 m*M* MgCl$_2$, 0.5 U
   of *Taq* DNA polymerase, 2.5 μL of DNA extract prepared as described under
   Subheading 3.3.
2. Perform the Trans-PCRs according to a modified touchdown protocol *(9)*:
   5 cycles consisting of denaturation at 94°C for 30 s, annealing at 66–61°C (the
   temperature was decreased 1°C between consecutive cycles) for 1 min and
   extension at 72°C for 1 min, and then 40 cycles of denaturation at 94°C for
   30 s, annealing at 61°C for 30 s, and extension at 72 °C for 1 min.
3. Load 10 μL of PCR product onto 1.5% agarose gel and run electrophoresis.
4. Stain the gel with ethidium bromide and visualize under UV transilluminator.

3.4.2. Evaluation of the Sensitivity of the PCR Assay (see **Note 4**)

In order to evaluate the sensitivity of PCR detection, three different types of
samples are used. Each sample should be tested in triplicate.

3.4.2.1. CELL SUSPENSION OF *C. BURNETII*

1. Prepare serial dilutions in PBS of one set of *C. burnetii* suspension containing
   between 10^8 and 10^1 *Coxiella* cells/mL.
2. Boil each dilution for 10 min or treat with proteinase K as described in **Subheading 3.3.2. step 2.**
3. Use 2.5 μL in the PCR assay (an example is shown in **Fig. 1**).

3.4.2.2. SPIKED MILK SAMPLES

1. Add 50 μL of *C. burnetii* Nine Mile strain (serial dilutions containing between
   10^8 and 10^1 *Coxiella* cells/mL) to 1 mL of *Coxiella*-free milk *(9)*. Commercial
   pasteurized milk, bovine, caprine, and ovine raw milk may be used.
2. Treat 1 mL of each dilution as described in **Subheading 3.3.3.**

3.4.2.3. SPIKED FECAL SAMPLES

1. Mix 50 μL of serially diluted *Coxiella* suspension (10^8 to 10^1 *C. burnetii* cells/
   mL) with 20 mg of *Coxiella*-free feces.
2. Extract the DNA as described in **Subheading 3.3.4.**

3.4.3. Evaluation of Specificity

Specificity can be examined by subjecting genomic DNA (10^6 template copies) of the following bacteria to Trans-PCR : *Chlamydophila psittaci,*
Chlamydophila pecorum, Brucella melitensis, Brucella abortus, Escherichia

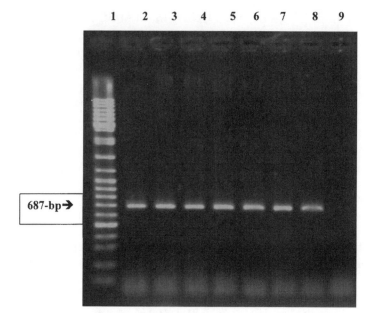

Fig. 1. Trans-PCR. Determination of the detection limit using a dilution series of *C. burnetti* cells (10^7, 10^6, 10^5, 10^4, 10^3, 10^2, and 10^1 cells per reaction, lanes 2–8; lane 1, 100-bp DNA ladder; lane 9, negative control (reagents without DNA).

coli, Klebsiella pneumoniae, Listeria monocytogenes, Salmonella abortus ovis, Salmonella typhimurium, Staphylococcus aureus, Staphylococcus chromogenese, Staphylococcus hominis, Streptococcus dysgalactiae, and *Streptococcus agalactiae.*

To verify the identity of the PCR amplicon obtained from milk and fecal material of naturally infected animals, PCR products from these samples and the reference strain can subsequently be subjected to *Taq*I and *Alu* restriction endonuclease digestion. *Taq*I digestion will result in the generation of 446-, 201-, and 10-bp fragments, whereas cleavage with *Alu*I will yield bands at 316-, 191-, and 179-bp *(9)*. The assay's specificity can be further evaluated by sequence analysis of the Trans-PCR product. In our laboratory, PCR products were cloned and sequenced using pCR2.1 primers, and in all cases, the sequence of the amplicon was identical to the sequence in the European Molecular Biology Laboratory (EMBL)/GenBank® database (Accession no. M80806) *(9)*.

3.4.4. Inhibitory Effect of Fecal Material

No *C. burnetii* DNA was amplified from the undiluted DNA solution purified from artificially contaminated fecal samples. Extraction using the QIAamp Tissue Kit procedure did not eliminate PCR inhibitors completely, suggesting

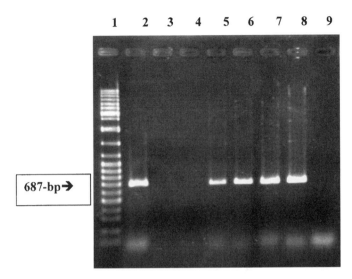

687-bp➔

Fig. 2. Inhibitory effect of fecal matrix on the sensitivity of the Trans-PCR assay. The DNA extract from a fecal sample spiked with 10^6 *C. burnetti* cells/mL was diluted 1:2, 1:5, 1:20, 1:50, and 1:100, respectively (lanes 3–8). and 2.5 μL were used as a template for PCR. Lane 1, 100-bp ladder; lane 2, positive control; lane 9, negative control (reagents without DNA).

they were co-isolated with DNA. To improve the sensitivity of PCR detection, DNA solution purified from fecal sample spiked with 10^6 *Coxiella* cells was diluted to 1:2, 1:5, 1:10, 1:20, 1:50, 1:100, and 2.5 μL were used as template for PCR (*see* **Fig. 2**). The *C. burnetii*-specific amplicon of 687-bp was detected at 1:10 dilution, and the sensitivity increased remarkably when the DNA solution was diluted 100-fold (*see* **Fig. 2**). This indicates that the fecal material completely inhibited PCR amplification, but the original sensitivity was restored simply by diluting the DNA extract. Thus, when DNA extracts of all spiked samples were diluted 10- or 100-fold and subjected to PCR, the 687-bp PCR product was obtained (*see* **Fig. 3**).

4. Notes

1. To prevent the risk of human infection, samples taken from abortions or suspected infected animals must be handled with all the precautions necessary to protect laboratory staff and to avoid contamination of the environment. Thus, as recommended, tissue manipulations and sample preparation must be performed in a confined level L3 laboratory.
2. Several DNA preparation procedures were compared to examine their detection levels of *C. burnetii* from ovine milk sample. The Trans-PCR sensitivity was

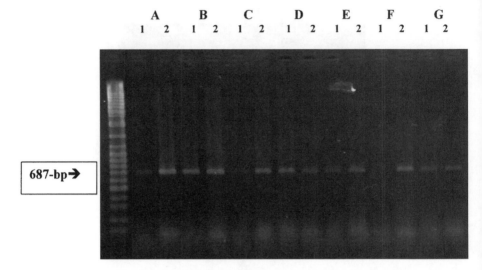

Fig. 3. Sensitivity of Trans-PCR with fecal samples. Serial decimal dilutions (10^8–10^2 *C. burnetti* particles/mL) were used to spike 20 mg of fecal samples (A to G), and DNA was extracted according to **Subheading 3.3.4.** The final DNA extract was then diluted 10-fold (lane 1) and 100-fold (lane 2) before being subjected to PCR.

determined by diluting a stock of *C. burnetii* reference strain of known titer with 1 mL of uncontaminated, i.e., *Coxiella*-free milk sample as described in **Subheading 3.4.2.2.**, and the DNA was prepared following three procedures *(9)*. The procedure used affected the detection limit of PCR, and amplified *C. burnetii* DNA was only detected from samples containing at least 10^6 organisms/mL using the DNA preparation procedure, which included dilution and proteinase K treatment of milk. A 100-fold increase of sensitivity was observed when an experimentally infected milk samples were centrifuged at 13,000*g* for 60 min. This DNA preparation procedure combined with a further DNA extraction using the QIAamp Tissue Kit (*see* **Subheading 3.3.**) increased the sensitivity and proved the most effective for the detection of *C. burnetii* by PCR *(9)*.

3. The QIAamp Tissue DNA Kit has been used very effectively in preparing and purifying *C. burnetii* DNA from vaginal swabs, milk, and fecal samples. However, other DNA extraction kits, such perfect gDNA Blood Mini Kit, also proved to be more effective regarding DNA preparation from bloody tissues, such placenta and spleen.

4. Although, the primers derived from superoxide dismutase enzyme gene (CB1-CB2 primers) are used routinely in many laboratories *(13)*, the Trans-PCR assay was demonstrated to be clearly more sensitive and specific *(9)*. The actual detection limit of the Trans-PCR assay is 10 particles/mL of bacterial suspension, 1 bacteria/mL of milk sample, or 1 *Coxiella*/mg of feces. The sensitivity was tested also on DNA templates extracted from genital swabs, milk, and fecal

samples taken from naturally infected animals *(9)*. The temperature-time profile of Trans-PCR can be modified using 60°C as annealing temperature and reducing the cycle number to 35. These modifications lead to a reduction of the time required and slight enhancement of the assay's sensitivity.

References

1. Baca, O. and Paretsky, D. (1983) Q fever and *Coxiella burnetii*: a model for host parasite interactions. *Microbiol. Rev.* **47,** 127–149.
2. Pinskey, R. L., Fishbein D. B., Greene, C. R., and Geinshemer, K. F. (1991) An outbreak of cat associated Q fever in the United States. *J. Infect. Dis.* **164,** 202–204.
3. Fournier, P-E. and Raoult, D. (1998) Diagnosis of Q fever. *J. Clin. Microbiol.* **36,** 1823–1834.
4. Frazier, M. E., Mallavia, L. P., Samuel, J. E., and Baca, O. G. (1990) DNA probes for the identification of *Coxiella burnetii* strains. *Ann. N.Y. Acad. Sci.* **590,** 445–457.
5. Willems, H., Thiele, D., Fröhlich-Ritter, R., and Krauss, H. (1994) Detection of *Coxiella burnetii* in cow's milk using the polymerase chain reaction (PCR). *J. Vet. Med. B.* **41,** 580–587.
6. Muramatsu, Y., Maruyama, M., Yanase, T., Ueno H., and Morita, C. (1996) Improved method for preparation of samples for the polymerase chain reaction for detection of *Coxiella burnetii* in milk using immunomagnetic separation. *Vet. Microbiol.* **51,** 179–185.
7. Yuasa, Y., Yoshiie, K., Takasaki, T., Yoshida H., and Oda, H. (1996) Retrospective survey of chronic Q fever in Japan by using PCR top detect *Coxiella burnetii* in paraffin-embedded clinical samples. *J.Clin. Microbiol.* **34,** 824–827.
8. Lorenz, H., Jäger, C., Willems, H., and Baljer, G. (1998) PCR detection of *Coxiella burnetii* from different clinical specimens, especially bovine milk, on the basis of DNA preparation with a silica matrix. *Appl. Environ. Microbiol.* **64,** 4234–4237.
9. Berri, M., Laroucau, K., and Rodolakis, A. (2000) The detection of *Coxiella burnetii* from ovine genital swabs, milk and faecal samples by the use of a single touchdown polymerase chain reaction. *Vet. Microbiol.* **72,** 285–293.
10. Berri, M., Souriau, A., Crosby, M., Crochet, D., Lechopier P., and Rodolakis, A. (2001) Relationship between *Coxiella burnetii* shedding, clinical signs and serological response of 34 sheep. *Vet. Rec.* **148,** 502–505.
11. Dufour, B., Moutou, F., and Rodolakis, A. (1998) Methods of sample selection and collection in *Manual for laboratory diagnosis of infectious abortions in small ruminants* (Rodolakis, A., Nettleton, P., and Benkirane, A., eds.), Espace Vétérinaire, Casablance, Morocco, pp. 29–36.
12. Houver, T. A., Vodkin, M. H., and Willems, J.C. (1992) A *Coxiella burnetii* repeated DNA element resembling a bacterial insertion sequence. *J. Bacteriol.* **174,** 5540–5548.
13. Stein, A., and Raoult, D., (1992) Detection of Coxiella burnetii by DNA amplification using polymerase chain reaction. *J. Clin. Microbiol.* **30,** 2462–2466.

11

Detection and Subtyping of Shiga Toxin-Producing *Escherichia coli* (STEC)

Peter Gallien

1. Introduction

When the Japanese microbiologist Shiga discovered a bacterium causing dysentery in humans in 1898, the organism was designated *Shigella dysenteriae* type 1. The toxin produced by the germ was found to have enterotoxic and neurotoxic properties. Later on, it became clear that, in most countries, *Shigella dysenteriae* type 1 does not play an important role in human infectious diseases. Nevertheless, this microorganism has spread a genetic message among other bacterial species, namely *Escherichia coli* and *Citrobacter freundii*.

Meanwhile, cases of diarrhea have been associated to an increasing extent with *E. coli*, a generally harmless occupant of the gut in humans and animals. Although these optionally anaerobic, Gram-negative, motile, rod-shaped bacilli only make up less than one percent of the total bacterial population in the gut, they play an important part in the symbiosis between host and intestinal flora as a consequence of their oxygen-consuming metabolism *(1)*. Alongside these nonpathogenic *E. coli*, there may be toxigenic representatives of the same species present in the host organism. Pathogenic *E. coli* can possess specific adhesins, which are genes encoding toxins and other factors of virulence. This group of pathogens is associated with intestinal and extra-intestinal diseases *(2)*.

Among the representatives residing in the intestinal tract are shiga toxin-producing *E. coli* (STEC), which can cause hemorrhagic colitis (HC), toxic extra-intestinal complications, such as the hemolytic-uremic syndrome (HUS) *(3–7)*, and thrombotic thrombocytopenic purpura (TTP) *(8)* in humans. In addition, the central nervous system can be affected *(9,10)*.

From: *Methods in Molecular Biology, vol. 216: PCR Detection of Microbial Pathogens: Methods and Protocols*
Edited by: K. Sachse and J. Frey © Humana Press Inc., Totowa, NJ

The first description of STEC as a source of HC was published in 1983 *(11)*. Meanwhile, STEC have been isolated from food, feces, stool, sewage, drinking water, and various other habitats connected with human infectious diseases. The toxins produced by STEC were shown to be cytotoxic for a number of tissue culture cell lines, e.g. kidney cells from the green African monkey known as Vero cells *(12)*. This led to the term verotoxin for shiga toxin and, consequently, verocytotoxin-producing *E. coli* (VTEC) as a synonym for STEC, with both terms being equally used at present.

Enterohemorrhagic *E. coli* (EHEC) form a subgroup of STEC (VTEC). However, it is still unclear which factors and/or mechanisms actually make an EHEC out of an STEC. Obviously, all EHEC strains possess shiga toxin genes (*stx*). The shiga toxins themselves exhibit RNA-glycosidase activity and inhibit the synthesis of proteins in eukaryotic cells *(13)*. The activity of the toxins is associated with subunit A, which consists of two parts (structural data of *stx* genes are given below). The A1 fragment harbors the specific N-glycosidase activity, whereas fragment A2 links A1 to five B subunits. In a eukaryotic host cell, the enzyme cleaves at the adenine at position 4324 in a loop structure of the 28S rRNA of the 60S ribosomal subunit *(13)*.

A schematic presentation of transmission routes to man is given in **Fig. 1**. Direct transmission of EHEC from ruminants as the main reservoir for STEC or from raw foods, such as milk, raw or undercooked meat, sausage containing beef, cheese produced from raw milk, to humans, as well as transmission from man to man are observed frequently *(14)*.

The classification of *E. coli* into serogroups O, K, and H is well established. Serogroup O157 occurs very often in human EHEC infections, but other serogroups such as O22, O26, O103, O111, and O145 also figure prominently *(15–18)*. As *stx* genes are located on temperent phages, many *E. coli* serotypes are subject to infection by these phages *(19)*. A World Health Organization (WHO) list contains more than 50 *E. coli* serotypes connected with EHEC infections in man *(20,21)*. Moreover, other bacteria, e.g., *Citrobacter freundii* are also known to harbor *stx* genes *(22)*.

Virulence genes of STEC/EHEC can be located on the DNA of temparent phages, on plasmids, or on the chromosome. The Shiga-toxin genes are located on the DNA of phages. At present, the following subtypes are known: *stx 1*; *stx 2*; *stx 2c*; *stx 2d*; *stx 2e*; and *stx 2f (23–30)*.

Pathogenicity islands like the locus of enterocyte effacement (LEE) or a high- pathogenicity-island (HPI) *(31)* and the *ast A* gene *(32)* are located on the chromosomal DNA of the cells. LEE contains a gene cluster coding for a type III secretion system. Some proteins expressed by genes of LEE form a channel to connect a bacterial cell and a host cell. Subsequently, other proteins encoded by genes of LEE can pass this channel.

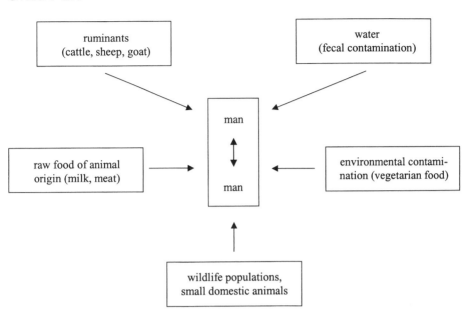

Fig. 1. Routes of transmission of STEC to humans

The *eae* gene (with subtypes α,β,γ,δ, and ε) coding for a protein called intimin completes a key function as it effects a closer connection between pathogen and host cell. A pedestrial cup formed beneath the bacterial cell is built by the effacing of microvilli so that the bacterial cell is able penetrate into the host.

The HPI also comprises a gene cluster. Some proteins derived from this locus are involved in iron uptake from the nutrient media or the host.

The *ast A* gene was found in enteroaggregative *E. coli*. The gene product is a heat-stable toxin, which may contribute to the disturbance of the electrolyte balance of host cells.

Virulence factors located on plasmids include the EHEC-hemolysin (EHEC-*hly*) *(33–36)*, the *kat P* gene, which encodes an enzyme with catalase-peroxidase-activity *(37)*, the *esp P* gene encoding a serin protease cleaving coagulation factor V of human blood *(38)*, the *etp* gene cluster encoding proteins of a type II secretion system *(39)*, the *col* D 157 gene involved in colicin production *(40)*, and the *ile X* gene, a tRNA gene related to *stx* 2 and subtypes only *(41)*.

In the present chapter, a cascade of methods for specific detection, isolation, and characterization of STEC from different habitats and sample matrixes is described. The general course of the procedure and individual steps involved are summarized in **Fig. 2** *(42,43)*.

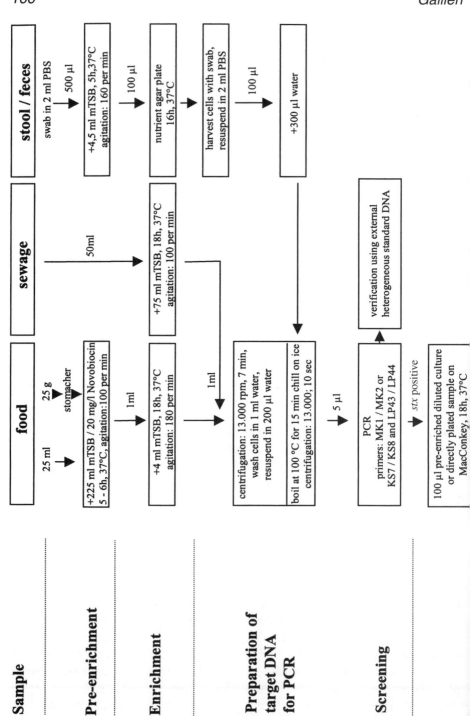

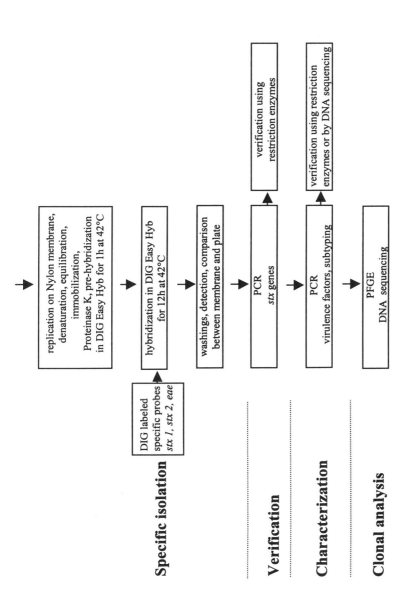

Fig. 2. Flow chart illustrating the general course of operation and the sequence of steps involved in the cascade of methods for specific detection, isolation, and characterization of STEC.

2. Materials

2.1. Bacterial Enrichment

1. Food samples: 25 g of food, e.g., raw minced beef, raw sausage containing beef, soft cheese produced from unpasteurized milk or 25 mL of raw milk.
2. Sewage samples: 50 mL are used.
3. Stool samples and feces: use swab samples.
4. Nutrient medium for enrichment (for preparation procedure, *see* **Subheading 3.1.4.**):
   a. Modified tryptose soy broth (mTSB) containing novobiocin pH 7.3: 33.0 g mTSB containing Bile salts No. 3 and dipotassium hydrogen phosphate (K_2HPO_4) (Mast Diagnostics, Merseyside, UK), 20 mg Novobiocin (Sigma, Deisenhofen, Germany), make up to 1 L with distilled water.
   b. MacConkey agar (Merck, Darmstadt, Germany): 17.0 g tryptone / peptone from casein, 3.0 g Bacto-Peptone from meat, 10.0 g lactose · H_2O, 1.5 g bile salts No. 3, 5.0 g sodium chloride, 0.03 g neutral red, 0.001 g crystal violet, 10 to 15 g of agar (depending on gelling properties), make up to 1000 mL with distilled water.

2.2. PCR, Electrophoresis, and Visualization

1. Water: autoclaved, free of DNase.
2. Primers: MK 1/MK 2, KS 7/KS 8, LP 43/LP 44, working solution 50 pmol/µL (*see* **Table 1**).
3. Primers HB10/HB11 (*see* **Note 1**):
   HB10 : 5' - att cca cac aac ata cga gcc g - 3', 50 pmol/µL,
   HB 11: 5' - gtt tcg cca cct ctg act tga g - 3', 50 pmol/µL.
4. Deoxynucleoside triphosphate mixture (dNTP Mixture) stock solutions consisting of: 10 mM 2'-Deoxy-adenosine-5'-triphosphate (dATP), 10 mM 2'-Deoxy-cytidine-5'-triphosphate (dCTP), 10 mM 2'-Deoxy-guanosine-5'-triphosphate (dGTP), 10 mM 2'-Deoxy-thymidine-5'-triphosphate (dTTP).
   Dilute 1:10 in water before use. The working solution contains 10 µL of each dNTP stock solution plus 360 µL of autoclaved distilled water.
5. Heat-stable DNA Polymerase: e.g., Gold Star, 5 U/µL (Eurogentech, Seraing, Belgium), AmpliTaq® LD, 5 U/µL (Perkin Elmer, Weiterstadt, Germany), or Dynazyme, 5 U/µL (Biometra, Göttingen, Germany).
6. PCR buffer: supplied 10-fold concentrated with or without $MgCl_2$. Those buffers containing $MgCl_2$ were optimized by the supplier for their particular polymerase.
7. $MgCl_2$: 25 mM.
8. TAE buffer: stock solution (50X) containing 242.0 g of Tris-[hydroxymethyl]-aminomethane (TRIS), 57.1 mL of glacial acetic acid, and 100.0 mL of 0.5 M ethylene diamine tetraacetic acid, disodium salt (EDTA-Na_2). Make up to 1000 mL with distilled water.
9. Agarose: type A 6013 (Sigma). For analytical separation of polymerase chain reaction (PCR) products prepare a 2% (w/v) suspension in TAE buffer and boil for a few min. The solution must become clear and transparent. Cool down to 50°C and pour the solution on a gel frame.

Table 1
PCR Conditions for Subtyping of STEC Based on *stx* Genes

Primers	Primer Sequences (5'-3')	Target Gene	PCR Denaturation	Annealing	Extension	Amplicon (bp)	References
Mk1 Mk2	ttt acg ata gac ttc tcg ac cac ata taa att att tcg ctc	*stx* A general	94°C for 60s	44°C for 60s 30 cycles	72°C for 90s	230	*23*
LP 30 LP 31	cag tta atg tgg tgg cga agg cac cag aca atg taa ccg ctg	*stx 1A*	94°C for 90s	64°C for 90s 30 cycles	72°C for 90s	348	*24*
LP 43 LP 44	atc cta ttc ccg gga gtt tac g gcg tca tcg tat aca cag gag c	*stx 2A*+ all subtypes	94°C for 90s	64°C for 90s 30 cycles	72°C for 90s	584	*24*
KS 7 KS 8	ccc gga tcc atg aaa aaa aca tta tta ata gc ccc gaa ttc agc tat tct gag tca acg	*stx 1 B*	94°C for 30s	52°C for 60s 30 cycles	72°C for 40s	285	*25*
GK 3 GK 4	atg aag aag atg ttt atg tca gtc att att aaa ctg	*stx 2 B ; stx 2c (B)*[a] *Hae III*-restriction[a]	94°C for 30s	52°C for 60s 30 cycles	72°C for 60s	260 128[a]+142[a]	*26,27*
slt2v start slt2v stop	atg aag aag atg ttt atg gcg tca gtt aaa ctt cac ctg ggc	*stx 2e B*	94°C for 30s	53°C for 60s 30 cycles	72°C for 60s	267	*28*
slt2v u sltsv d	cca cca gga agt tat att tcc ttc acc agt tgt ata taa aga	*stx 2e A*	94°C for 60s	55°C for 60s 30 cycles	72°C for 120s	759	*28*
VT2-cm VT2-f	aag aag ata ttt gta gcg g taa act gca ctt cag caa at	*stx 2d A*	94°C for 30s	55°C for 60s 30 cycles	72°C for 40s	256	*29*
VTAM-I VTAM-II	agg gcc cac tct tta aat aca tcc cgt cat cct gt taa ctg tgc g	*stx 2d B*	94°C for 30s	62°C for 30s 30 cycles	72°C for 30s	248	Bülte et al. personal communication
128-1 128-2	aga ttg ggc gtc att cac tgg ttg tac ttt aat ggc cgc cct gtc tcc	*stx 2f A*	94°C for 30s	57°C for 60s 30 cycles	72°C for 60s	428	*30*

[a] For screening PCR.

10. Ethidium bromide: 1% (w/v) stock solution (Roth, Karlsruhe, Germany). Dilute 1:1000 with distilled water to a final concentration of 10 mg/L. Alternatively, Gelstar staining solution (Biozym) can be used. Dilute the stock solution 1:100 in distilled water.

11. DNA molecular weight markers II, V, and IX (Roche Diagnostics, Mannheim, Germany) containing the following fragments (in bp): Marker II : 125; 564; 2027; 2322; 4361; 6557; 9416; 23130; and Marker V: 8; 11; 18; 21; 51; 57; 64; 80; 89; 104; 123; 124; 184; 192; 213; 234; 267; 434; 458; 504; 540; 587; and Marker IX: 72; 118; 194; 234; 271; 281; 310; 603; 872; 1078; 1353.

12. Gel loading buffer: 62.5 mg bromophenol blue, 62.5 mg xylene cyanol, 6.25 g Ficoll® (Type 400; Sigma), make up to 25 mL with distilled water.

2.3. DNA Hybridization and Digoxigenin Labeling of Probes

2.3.1. Preparation of Agar Plates and Nylon Membranes

1. Solution for denaturation: 0.5 M sodium hydroxide, 1.5 M sodium chloride.
2. Solution for neutralization, pH 7.4: 1 M Tris-HCl, 1.5 M sodium chloride.
3. Solution for equilibration (20X SSC): 3 M sodium chloride, 0.3 M sodium citrate, pH 7.0.
4. Proteinase K: 1 mg/mL, solution in 2X SSC.

2.3.2. Digoxigenin Labeling of Probes

1. Water: autoclaved, free of DNase.
2. Primers: MK1/MK2, KS7/KS8, LP43/LP44; SK1/ SK2, working solution 50 pmol/μL (*see* **Table 1**).
3. dNTP mixture: *see* **Subheading 2.2.**, **step 4**
4. Digoxigenin-11-2'-deoxy-uridine-5'-triphosphate (DIG-11-dUTP): 1 m*M*.
5. DNA polymerase, PCR buffer, MgCl$_2$: *see* **Subheading 2.2.**
6. Agarose: low-melting grade (Serva, Heidelberg, Germany).
7. TAE buffer (50X): *see* **Subheading 2.2.**, **item 8.**

2.3.3. Prehybridization and Hybridization

1. DIG Easy Hyb® solution (Roche Diagnostics) (*see* **Note 2**).
2. DIG-labeled probes for *stx 1*, *stx 2*, and *eae* (at least 25 ng/mL for 100 mm^2 of membrane disc).
3. Reference strains: *E. coli* C 600 (*stx* and *eae* negative), *E. coli* C 600 J1 (*stx 1* positive), *E. coli* C 600 W 34 (*stx 2* positive), *E. coli* 161 - 84 (*stx 1* , *stx 2*, and *eae* positive).

2.3.4. Washings and Detection

1. Washing solution 1: sodium dodecyl sulfate (SDS) 1 g/L, 2X SSC (*see* **Subheading 2.3.1.**, **item 3**).
2. Washing solution 2: SDS 1g/L, 0.5X SSC.
3. DIG Nucleic Acid Detection Kit (Roche Diagnostics).
4. Buffer 1 (maleic acid buffer): 0.1 M maleic acid (11.6 g/L), 0.15 M NaCl (8.77 g/L).

5. Washing buffer: buffer 1 with 0.3% Tween®20 (Sigma).
6. Blocking stock solution (10X): 10% solution of blocking reagent (Roche Diagnostics) in buffer 1.
7. Buffer 2 blocking solution: 1:10 dilution of blocking stock solution in buffer 1 (final concentration 1%).
8. Buffer 3 (detection buffer): 0.1 M Tris (15.76 g/L), 0.1 M NaCl (5.85 g/L).
9. TE buffer: 100 mM Tris (12.1 g/L), 1 mM EDTA-Na$_2$ (0.377 g/L).
10. Anti-digoxigenin-AP-conjugate: dilute reagent (Roche Diagnostics) 1:5,000 in buffer 2. (Centrifuge vessel containing the stock solution at 13,000g for 15 min to sediment agglomerates. These could react later and give false positive results on the membrane disc. The diluted AP-conjugate is only stable for 12 h at 4°C.)
11. Color substrate solution: pipet 45 µL of 4-nitroblue tetrazolium chloride (NBT) solution as supplied ready to use from Roche Diagnostics (75 mg/mL in 70% dimethyl formamide), and 35 µL of 5-bromo-4-chloro-3-indolyl-phosphate (BCIP or X-phosphate) from the same source (50 mg/mL in dimethylformamide) into a plastic tube. Make up to 10 mL with buffer 3.
12. Phosphate-buffered saline (PBS) pH 7.5: 8.5 g NaCl, 1.14 g disodium hydrogen phosphate (Na$_2$HPO$_4$ × 12 H$_2$O), 0.5 g potassium dihydrogen phosphate (KH$_2$PO$_4$), make up to 1000 mL with distilled water. (Dissolve all salts in approx 900 mL of distilled water and adjust to pH 7.5 with 2 N NaOH. Store at 4°C and use the solution for no longer than 4 wk.

2.4. Characterization of Isolates

Reagents for characterization of STEC/EHEC isolates (*stx* subtyping, detection of virulence factors, *eae* subtyping) are given in **Tables 1–4**.

3. Methods

The general course of operation is given in **Fig. 2**.

3.1. Bacterial Enrichment for Examination of Different Sample Matrixes for STEC / EHEC

3.1.1. Foods

1. Mix 25 mL (e.g., raw milk) or 25 g (e.g., ground beef, soft cheese, raw sausage containing beef) and 225 mL of mTSB containing Novobiocin in a stomacher and cultivate with agitation (frequency: 100/min) at 37°C for 5 to 6 h.
2. Mix 1 mL of the pre-enriched culture and 4 mL of mTSB and cultivate with agitation frequency: 180/min) at 37°C for 18 h.

3.1.2. Sewage

1. Mix 50 mL sewage and 75 mL mTSB and cultivate at 37°C with agitation (frequency: 100/min) for 18 h.

Table 2
Components (in µL) of 25-µL Reaction Mixtures for PCR of STEC and Subtyping Based on *stx* Genes

Target Gene	Primers	Water	PCR Buffer (10 X)	MgCl$_2$ (25 mmol/L)	dNTP Mixture	Primer 1 (50pmol/µL)	Primer 2 (50pmol/µL)	DNA Polymerase (5U/µL)	Template
stx A general (except *stx 2f* A)	MK 1/Mk2	15.25/12.75[a]	2.5	2	2	0.5	0.5	0.25 Goldstar	2.5 / 5[a]
stx 1 A	LP 30/LP 31	14.75	2.5	2	2	0.5	0.5	0.25 Goldstar	2.5
stx 2A general (except *stx 2f*)	LP 43/LP 44	14.75/12.25[a]	2.5	2	2	0.5	0.5	0.25 Goldstar	2.5 / 5[a]
stx 1 B	KS 7/KS 8	16.85/14.35	2.5	—	2	0.5	0.5	0.15 Amplitaq LD	2.5 / 5[a]
stx 2 B ; stx 2c	GK 3/GK 4	14.75	2.5	2	2	0.5	0.5	0.25 Goldstar	2.5
stx 2e B	slt 2v start/stop	16.50[b]	2.5	—	2	0.5	0.5	0.5 Dynazyme	2.5
stx 2e A	slt 2v u/slt 2v d	16.50[b]	2.5	—	2	0.5	0.5	0.5 Dynazyme	2.5
stx 2d A	VT 2-cm/VT 2-f	16.85[b]	2.5	—	2	0.5	0.5	0.15 Amplitaq LD	2.5
stx 2d B	VTAMI/VTAMII	15.25	2.5	1.5	2	0.5	0.5	0.25 Goldstar	2.5
stx 2f A	128 - 1/128 - 2	15.25	2.5	1.5	2	0.5	0.5	0.25 Goldstar	2.5

[a] For screening PCR.
[b] Optimized buffer containing 1.5 mmol/l MgCl$_2$; (concentrations given refer to stock solutions).

Table 3
Conditions of PCR Assays for the Detection of Different Target Genes of STEC

Primers	Primer Sequences (5'-3')	Target Gene	Denaturation	Annealing	Extension	Amplicon (bp)	References
				PCR			
hly A1 hlyA4	ggt gca gca gaa aaa gtt gta g tct cgc ctg ata gtg ttt ggt a	EHEC-*hlyA*	94°C for 30s	57°C for 60s 30 cycles	72°C for 90s	1551	*34*
SK1 SK2	ccc gaa ttc ggc aca agc ata agc ccc gga tcc gtc tcg cca gta ttc g	*eae*	94°C for 30s	52°C for 60s 30 cycles	72°C for 60s	863	*44*
AstA1 AstA2	gcc atc aac aca gta tat ccg gcg agt gac ggc ttt gta gt	*ast A*	94°C for 30s	53°C for 60s 30 cycles	72°C for 50s	140	*32*
wkat-B wkat-F	ctt cct gtt ctg att ctt ctg g aac tta ttt ctc gca tca tcc c	*kat P*	94°C for 30s	56°C for 60s 30 cycles	72°C for 40s	2125	*37*
Esp A Esp B	aaa cag cag gca ctt gaa cg gga gtc gtc agt cag tag at	*esp P*	94°C for 30s	56°C for 60s 30 cycles	72°C for 150s	1830	*38*
D1 D13R	cgt cag gag gat gtt cag cga ctg cac ctg ttc ctg att a	*etp D*	94°C for 30s	52°C for 60 30 cycles	72°C for 70s	1062	*39*
356 595	gca gga tga ccc tgt aac gaa g ccg aag aaa aac cca gta aca g	*ile X*	94°C for 30s	52°C for 60s 30 cycles	72°C for 60s	640	*41*
col D1 col D2	gta aat ctg cct gtt cgt gga c cct ttt tct ctt cgg tat gtt c	*col D 157*	94°C for 60s	57°C for 60s 30 cycles	72°C for 60s	587	*40*
irp 2FP irp 2RP	aag gat tcg ctg tta ccg gac tcg tcg ggc agc gtt tct tct	*irp 2*	94°C for 60s	60°C for 60s 30 cycles	72°C for 60s	280	*31*
fyu A FP fyu A RP	gcg acg gga agc gat tta cgc agt agg cac gat gtt gta	*fyu A*	94°C for 60s	57°C for 60s 30 cycles	72°C for 60s	780	*31*
ANK 49 ANK 50	atg tta tcc tca tat aaa ata aac tta ata cga cag tgg aat atg	*pas*	94°C for 20s	50°C for 60s 30 cycles	72°C for 120s	1221	*44*

Table 4
Components (in μL) of 25-μL Reaction Mixtures for PCR to Detect Different Target Genes of STEC

Target	Gene Primers	Water	PCR buffer (10X)	MgCl$_2$ (25 mmol/L)	dNTP Mixture	Primer 1 (50 pmol/μL)	Primer 2 (50 pmol/μL)	DNA-Polymerase	Template
EHEC *hly A*	hly A1/hlyA4	14.6	2.5	2	2	0.5	0.5	0.4 Hifi[a]	2.5
eae general	SK1/SK2	14.6	2.5	2	2	0.5	0.5	0.4 Hifi[a]	2.5
pas	ANK49/ANK 50	14.6	2.5	2	2	0.5	0.5	0.4 Hifi[a]	2.5
ast A	Ast A1/Ast A2	16.5	2.5[b]	—	2	0.5	0.5	0.5 Dynazyme[c]	2.5
kat P	wkat-B/wkat-F	17.0	2.5[b]	—	2	0.5	0.5	0.4 Hifi[a]	2.5
esp P	Esp A/Esp B	16.6	2.5[b]	—	2	0.5	0.5	0.4 Hifi[a]	2.5
etp D	D1/D13R	16.6	2.5[b]	—	2	0.5	0.5	0.4 Hifi[a]	2.5
ile X	356/595	14.6	2.5	2	2	0.5	0.5	0.4 Hifi[a]	2.5
col D O157	col D1/col D2	16.6	2.5[b]	—	2	0.5	0.5	0.4 Hifi[a]	2.5
irp 2	irp 2 FP/irp 2 RP	15.25	2.5	2	2	0.5	0.5	0.25 Goldstar[d]	2.5
fyu A	fyu A FP/fyu A RP	14.6	2.5	2	2	0.5	0.5	0.4 Hifi[a]	2.5

[a]3.5 U/μL.
[b] Optimized buffer containing 1.5 mmol/L MgCl$_2$.
[c]5 U/μL.
[d] 5 U/μl (concentrations given refer to stock solutions).

3.1.3. Stool Samples and Feces

1. Dip a small swab into the sample and stir it in 2 mL PBS.
2. Mix 0.5 mL of the suspension with 4.5 mL of mTSB and cultivate with agitation (frequency: 160/min) at 37°C for 5 h.
3. Plate 100 μL on a nutrient agar plate and cultivate overnight at 37°C.
4. Rinse the plate with 2 mL of physiological saline and prepare appropriate dilutions (e.g. 1:4; 1:8; 1:10) in water.

3.1.4. Preparation of Nutrient Medium (see **Subheading 2.1.**)

1. For modified tryptose soy broth (mTSB) containing Novobiocin, dissolve 33.0 g of mTSB containing Bile salts No. 3 and K_2HPO_4 in 1l distilled water. Autoclave the solution at 121°C for 15 min. The pH value should be 7.3 ± 0.2. Add Novobiocin as aqueous solution to a final concentration of 20 mg/L.
2. MacConkey agar: dissolve all components in 1l of distilled water and autoclave the solution at 121°C for 15 min. Cool down to 45°C and prepare agar plates.

3.2. Screening PCR

3.2.1. Pre-PCR Treatment of Samples

1. Sediment cells from 1 mL of enriched culture or from an appropriate dilution (stool samples and feces) by centrifugation at 13,000g for 10 min.
2. Discard the supernatant and resuspend the sediment in 1 mL of physiological saline. Repeat the washing step.
3. Resuspend the sediment in 200 μL of water.
4. Heat the suspension at 100°C for 15 min.
5. Chill on ice to 0°C.
6. Centrifuge the sample for 10 s at 13,000g.
7. Use 5 μL for a 25-μL PCR mixture.

3.2.2. Amplification by PCR

Prepare reaction mixes for amplification according to **Table 2**. Temperature-time profiles are given in **Table 1** (*see* **Notes 1** and **3**).

3.2.3. Agarose Gel Electrophoresis and Visualization

1. Mix 25 μL of PCR product with 2.5 μL of gel loading buffer. If Gelstar staining solution is used instead of ethidium bromide, add 1μL of a 1:100 dilution.
2. Pipet 15–20 μL of this mixture into the wells of a 2% agarose gel.
3. Run electrophoresis in 1X TAE buffer.
4. Stop the run when the bromophenol blue front has reached the lower quarter of the gel.
5. Identify specific bands of PCR amplicons under UV light and estimate their size in bp in relation to marker bands.

3.3. Specific Isolation of STEC / EHEC using DNA Hybridization and Digoxigenin-Labeled Probes for Detection of stx 1, stx 2, or eae

3.3.1. Preparation of Plates and Membrane Discs

1. Use a suitable dilution of the pre-enriched or enriched culture in physiological saline.
2. Plate approx 100 µL of the diluted culture onto a MacConkey agar plate.
3. Cultivate overnight at 37°C. Only single colonies should be grown on the plate. Do not use plates with more than 1000 colonies for replica blots.
4. Pre-cool colonies on agar plates for approx 30 min at 4°C.
5. Use tweezers to place a membrane disc for colony and plaque hybridization onto the MacConkey plate congruently. Avoid air bubbles.
6. Mark the membrane disc and the plate at least on two spots. This is necessary for the identification of STEC/EHEC by comparing the hybridized and developed membrane disc with the master plate (*see* **Subheading 3.3.5.**).
7. Use a spatula to press the membrane gently on the agar plate.
8. Remove the membrane disc carefully with filter tweezers and blot briefly (colonies upside) on Whatman 3 MM paper. Cultivate the plate for another 2 to 3 h at 37°C to create a master plate for subsequent specific isolation.
9. Place membrane disc (colonies upside) for 15 min on a prepared filter paper soaked with denaturation solution.
10. Place membrane disc for 15 min onto a prepared filter soaked with neutraliza tion solution.
11. Place membrane disc for 5 min onto prepared filters soaked with 2X SSC for equilibration.
12. Bake the air-dried membrane disc at 80° C for 1 h.
13. Treat colony lifts with proteinase K to remove cell debris. Place membrane disc on a foil, pipet 1 mL on it, and distribute the solution.
14. Incubate at 37°C for 1 h.
15. Blot membrane disc under pressure (use a tube) between filter paper, soaked in distilled water and remove debris. The cell debris will stick to the filter paper. The membrane is now ready for hybridization.

3.3.2. Preparation of DIG-Labeled Probes

1. Mix the following substances for PCR to prepare DIG-labeled probes. Do not mix all dNTPs before use in the master mixture, use a reduced amount of dTTP (0.65 µL) and, as an additional dNTP, add 0.35 µL of DIG-11-dUTP.
2. Each 50-µL amplification mixture should contain: 27.35 µL of water, 5.0 µL of 10X PCR buffer (without $MgCl_2$), 3.0 µL of 25 mM $MgCl_2$, 1.0 µL of 1 mM dATP, 1.0 µL of 1 mM dCTP, 1.0 µL of 1 mM dGTP, 0.65 µL of 1 mM dTTP, 3.5 µL of DIG-11-dUTP (1 nmol/µL), 1.0 µ of primer 1 (MK1 or KS7 or SK1) (50 pmol/µL), 1.0 µL of primer 2 (MK2 or KS8 or SK2) (50 pmol/µL), 0.5 µL Goldstar DNA Polymerase (5U/µL), and 5.0 µL of template DNA.
3. Pick 2 or 3 colonies from an agar plate and suspend them in 100 µL of water. Use 5 µL of this suspension as template in a 50-µL amplification reaction.

Table 5
Combinations of Reference Strains and Primer Pairs
for the Preparation of DIG-Labeled Probes

Virulence Factor	Reference Strain	PCR Primers
stx 1	C 600 J 1	MK 1 / MK 2
stx 1	C 600 J 1	KS 7 / KS 8
stx 2	C 600 W 34	MK 1 / MK 2
stx 2	C 600 W 34	LP 43 / LP 44
eae	16–84	SK1 / SK 2

4. Choose a suitable combination of primers and reference strains according to **Table 5**.
5. Run the specific temperature-time profile given in **Tables 1** and **3**.
6. Prepare 1.8% low-melting agarose gel in TAE buffer.
7. After finishing PCR, follow the steps given in **Subheading 3.2.3.**
8. Excise the specific bands under UV light using a scalpel.
9. Dissolve one plug (vol approx 80 µL) by boiling in 200 µL of water.
10. Chill on ice to 0°C. The amount is enough for hybridization of one membrane disc. The solution can be stored at −20°C for several mo.

3.3.3. Prehybridization and Hybridization

1. Place at most 3 membrane discs into a roller bottle and fill in 60 mL of DIG Easy Hyb. Prehybridize for 1 h at 42°C in a hybridization oven for roller bottles. Make certain that the membranes do not stick to each other and are sufficiently covered with the solution.
2. Denature the labeled probe by boiling for 5 min at 95 to 100°C and chill on ice.
3. Mix a suitable amount of denatured labeled probe and 6 mL of prewarmed DIG Easy Hyb.
4. Remove the prehybridization solution and add approx 6.6 mL of the probe/DIG Easy Hyb mixture.
5. Hybridize overnight at 42°C. The hybridization solution with the DIG-labeled probe is stable at −20°C for more than 12 mo and can be reused several times when freshly denatured.

3.3.4. Washings and Detection

1. For stringent washes, wash membrane discs twice in ample washing solution 1 for 5 min each at room temperature.
2. Transfer to washing solution 2 and wash twice for 15 min at 68°C with gentle agitation.
3. Before starting with the detection procedure, prepare all the buffers and solutions for STEC detection using the DIG Nucleic Acid Detection Kit as given in **Subheading 2.3.4.**

Table 6
PCR Conditions for Subtyping Based on *eae* Genes

Primers	Primer Sequences (5'-3')	Target Gene	PCR Denaturation	PCR Annealing	PCR Extension	Amplicon (bp)	References
SK1 SK2	ccc gaa ttc ggc aca agc ata agc ccc gga tcc gtc tcg cca gta ttc g	*eae* cons.	94°C for 30s	52°C for 60s 30 cycles	72°C for 60s	863	*44*
SK1 LP2	ccc gaa ttc ggc aca agc ata agc ccc gaa ttc ttt tac aca agt ggc	*eae*-a	94°C for 30s	55°C for 60s 30 cycles	72°C for 120s	2807	*45,46*
SK1 LP4	ccc gaa ttc ggc aca agc ata agc ccc gtg ata cca gta cca att acg gtc	*eae*-b	94°C for 30s	55°C for 60s 30 cycles	72°C for 120s	2287	*45,46*
SK1 LP3	ccc gaa ttc ggc aca agc ata agc ccc gaa ttc tta ttc aca aac cgc	*eae*-g	94°C for 30s 94°C for 30s	45°C for 60s 3 cycles 52°C for 60s 28 cycles	72°C for 120s 72°C for 120s	2792	*45,46*
Int -d Int - Ru	tac gga ttt tgg ggc at ttt att tgc agc ccc cca t	*eae*-d	95°C for 20s	45°C for 60s 30 cycles	74°C for 60s	544	*34*
SK1 LP5	ccc gaa ttc ggc aca agc ata agc agc tca ctc gta gat gac ggc aag cg	*eae*-e	94°C for 30s	55°C for 60s 30 cycles	72°C for 120s	2608	*45,46*

Note: All reaction mixtures should contain the following ingredients: 14.6 μL of water, 2.5 μL of 10X PCR buffer, 2 μL of $MgCl_2$, 2 μL of dNTP mix, 0.5μL of each primer, 0.4μL of DNA polymerase, 2.5 μLof template.

4. Wash membrane discs briefly (5 min) in washing buffer at room temperature.
5. Block for 30 min in 40 mL of buffer 2 at room temperature.
6. Dilute the appropriate Anti-DIG-AP conjugate in buffer 2.
7. Incubate the membrane for 30 min in 15 mL of antibody conjugate solution.
8. Wash 2 for 15 min in 40 mL of washing buffer 3.
9. Equilibrate for approx 5 min in 20 mL of buffer 3.
10. Add 4 mL of color substrate solution to each membrane disc and incubate in the dark. (Membranes should be processed separately. Do not shake or move during incubation.)
11. The color reaction is usually completed after 1 to 2 h.
12. When the expected signals appear, stop the reaction by rinsing in approx 50 mL of TE buffer.

3.3.5. Specific Isolation of STEC/EHEC

1. Use a needle to get through the brown violet spots on the membrane disc.
2. Put the membrane disc on a light box. Place the master plate (*see* **Subheading 3.3.1., step 8**) onto the membrane disc congruently. Note the marks on the membrane disc and the MacConkey plate made before creating the replica. The developed membrane disc is a mirror image of the master plate.
3. Compare spots on the membrane disc and colonies on the master plate. Isolate single colonies by using a sterile glass stick and cross out on a new agar plate. (In case of a mixture of STEC and other microorganisms, or when the number of colonies on the master plate is very high, repeat the hybridization, and use this newly prepared agar plate for replica. **Figure 3** illustrates the relationship between master plate and developed membrane disc.

3.4. Characterization of Isolates

1. For detection of specific virulence factors by PCR, prepare reaction mixtures given in **Tables 2** and **4**. Then run the specific temperature-time profile (*see* **Tables 1, 3** and **6**).
2. Detect specific amplicons by gel electrophoresis as described in **Subheading 3.2.3.** and estimate their size (in bp) in relation to marker bands (*see* **Note 4**).

4. Notes

1. It is necessary to check the sample DNA preparation by an alternative PCR system to verify the PCR result. For this purpose, a system based on a pUC19 target sequence is recommended as external amplification control.
   Prepare a 25-µL PCR mixture containing components given in **Table 7**. Run the following temperature-time profile: initial denaturation at 95°C for 5 min, 30 cycles of denaturation at 95°C for 0.5 min, annealing at 65°C for 0.5 min, extension at 72°C for 1 min, and final extension at 72°C for 7 min. The size of the correct amplicon is 429 bp.
   In case of failure of amplification controls, plate a suitable dilution of the enriched culture on a nutrient agar plate. Cultivate the plate overnight at 37°C. Then rinse

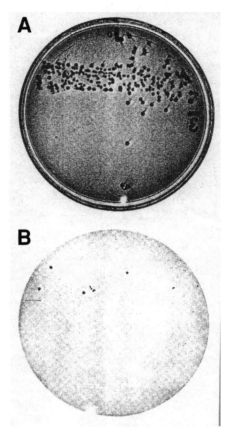

Fig. 3. Photographic image showing a master plate in the upper part and a replica nylon membrane after hybridization in the lower part (*see* **Subheadings 3.3.1.** and **3.3.5.**).

the plate with 2 mL of physiological saline and prepare appropriate dilutions in water. Use these suspensions as template in a screening PCR again.

2. It is possible to use another solution (called solution B) for prehybridization and hybridization. It contains the following components: 5 X SSC solution (*see* **Subheading 2.3.1.**), 0.1 % (w/v) N-lauroylsarcosine, 0.1% (w/v) SDS, Blocking reagent® (Roche Diagnostics). To prepare the solution, dissolve N-lauroylsarcosine and SDS in 5X SSC. Calculate and add the amount of blocking reagent. Heat to 70°C and stir the solution for 1 h. The solution remains opaque. When applying solution ('B'), only use the following solution for all washing steps: 0.1% (w/v) SDS in 0.04X SSC. It is possible to use the solution containing labeled probes 2 or 3 irrespectively of the kind of hybridization solution (DIG Easy Hyb or B). For reuse, heat the solution to 70°C and chill on ice.

Table 7
Reaction Mix for Amplification Control Using pUC 19 DNA as Target
(*see* Note 1)

Component	Volume (µl)	Final Concentration in 25-µL Reaction Mix
Water	11.8	—
10X buffer	2.5	1X buffer
MgCl₂ solution (25 mmol/L)	1.5	1.5 mmol/L
dNTP-Mix (10 mmol/l)	2.0	each dNTP 200 µmol/L
Primer HB 10 (20 µmol/L)	0.5	0.4 µmol/L
Primer HB 11 (20 µmol/L)	0.5	0.4 µmol/L
DNA polymerase	0.2	at least 1 U per mix
pUC 19 DNA (1 pg/µL)	1	0.04 pg/µL
Template DNA	5	—

Table 8
Cleavage of PCR Amplicons by Restriction Endonucleases

Virulence Factor	Restriction Enzyme	Fragments in bp
stx 1	*Acc* I	91, 191
stx 2	*Hae* III	248, 336
eae	*Pst* I	241, 248, 374

3. Always include a positive, a negative, and a water control. Use the strains given in **Table 6** as positive controls and strain C 600 or another *stx*-negative strain as negative control.

4. The most important virulence factors of STEC/EHEC are *stx 1*, *stx 2*, and *eae*. Positive PCR results should be verified by restriction enzymes cleaving within the amplified fragment. More data are given in **Table 8**. In a standard procedure, the following mix should be used: 16 µL of PCR product (after running the specific temperature-time profile), 2 µL of 10X restriction buffer depending on the endonuclease (e.g., SuRE/cut buffer, Roche Diagnostics), and 2 µL (at least 20 U) of restriction enzyme. Incubate for 2 h at 37°C. After finishing, add 2 µL of gel loading buffer and start electrophoretic separation in a 2% agarose gel. Detect specific bands as described in **Subheading 3.2.3.**

References

1. Savage, D. C. (1977) Microbial ecology of the gastrointestinal tract. *Ann. Rev. Microbiol.* **31**, 107–133.
2. Nataro, J. P., and Kaper, J. B. (1998) Diarrheagenic *Escherichia coli*. *Clin. Microbiol. Rev.* **11**, 142–201.

3. Boyce, T. G., Swerdlow, D. L., and Griffin, P. M. (1995) *Escherichia coli* O157:H7 and the hemolytic–uremic syndrome. *N. Engl. J. Med.* **333**, 364–368.
4. Karmali, M. A. 1989. Infection by verocytotoxin - producing *Escherichia coli*. *Clin. Rev.* **2**, 15–38.
5. Karmali, M.A., Steele, B.T., Petric, M., and Lim, C. (1983) Sporadic cases of haemolytic uremic syndrome associated with faecal cytotoxin and cytotoxin- producing *Escherichia coli* in stools. *Lancet* **1**, 619–620.
6. Moake, J. L. (1994) Haemolytic uraemic syndrome: basic science. *Lancet* **343**, 393–397.
7. Remuzzi, G. and Ruggenenti, P. (1995) The hemolytic uremic syndrome. *Kidney Inter.* **47**, 2–19.
8. Keusch, G.T. and Acheson, W.K. (1997) thrombotic thrombocytopenic Purpura associated with shiga toxins. *Seminars in Hematology* **34**, 106–116.
9. Kaper, J. B., and O' Brien, A. D. (ed.) (1998) *Escherichia coli* O157 and other Shiga toxin-producing *E. coli* strains. ASM Press, Washington DC, USA.
10. Karch, H., Gunzer, F., Schwarzkopf, A. and Schmidt, H. (1993) Molekularbiologie und pathogenetische Bedeutung von Shiga- und Shiga-like Toxinen. *Bio Engineering* **3**, 39–45.
11. Riley, L. W., Remis, R. S., Helgerson, S. D., et al. (1983) Hemorrhagic colitis associated with a rare *Escherichia coli* serotype. *New Engl. J. Med.* **308**, 681–685.
12. Konowalchuk, J., Speiers, J.I., and Stavrik., S. (1977) Verocell response to a cytotoxin of *Escherichia coli*. *Infect.Immun.* **18**, 775–779.
13. Hovde, C. J., Calderwood, S. B., Mekalanos, J. J., and Collier, R. J. (1988) Evidence that glutamic acid 167 is an active-site residue of Shiga-like toxin 1. *Proc. Natl. Acad. Sci. USA* **85**, 2568–2572.
14. Griffin, P. M., and Tauxe, R. V. (1991) The epidemiology of infections caused by *Escherichia coli* , and the associated hemolytic uremic syndrome. *Epidemiol. Rev.* **13**, 60–98.
15. Scotland, S. M., Willshaw, G. A., Smith, H. R., and Rowe, B. (1990) Properties of strains of *Escherichia coli* O26:H11 in relation to their enteropathogenic or enterohemorrhagic classification. *J. Infect. Dis.* **162**, 1069–1074.
16. Scheutz, F. (1995) Epidemiology of non-O 157 VTEC. Proceedings of WHO Consultation on Emerging Food-borne Diseases, Berlin , March 20–24, 1995.
17. Tarr, P. I., and Neill, M. A. (1996) Perspective: the problem of non-O157 : H7 shiga toxin (verocytotoxin) - producing *Escherichai coli*. *J. Infect. Dis.* **174**, 1136–1139.
18. Willshaw, G. A., Smith, H. R., Roberts, D., Thirlwell, J., Cheasty, T., and Rowe, B. (1993) Examination of raw beef products for the presence of vero cytotoxin producing *Escherichia coli* , particularly those of serogroup O157. *J. Appl. Bacteriol.* **75**, 420–426.
19. Datz, M., Janetzki-Mittmann, C., Franke, S., Gunzer, F., Schmidt, H., and Karch, H. (1996) Analysis of the enterohemorrhagic *Escherichia coli* O157 DNA region containing lambdoid phage gene p and shiga - like - toxin structural genes. *Appl. Environml. Microbiol.* **62**, 791–797.

20. Jackson, M. P., Neill, R.J., O'Brien, A. D., Holmes, R. K., and Newland, J. W. (1987) Nucleotide sequence analysis and comparison of the structural genes for shiga-like toxin I and shiga-like toxin II encoded by bacteriophages from *Escherichia coli* 933. *FEMS Microbiol. Lett.* **44**, 109–114.

21. WHO Consultation on Selected Emerging Food-borne Diseases (1995). Berlin, Germany, March 20–24, WHO/CDS/VPH / 95.142.

22. Schmidt, H., Montag, M., Bockemühl, J., Heesemann, J., and Karch, H. (1993) Shiga-like toxin II related cytotoxins in *Citrobacter freundii* strains from humans and bovine beef samples. *Infect. Immunol.* **61**, 534–543.

23. Karch, H. and Meyer, T. (1989) Single primer pair for amplifying segments of distinct Shiga-like-toxin genes by polymerase chain reaction. *J. Clin. Microbiol.* **27**, 2751–2757.

24. Cebula, T. A., Payne, W. I., and Feng, P. (1995) Simultaneous identification of strains of *Escherichia coli* serotype O157:H7 and their shiga-like toxin type by mismatch amplification mutation assey- multiplex PCR. *J. Clin. Microbiol.* **33**, 248–250.

25. Rüssmann, H., Kothe, E., Schmidt, H., et al. (1995) Genotyping of Shiga-like toxin genes in non-O157 *Escherichia coli* strains associated with haemolytic uraemic syndrome. *J. Med. Microbiol.* **42**, 404–410.

26. Schmidt, H., Rüssmann, H., Schwarzkopf, A., Aleksic', S., Heesemann, J., and Karch, H. (1994). Prevalence of attaching and effacing *Escherichia coli* in stool samples from patients and controls. *Zbl. Bact.* **281**, 201–213.

27. Schmitt, C. K., McKee, M. L., and O'Brien, A. D. (1991) Two copies of shiga-like-toxin II-related genes common in enterohe morrhagic *Escherichia coli* strains are responsible for the antigenic heterogeneity of the O157:H- strain E32511. *Infect.Immun.* **59**, 1065–1073.

28. Weinstein, D. L., Jackson, M. P., Samuel, J. E., Holmes, R. K., and O'Brien, A. D. (1988) Cloning and sequencing of a shiga-like toxin type II variant from an *Escherichia coli* strain responsible for Edema disease of swine. *J. Bacteriol.* **170**, 4223–4230.

29. Pierard, D., Muyldermans, G., Moriau, L., Stevens, D., and Lauwers, S. (1998) Identification of new Verocytotoxin type 2 variant B-subunit genes in human and animal *Escherichia coli* isolates. *J. Clin.Microbiol.* **36**, 3317–3322.

30. Schmidt, H., Scheef, J., Morabito, St., Caprioli, A., Whieler, L., and Karch, H. (2000) A new Shiga toxin 2 variant (Stx 2f) from *Escherichia coli* isolated from pigeons. *Appl. Environ. Microbiol.* **66**, 1205–1208.

31. Karch, H., Schubert, S., Zhang, D., Schmidt, H., Ölschläger, T., and Hacker, J. (1999) A genomic island, termed High-Pathogenicity Island, is present in certain Non O157 Shiga toxin-producing *Escherichia coli* clonal lineages. *Infect. Immunol.* **67**, 5994–6001.

32. Savarino, S. J., Fasano, A., Watson, J., et al. (1993) Enteroaggregative *Escherichia coli* heat-stable enterotoxin I represents another subfamily of *E. coli* heat-stable toxin. *Proc. Natl. Acad. Sci. USA* **90**, 3093–3097.

33. Beutin, L., Prada, J., Zimmermann, S., Stephan, R., Orskov, I., and Orskov, F. (1988) Enterohemolysin, a new type of hemolysin produced by some strains of

enteropathogenic *E. coli* (EPEC). *Zentralbl. Bacteriol. Microbiol. Hyg. A* **267**, 576–588.

34. Adu-Bobie, J., Frankel, G., Bain, Ch., et al. (1998) Detection of intimins α, β, γ, and δ, four intimin derivatives expressed by attaching and effacing microbial pathogens. *J. Clin. Microbiol.* **36**, 662–668.

35. Schmidt, H., and Karch, H. (1996) Enterohemolytic phenotypes and genotypes of shiga toxin-producing *Escherichia coli* O111 strains from patients with diarrhoea and hemolytic-uremic syndrome. *J. Clin. Microbiol.* **34**, 2364–2367.

36. Schmidt, H., Kernbach, C., and Karch, H. (1996). Analysis of the EHEC hly operon and its location in the physical map of the large plasmid of enterohaemorrhagic *Escherichia coli* O157 : H7. *Microbiology* **142**, 907–914.

37. Brunder, W., Schmidt, H., and Karch, H. (1996) *katP*, a novel catalase-peroxidase encoded by the large plasmid of enterohaemorrhagic *Escherichia coli* O157:H7. *Microbiology* **142**, 3305–3315.

38. Brunder, W., Schmidt, H., and Karch, H. (1997) *EspP*, a novel extracellular serine protease of enterohaemorrhagic *Escherichia coli* O157:H7 cleaves human coagulation factor V. *Molec. Microbiol.* **24**, 767–778.

39. Schmidt, H., Henkel, H., and Karch, H. (1997) A gene cluster closely related to type II secretion pathway operons of Gram-negative bacteria is located on the large plasmid of enterohemorrhagic *Escherichia coli* O157 strains. *FEMS Microbiol. Lett.* **148**, 265–272.

40. Hofinger, C., Karch, H. , and Schmidt, H. (1998) Structure and function of plasmid *pColD 157* of enterohemorrhagic *Escherichia coli* O157 and its distribution among strains from patients with diarrhea and hemolytic-uremic syndrome. *J. Clin. Microbiol.* **36**, 24–29.

41. Schmidt, H., Scheef, J., Janetzki-Mittmann, C., Datz, M., and Karch, H. (1997) An *ileX* tRNA gene is located close to the shiga toxin II operon in enterohemorrhagic *Escherichia coli* O157 and non-O157 strains. *FEMS Microbiol. Lett.* **149**, 39–44.

42. Gallien, P., Klie, H., Perlberg, K-W., and Protz, D. (1996) Einsatz von Nylon Membranen zur gezielten Isolierung und Charakterisierung Verotoxin-bildender *Escherichia coli* mittels DNA-Sonden. *Berl. Münch. Tierärtzl. Wschr.* **109**, 431–433.

43. Gallien, P., Much, Ch., Perlberg, K.-W. und Protz, D. (1999) Subtypisierung von *stx*-Genen in Shigatoxin-produzierenden *Escherichia coli* (STEC). *Fleischwirtschaft* **6**, 99–103.

44. Schmidt, H., Rüssmann, H., and Karch, H. (1993) Virulence determinants in nontoxinogenic *Escherichia coli* O157 strains that cause infantile diarrhoea. *Infect. Immunol.* **61**, 4894–4898.

45. Oswald, E., Schmidt, H., Morabito, S., Karch, H., Marches, O., and Caprioli, A. (2000) Typing of intimin genes in human and animal enterohemorrhagic and enteropathogenic *Escherichia coli*: Characterization of a new intimin variant. *Infect. Immunol.* **68**, 64–71.

46. Gallien, P., Karch, H., Much, Ch., et al. (2000) Subtypisierung von eae-Genen in Shigatoxin-produzierenden *Escherichia coli* (STEC). *Fleischwirtschaft* **2**, 84–89.

12

Detection of *Listeria monocytogenes* Using a PCR/DNA Probe Assay

Louise O'Connor

1. Introduction

It is only in recent years that *Listeria monocytogenes* has become regarded as a significant food-borne pathogen. Interest in the organism arose due to several food-borne outbreaks in the early 1980s of listeriosis. The high mortality rate associated with the illness prompted widespread public concern about the pathogen and resulted in health authorities and the food industry initiating programs to control the organism and the disease. Listeriosis is not characterized by a unique set of symptoms, since its course depends on the state of the host. Approximately one-third of human *L. monocytogenes* infections are associated with pregnant women and their unborn infants, and the other two thirds occur in non-pregnant immunocompromised individuals of all ages. Symptoms caused by the infection during pregnancy are generally only a mild fever in the mother, with or without gastroenteritis or flu-like symptoms, but the consequences for the fetus or newborn are often major or fatal.

Many food products pose a risk of causing an outbreak of listeriosis because the organism can be present in many raw materials such as raw milk. In most instances the outbreaks have resulted due to a lack of adequate control and monitoring systems. The availability of methods to detect the organism is, therefore, of extreme importance in determining the status of a particular food product prior to release for sale and consumption. In addition, reliable detection methods are of vital importance in determining the source of outbreaks. In recent years the advent of nucleic acid-based technologies has allowed the development of tests enabling pathogens to be rapidly identified without the need to isolate pure cultures.

From: *Methods in Molecular Biology, vol. 216: PCR Detection of Microbial Pathogens: Methods and Protocols*
Edited by: K. Sachse and J. Frey © Humana Press Inc., Totowa, NJ

This chapter will describe a polymerase chain reaction (PCR)/DNA probe membrane-based method developed in our laboratory for the rapid detection of *L. monocytogenes* in food samples. DNA probes are now commonly used to detect and identify microorganisms, particularly pathogenic microorganisms in a variety of areas including medicine, the food industry, and the environment. The ribosomal operon represents a good target for development of diagnostic assays for microbial pathogens, since it is present in multiple copies in most organisms. This region contains conserved sequence interspersed with variable sequence regions allowing the scope to design broad ranging or specific detection assays as appropriate *(1)*. These DNA probes can be employed directly in hybridization assays or they may be combined with an in vitro amplification step such as PCR for pathogen detection. PCR is currently one of the most widely employed techniques to complement classical microbiological methods for the detection of pathogenic microorganisms. Detection of PCR amplification products can be carried out by gel electrophoresis combined with ethidium bromide staining and visual examination of the gel using UV light. Southern blotting and hybridization with a specific labeled probe can follow electrophoresis, thus allowing confirmation of the identity of the PCR product and increasing the sensitivity of the assay. Colorimetric or fluorometric hybridization of amplification products can also be carried out using specific DNA probes bound to solid phases, such as microtiter plates, allowing rapid and simplified detection. The detection format adapted in our laboratory is a colorimetric reverse hybridization assay *(2–5)*. This membrane-based detection of PCR products was first developed by Saiki et al. *(6)*, later modified by Zhang et al. *(7)*, and adopted by Chehab and Wall *(8)* for screening cystic fibrosis mutations. The method incorporates biotin into the PCR product by including a biotinylated primer in the PCR. This biotinylated PCR product can hybridize to a capture probe, which is immobilized on a chemically modified nylon membrane. The PCR/DNA probe hybrid can be detected colorimetrically using a conjugate such as streptavidin alkaline phosphatase and a chromogenic substrate (*see* **Fig. 1A** and **1B**). A panel of capture probes to different pathogenic microorganisms can be immobilized on a single membrane, allowing multipathogen detection.

2. Materials

2.1. Enrichment of Food Samples

1. *Listeria*-enrichment broth (Merck, Darmstadt, Germany; Cat. no. 1.11951).
2. *Listeria*-enrichment broth supplement (Merck; Cat. no. 1.11883).
3. Stomacher (Seward; Cat. no. 400).
4. Sterile stomacher bags.
5. 37°C incubator.

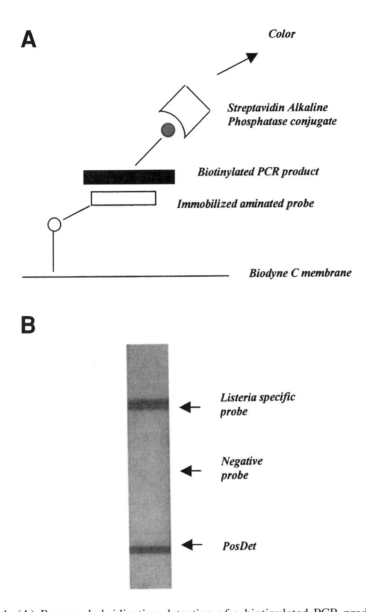

Fig. 1. (**A**) Reverse hybridization detection of a biotinylated PCR product. The probe on the membrane specifically captures the biotinylated PCR product. The probe product hybrid is then detected via a streptavidin alkaline phosphate conjugate and conversion of the substrate to an insoluble product, yielding a purple color on the membrane. (**B**) Actual Biodyne C membrane strip following colorimetric detection using an *L. monoctogenes*-specific probe in the membrane. Also on the strip is the positive detection control PosDet and the unrelated negative control.

2.2. PCR of Enriched Food Samples

1. Oligonucleotide primers: LGS2 5'-ccgtgcgccctttctaactt-3', LMONOREV 5'-tttgttcagttttgagaggt-3' (Genosys Biotechnologies, Ltd., London Road, Pampisford, Cambridgeshire).
2. Nuaire 425 Class II Biological Safety Cabinet.
3. Thermal cycler.
4. PCR tubes, 0.5 mL.
5. PCR-grade H_2O (Sigma UK).
6. Proteinase K (10 mg/mL) (Sigma).
7. *Taq* DNA polymerase, $MgCl_2$, and reaction buffer (Promega, Madison, WI, USA).
8. DU:dNTPs (Promega): Prepare from 100 mM stocks by combining 12.5 µL dATP, dGTP, dCTP, 4.45 µL dTTP, and 8 µL dUTP.
9. Uracil-N-DNA glycosylase (UNG) (Roche Molecular Biochemicals, UK)

2.3. Reverse Hybridization Assay

1. Biodyne C membrane (Pall Biosupport; Pall Europe, Ltd. Portsmouth UK).
2. EDAC-HCl (Sigma) 16% (w/v) in distilled water (dH_2O).
3. Oligonucleotide probes (LMONO20 5'-gatgcttcaaggcatagtgcc-3'), including PosDet a positive detection control probe (Genosys Biotechnologies Ltd.). PosDet consists of a random stretch of oligonucleotide bases modified with a biotin group at the 3'-end and an amine group at the 5'-end.
4. $NaHCO_3$: 0.5 M, pH 8.4.
5. Biodot liquid dispensing machine (xyz 3000) (Biodot, Irvine, CA, USA).
6. NaOH: 0.1 M.
7. Biodot cutter 3000.
8. Heating block.
9. Hybridization buffer: 5X SSPE (20X SSPE: 175.3 g NaCl, 27.6 g $NaH_2PO_4 \cdot H_2O$, 7.4 g EDTA/L H_2O), 0.1% (w/v) sodium dodecyl sulfate (SDS).
10. TBS: 50 mM Tris-HCl, pH 7.5, 0.15 M NaCl.
11. Shaking platform.
12. Strepatvidin alkaline phosphatase conjugate (Calbiochem-Novabiochem UK, Ltd., Nottingham, UK).
13. TBS/Tween® 20: 50 mM Tris-HCl, pH 7.5, 0.15 M NaCl, 1% Tween.
14. Color development solution: 0.1 M Tris-HCl, pH 9.5, 50 mM $MgCl_2$, 0.1 M NaCl.
15. Chromogenic substrates: nitroblue tetrazolium chloride (NBT) and 5-bromo-4-chloro-3-indolyl-phosphate, 4-toluidine salt (BCIP) (Roche Molecular Biochemicals). To 10 mL of color development solution, 50 µL of NBT and 37.5 µL BCIP are added.

3. Methods

3.1. Enrichment of Food Samples (see Notes 1 and 2)

1. Weigh out 25 mL/25 g of sample.
2. Place sample on sterile stomacher bag and add 225 mL of enrichment broth.
3. Homogenize the sample, seal the top of the bag, and incubate at 37°C overnight.

3.2. PCR Amplification of Enriched Food Samples (see Notes 3 and 4)

1. Remove 5 µL of the enriched sample from the stomacher bag and place in 0.5 mL PCR reaction tube containing 1 µL proteinase K and add sterile PCR-grade H_2O to 50 µL.
2. Heat the reaction to 55°C for 10 min to lyse the bacterial cells. The proteinase K is inactivated by following this step with a 95°C step for 10 min.
3. Prepare the PCR master mixture in a class II safety cabinet. For one reaction, add 10 µL reaction buffer, 16 µL $MgCl_2$ (25 mM), 12.5 µL dU:dNTPs, 1 µL of each primer, 0.3 µL *Taq* DNA polymerase and 9.1 µL H_2O, 0.1 µL UNG, and vortex mix well (*see* **Note 5**).
4. Add 50 µL of this mixture to the 50 µL of lysed cell reaction.
5. Cycle through 1 cycle of 37°C for 10 min, 35 cycles of denaturing at 94°C, annealing at 55°C and extension at 72°C for 1 min.
6. Include a no-template negative control in all experiments.

3.3. Reverse Hybridization Assay (see Notes 6–8)

1. Cut the Biodyne C membrane into a strip 35 mm × 4 cm and activate by soaking in 16% EDAC HCl for 20 min.
2. Wash the membrane several times in distilled water and allow to dry at room temperature.
3. Prepare 40 pmol/ µL solution of the probes in 0.5 M $NaCO_3$ pH 8.4.
4. Apply the probes using the Biodot dispensing machine. Alternatively, if a liquid dispensing machine is not available the probes can be applied using a hand-held pipet, and 1 µL dots are adequate (*see* **Note 9**).
5. Quench all active carboxyl groups on the membrane by soaking it in a solution of 0.1 M NaOH for 5 min followed by several rinses with dH_2O.
6. Remove a 50 µL aliquot of the PCR product from the reaction mixture and denature at 95°C for 10 min in a heating block. Quench immediately on ice.
7. To this denatured PCR product add 1 mL of hybridization buffer.
8. Immerse the membrane strip containing the probes in this buffer/PCR product mixture in a trough and incubate on a shaking platform for 1 h at 37°C.
9. Following hybridization, wash the membrane 2× for 10 min in hybridization buffer and once in TBS.
10. Prepare a 1/2000 dilution of the streptavidin alkaline phosphatase conjugate in TBS/Tween solution (*see* **Note 10**).
11. Incubate the washed membrane in the conjugate for 20 min at room temperature on a shaking platform.
12. Wash the membrane 2× for 10 min in TBS and once in color development solution.
13. Add 50 µL of NBT and 37.5 µL of BCIP to 10 mL of color development solution.
14. Immerse the membrane in this chromogenic substrate until the Posdet purple line on the membrane becomes visible (1 to 2 min).
15. Terminate color development by soaking the membrane in dH_2O.

4. Notes

1. In enriching food samples, which are particulate in nature, such as meat, it is a good idea to use stomacher bags containing an inner gauze insert. This retains the gritty material, allowing the enrichment broth to remain free from particulate material. The samples are usually incubated without shaking because of the nature of the stomacher bags, which will leak if not sealed properly. Liquid samples such as milk can be enriched in sealed culture flasks in a shaking incubator.

2. One of the limitations of using PCR assays for the detection of microbial pathogens is that PCR will detect both viable and nonviable organisms if the target sequence is present in the sample. The inclusion of a pre-enrichment step prior to PCR increases the number of target cells in the media, dilutes the inhibitory effects of the matrix, and confines detection to viable culturable cells *(9)*. It is, therefore, important to consider carefully the type of enrichment medium chosen, which may in itself contain inhibitors. The length of the enrichment step should also be taken into consideration. This should be optimized to allow maximum growth of the target organism, while at the same time fitting the time-frame for the assay. This allows detection of the organism in the shortest time frame possible.

3. One of the main problems associated with the use of PCR assays for food samples is the presence of inhibitors in the food. False negative results can occur for many reasons, including nuclease degradation of the target nucleic acid sequences or primers, the presence of substances that chelate divalent magnesium ions *(10)* essential for the PCR reaction, and lastly, inhibition of the DNA polymerase *(11)*. Inhibition of the PCR varies with the type of food under analysis. High levels of oil, salt, carbohydrate, and amino acids have been shown to have no inhibitory effect, while casein hydrolysate, calcium ions, and certain components of some enrichment broths are inhibitory for PCR. The removal of inhibitory substances from the target to be amplified can be an important step prior to PCR amplification.

   A method that we have used to reduce the amount of PCR inhibitors is immunomagnetic separation (IMS). This involves the use of magnetic particles coated with antibodies to *Listeria* to capture the organism from the enrichment culture. These magnetic particles can then be included directly in the PCR. The antibody must have a high affinity for the target organism to achieve effective capture. Also, if the specificity of the antibody for the target organism is high, use of IMS introduces an extra degree of specificity into the assay. The problem associated with antibody-based separation is that certain food components, such as fats, can interfere with the antibody–organism interaction, as can food debris.

4. It is recommended that, if possible, the methods outlined here should be performed in biological cabinets, which allow product protection by providing an additional safeguard against PCR contamination.

5. The inclusion of UNG in the reaction is important for the control of contamination from PCR products. UNG catalyses the destruction of dUTP containing DNA, but not dTTP containing DNA. When dUTP is used together with dTTP, the generation of a PCR product that is susceptible to destruction by UNG is

facilitated. Deoxyuridine is not present in the target DNA, but is always present in the PCR product. Incubation of the reaction mixture and the target DNA at 37°C, allows the destruction of any dUTP-containing product. Increasing the temperature to 94°C in the first round of amplification inactivates the UNG ensuring that real product is not destroyed.

6. The reverse hybridization assay offers a number of advantages in comparison to conventional methods. It is rapid and convenient to use and is therefore ideal for routine laboratory use. Following PCR amplification, detection of the PCR products takes in the region of 2 h to perform, which represents a significant reduction in time when compared to classical methods of pathogen detection. The detection limit of this technology is comparable to that attainable using Southern blot analysis and therefore, represents a sensitive assay format.

7. When designing primer pairs and probes for use in this reverse hybridization format, it is advisable to design more than one set, since the dynamics of the reaction are different to Southern blot analysis in which the PCR product is denatured and immobilized on the membrane. In the reverse hybridization format, the PCR product is free in solution where secondary structures can form more easily. Secondary structures can have a pronounced effect in the reverse hybridization format, particularly where a large PCR product is being hybridized to a membrane-bound probe. It is also important to ensure that secondary structures are not present in the sequence from which probes are designed. A DNA folding program, such as the Zuker folding program can be used for determining secondary structure and is available on the Internet at (http://bioinfo.math.rpi.edu/).

8. One point to note with this reverse hybridization method is that the pH of the $NaHCO_3$ is critical. This is the buffer used to dilute the probe that is to be attached to the membrane. The 5'-amine on the probe will not bind to the carboxyl group on the membrane if this pH is not at 8.4.

9. It is not necessary to apply the probes using a Biodot liquid dispensing machine. They can be applied using a pipet. However, for uniformity and production of large batches of membrane strips, a liquid dispensing machine is useful.

10. For each individual assay, it may be necessary to titer the conjugate and use a dilution that will give optimal color with low background.

References

1. Barry, T. G., Colleran, G., Glennon, M., Dunican, L. K., and Gannon F. (1991) The 16S-23S ribosomal spacer region as a target for DNA probes to identify eubacteria. *PCR Method Appl.* **1**, 51–56.
2. Martin, C., Roberts, D., van der Weide, M., et al. (2000) Development of a PCR-based line probe assay for identification of fungal pathogens. *J. Clin. Microbiol.* **38**, 3735–3742.
3. O'Connor, L., Joy, J., Kane, M., Smith, T., and Maher, M. (2000) Rapid PCR/DNA probe membrane-based colorimetric assay for the detection of *Listeria* and *Listeria monocytogenes* in foods. *J. Food Prot.* **63**, 337–342.

4. O'Sullivan, N., Fallon, R., Carroll, C., Smith, T., and Maher, M. (2000) Detection and differentiation of *Campylobacter jejuni* and *Campylobacter coli* in broiler chicken samples using a PCR/DNA membrane based colorimetric detection assay. *Mol. Cell. Probes* **14,** 7–16.

5. Smith, T., O'Connor, L., Glennon, M., and Maher, M. (2000) Molecular diagnostics in food safety: rapid detection of food-borne pathogens. *Irish J. Agr. Res.* **39,** 309–319.

6. Saiki, R .K., Walsh, P. S., Levenson, C. H., and Erlich, H. A. (1989) Genetic analysis of amplified DNA mobilised sequence-specific probes. *Proc. Natl. Acad. Sci. USA* **86,** 6230–6234.

7. Zhang, Y., Coyne, M. Y., Will, S. G, Levenson, C. H., and Kawaski, E. S. (1991) Single base mutational analysis of cancer and genetic diseases using membrane bound modified oligonucleotides. *Nucleic. Acids Res.* **19,** 3929–3933.

8. Chehab, F. F., and Wall, J. (1992) Detection of multiple cystic fibrosis mutations by reverse dot blot hybridization: a technology for carrier screening. *Hum. Genet.* **89,** 163–168 .

9. Josephson, K.L., Gerba, C.P., and Pepper, I.L. (1993) Polymerase chain reaction: detection of non-viable bacterial pathogens. *Appl. Environ. Microbiol.* **59,** 3513–3515.

10. Bickley, J., Short, J. K., Mc Dowell, D. G., and Parker, H. C. (1996) Polymerase chain reaction (PCR) detection of *Listeria monocytogenes* in diluted milk and reversal of PCR inhibition caused by calcium ions. *Letts Appl. Microbiol.* **22,** 153–158.

11. Demeke, T., and Adams, P. (1992) The effect of plant polysaccharides and buffer additives on PCR. *BioTechniques* **12,** 332–334.

13

Detection of *Leptospira interrogans*

John W. Lester and Rance B. Lefebvre

1. Introduction

Leptospirosis, which is caused by pathogenic spirochetes of the genus *Leptospira*, is an important zooanthroponosis, both from a clinical and economic standpoint. The wide variety of potential disease reservoirs, coupled with the ability of *Leptospira* to survive for long periods in soils and surface water combine to make leptospirosis the most common zoonosis in the world *(1)*. Most leptospiral infections are either subclinical or self-limiting and are characterized by flu-like symptoms. If left untreated, though, more serious forms of the illness can result in spontaneous abortion, uveitis, jaundice, damage and failure of the heart, liver, and kidneys, and even a severe hemorrhagic condition resembling viral hemorrhagic fevers *(1)*.

The taxonomy of the genus *Leptospira* is complex. The genus was historically divided into two main species designations depending on the pathogenicity of the organism. The pathogenic leptospires, designated *L. interrogans*, were originally subclassified into taxa called serovars on the basis of a comparison of cross-absorption of rabbit sera against each serovar. Related serovars were assigned serogroups *(2)*. Recently however, genetic taxonomic analyses of the genus have led to the establishment of several new species within *L. interrogans* sensu lato, but these new species designations do not correlate with the earlier serologic classification *(3)*.

Detection of this diverse group of bacteria in clinical and environmental samples can be accomplished by serology (microscopic agglutination test), culture, direct examination by darkfield microscopy, or polymerase chain reaction (PCR). Serological methods may fail to detect infections in which leptospires do not elicit an immune response that reacts to known standards, and many animals that act as reservoir hosts of *L. interrogans* can have very low serum antibody

From: *Methods in Molecular Biology, vol. 216: PCR Detection of Microbial Pathogens: Methods and Protocols*
Edited by: K. Sachse and J. Frey © Humana Press Inc., Totowa, NJ

titers *(4)*. Furthermore, since a serological titer can persist for as long as 7 yr following an infection, a positive serological reaction does not necessarily indicate a current infection *(5)*. Culture methods are slow, susceptible to contamination, and will fail to detect leptospires that grow poorly or are uncultivable. Direct demonstration of the leptospires will only be positive during certain stages of infection. PCR, on the other hand, is capable of detection of minute amounts of leptospiral DNA in most environmental, culture, or clinical samples and is thus a much more reliable tool than the previous methods.

Frequently, it is desired to specifically identify *L. interrogans* in a sample that may also contain saprophytic leptospires. While there are a number of targets for PCR capable of this level of discrimination, the genes for ribosomal RNA are particularly well suited as PCR targets for the identification and speciation of bacteria, because they are present in all organisms. Furthermore, they possess a variety of regions with differing amounts of homology within and between species, which allows their phylogenetic classification *(6)*. The primers used in this protocol were designed by making a comparison of 16S ribosomal gene sequences from known pathogenic leptospires (*see* **Note 1**) *(7)*. The resulting set of 4 nested primers is specific for pathogenic *Leptospira* spp.

The most important aspect of this protocol is the sample preparation. Samples that are very dilute, such as large vol of urine, will require concentration, while other samples, such as semen, will require extensive removal of cell debris and polysaccharide from the DNA prior to amplification. Three basic methods are described here, a simple boiling method, a concentration method, and a hexadecyltrimethylammonium bromide (CTAB)/organic extraction method, which will yield the purest DNA, but which is also the most time-consuming.

2. Materials

2.1. Sample Preparation

2.1.1. Sample Concentration Method (see **Note 2**)

1. EDTA: 0.5 mM, pH 8.0.

2.1.2. CTAB/Organic Extraction Method (see **Note 3** and **ref. 8**)

1. Water bath 65°C.
2. TE Buffer: 10 mM Tris-HCl, pH 8.3, 1 mM EDTA containing 2% Triton® X-100.
3. Phenol/chloroform/isoamyl alcohol: 25:24:1 (v/v/v).
4. NaCl: 5% (w/v) in water.
5. CTAB: 10 % (w/v) in water.
6. Chloroform/isoamyl alcohol: 24:1 (v/v).
7. Isopropanol.

Table 1
Oligonucleotide Primers for PCR of Pathogenic Leptospira

Name and Position[a]	Nucleotide sequence, 5' to 3'
Outer primer set	
428–450	5'-AGGGAAAAATAAGCAGCGATGTG-3'
981–999c[b]	5'-ATTCCACTCCATGTCAAGCC-3'
Product size: 571 bp	
Nested primer Set	
552–570	5'-GAAAACTGCGGGCTCAAAC-3'
925–940c[b]	5'-GCTCCACCGCTTGTGC-3'
Product size: 370 bp	

[a] Corresponds to the *L. interrogans* sensu stricto 16S gene sequence (Accession no. X17547)
[b] Primers labeled "c" are complementary to those in the sequence.

2.2. PCR Amplification

1. Sterile double distilled H_2O.
2. *Taq* DNA Polymerase (*see* **Note 4**).
3. PCR buffer (10X): 200 mM Tris-HCl, pH 8.4, 500 mM KCl, 15 mM $MgCl_2$, or as supplied by manufacturer with *Taq* DNA polymerase (*see* **Note 5**).
4. dNTP mixture (dATP, dTTP, dCTP, and dGTP): 1.3 mM each.
5. Oligonucleotide primers (*see* **Table 1**), concentration adjusted to 50 pmol/µL.
6. Sterile mineral water.
7. Thermal cycler.
8. Agarose (molecular biology-grade).
9. Tris-Borate-EDTA buffer (TBE): To make a stock solution of 5X TBE, dissolve 54 g of Tris-base and 27.5 g of boric acid in about 750 mL of purified water, add 20 mL of 0.5 M EDTA, pH 8.0, and adjust the vol of the solution to 1 L with purified water. Dilution of the buffer 1 in 10 with purified water will give a final 0.5X running buffer.
10. Ethidium bromide solution: 0.5 µg/mL in 0.5X TBE buffer.
11. Loading buffer: 20% Ficoll® 400, 0.1 M Na_2 EDTA pH 8.0, 1.0% sodium dodecyl sulfate (SDS), 0.25% bromophenol blue.

3. Methods
3.1. Sample Preparation
3.1.1. Simple Boiling Method (see **Note 6**)

1. Boil the sample for 10 min in a microcentrifuge tube.
2. Centrifuge at 14,000g for 5 min.
3. Transfer supernatant to a fresh tube and store at −20°C.

3.1.2. Sample Concentration Method (see **Note 2**)

1. Transfer the sample from collection vessel to 50-mL centrifuge tubes and centrifuge at 14,000g for 30 min.
2. Decant supernatant and resuspend pellet in 1.5 mL 0.5 mM EDTA, pH 8.0.
3. Transfer sample to a microcentrifuge tube and centrifuge at 14,000g for 10 min.
4. Resuspend pellet in 60 µL double distilled sterile water (*see* **Note 7**).
5. Boil the sample for 10 min.
6. Centrifuge at 14,000g for 5 min and transfer supernatant to a fresh tube. Store sample at –20°C.

3.1.3. CTAB/Organic Extraction Method (see **Note 3** and **refs. 4** and **8**)

1. Add an equal volume of TE buffer with 2% Triton X-100 to 600 µL sample in a microcentrifuge tube and vortex mix.
2. Heat the sample at 94°C for 10 min.
3. Centrifuge at 14,000g for 5 min and transfer supernatant to a fresh tube.
4. Extract with 25:24:1 phenol/chloroform/isoamyl alcohol, and transfer aqueous phase to a fresh tube.
5. Add 75 µL of 5% NaCl and 75 µL of 10% CTAB and mix thoroughly.
6. Incubate at 65°C for 10 min.
7. Extract with 750 µL of 24:1 chloroform/isoamyl alcohol.
8. Centrifuge at 14,000g for 5 min and transfer the supernatant to a fresh tube (*see* **Note 8**).
9. Extract with 750 µL of 25:24:1 phenol/chloroform/isoamyl alcohol and transfer supernatant to a fresh tube.
10. Precipitate the DNA with 750 µL of isopropanol.
11. Centrifuge for 2 min at 14,000g and wash with 70% ethanol.
12. Resuspend DNA pellet in 5–10 µL of TE buffer.

3.2. PCR Amplification (see **Note 9**)

1. Prepare the PCR master mixture for 10 reactions by adding the following solutions to a microcentrifuge tube (*see* **Note 10**): 110 µL 10X PCR buffer, 110 µL dNTP mixture, 11 µL primer 428–450 (total of 500 pmol), 11 µL primer 981–999c (total of 500 pmol), 523 µL sterile double distilled water, and 5.5 µL (5–25 U) *Taq* DNA Polymerase.
2. Add 70 µL of the master mixture to each of ten 0.5-mL thin-walled microcentrifuge tubes (*see* **Note 11**).
3. To each tube add 30 µL of a prepared sample from **Subheading 3.1.**, and overlay with 1 or 2 drops (30–50 µL) of sterile mineral oil (*see* **Note 12**).
4. Place the PCR tubes into the thermal cycler and amplify with the following parameters (*see* **Note 13**): 94°C for 4 min (initial denaturation), then 35 cycles of 94°C for 1 min (denaturation), 55°C for 1 min (annealing), 72°C for 1 min (extension), followed by 72°C for 7 min (final extension), and then hold at 4°C until collection.

5. Remove the aqueous phase from each reaction and store in fresh tube. Aliquot 10 µL of each PCR product into a fresh tube for the nested reaction.
6. Prepare the PCR master mixture below for the nested amplification reactions, and add 90 µL of it to each of the 10 µL of templates produced in the previous PCR. The PCR master mixture contains: 110 µL 10X PCR buffer, 110 µL dNTP mixture, 11 µL primer 552–570 (total of 500 pmol), 11 µL primer 925–940c (total of 500 pmol), 743 µL sterile double distilled water, and 5.5 µL (5–25 U) *Taq* DNA Polymerase.
7. Overlay each with a drop of mineral oil and amplify again as in **step 4**.
8. Remove the aqueous phase from each reaction on completion, and store at 4°C.
9. Add 1 µL of loading buffer to 10 µL of each nested PCR product.
10. Perform electrophoresis using 1% agarose and 0.1 µg/mL ethidium bromide in TBE and visualize the PCR products using UV light. An example illustrating the sensitivity of the present method is shown in **Fig. 1**.

4. Notes

1. The species used were *L. interrogans* sensu stricto (GenBank® Accession no. X17547), *L. borgpetersenii* serovars hardjobovis and sponselee (Accession no. U12670), *L. borgpeterseni* 1627 serovar burgas (Accession no. U12669), *L. pomona* serovar kennewicki (Accession no. 71241), and *L. interrogans* serovars canicola, moulton (Accession no. 17547), fainei (Accession no. U60594), and inadai (accession no. Z21634) (**7**).
2. Samples with large vol such as bovine urine samples or water samples may require concentration prior to sample preparation for PCR. This method of sample concentration can also be combined with the method described in **Subheading 3.1.3.** if purification of DNA is desired prior to PCR amplification. Filtration cannot be used as a method of concentration for *Leptospira* because they are thin enough to pass through 0.22-µ*M* filters (**1**). Previous studies have found that following filtration with either cellulose or polycarbonate filters, *Leptospira* can be found both on the filter and in the filtrate (**8**).
3. Substances such as blood or semen that contain polysaccharides, cell lysis products, or other substances interfering with the amplification reaction usually require at least a phenol/chloroform extraction of the DNA. This method also uses CTAB, which removes carbohydrates from the aqueous phase.
4. *Taq* DNA polymerase (along with other thermostable polymerases) is available from many manufacturers. We used Promega (Madison, WI, USA) products with this protocol, but generally speaking, most products will perform similarly well. The *Taq* DNA polymerase and proprietary buffer that comes with it should always be used together, though, and never mixed.
5. The buffer that comes with the *Taq* DNA polymerase is often sufficient. Occasionally, however, you may need to adjust the $MgCl_2$ concentration. Many *Taq* DNA polymerase kits also come with a buffer lacking $MgCl_2$, and a separate tube of $MgCl_2$ that can be used if this is the case.

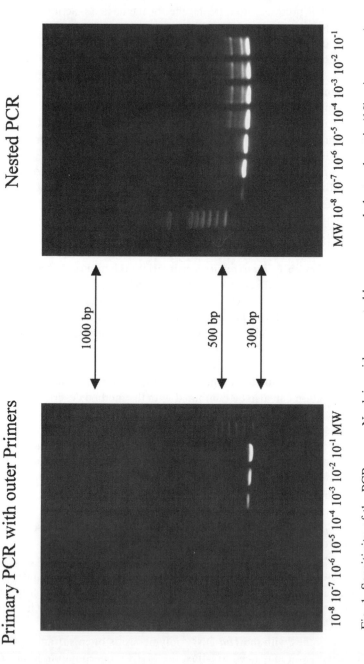

Fig. 1. Sensitivity of the PCR assay. Nucleic acids separated by agarose gel electrophoresis (1% w/v agarose) and stained with ethidium bromide. MW, molecular weight marker ladder (Promega). Ten milliliters of a 3-wk *Leptospira icterohemorrhagiae* culture (5×10^8 cells/mL) was processed by the simple boiling method (*see* **Subheading 3.1.1.**). Serial 10-fold dilution of the prepared DNA was made to 10^{-8}, and processed by PCR.

6. This method can be used for small vol of relatively "clean" samples, which do not require concentration and do not contain large amounts of substances, such as polysaccharides that may interfere with the amplification reaction. This may include small water or urine samples, cerebrospinal fluid (CSF) samples, and aqueous humor samples.
7. At this point the Sample Concentration Method (*see* **Subheading 3.1.2.**) is completed as for the Simple Boiling Method (*see* **Subheading 3.1.1.**), but the CTAB method (*see* **Subheading 3.1.3.**) may be used to prepare the sample if required (*see* **Note 4**).
8. At this point, if a layer of CTAB-carbohydrate is still visible at the interface, additional CTAB can be added to the supernatant, and the extraction can be repeated until no carbohydrate layer is visible.
9. Getting accurate results from PCR requires a high degree of cleanliness. We use a clean UV hood for preparing the PCR in an area of the laboratory separate from that used for sample preparation. The area used for preparation of the PCR should be thoroughly cleaned with bleach before use, and filter-tipped pipets should be used in order to prevent cross-contamination of samples. Adequate use of both negative and positive controls will also insure the accuracy of the results.
10. The master mixture should be made fresh as required and not prepared in advance. While this amount of master mixture is sufficient for 11 samples, it is advisable to make more master mixture than needed as this will allow for small pipeting mistakes. Additionally, by preparing fresh master mixture for every run, the risk of contamination is lessened.
11. Thin-walled microcentrifuge tubes designed for PCR allow faster heat transfer, which allows shorter PCR cycles to be used.
12. This prevents evaporation of the solution during heating. Fresh molecular biology-grade mineral oil should be used, as old mineral oil that has been damaged by UV radiation can inhibit PCR *(9)*. Other substances such as wax beads can also be used.
13. We use an MJ thermal cycler (MJ Research, Waltham, MA, USA), but any thermal cycler can be used. Other thermal cyclers can allow different (and in some cases faster) amplification conditions.

References

1. Faine, S. (1994) *Leptospira* and Leptospirosis. CRC Press, Boca Raton.
2. Kmety, E., and Dikken, H. (1993) Classification of the Species *Leptospira interrogans* and history of its serovars. University Press, Groeningen, The Netherlands.
3. Zuerner, R. L., Alt, D., and Bolin, C. A. (1995) IS1533-based PCR assay for identification of *Leptospira interrogans* sensu lato serovars. *J. Clin. Microbiol.* **33,** 3284–3289.
4. Masri, S. A, Nguyen, S., Gale, P., Howard, C. J., and Jung, S. (1997) A polymerase chain reaction assay for the detection of *Leptospira* spp. in bovine semen. *Can. J. Vet. Res.* **61,** 15–20.

5. Swart, K. S., Calvert, K., and Meney, C. (1982) The prevalence of antibodies to serovars of *Leptospira interrogans* in horses. *Aust. Vet. J.* **59,** 25–27.
6. Woese, K. H. (1987) Bacterial evolution. *Microbiol. Rev.* **51,** 221–271.
7. Faber, N. A., Crawford, M., LeFebvre R. B., Buyukmihci, N. C., Madigan, J. E., and Willits, N. H. (2000) Detection of *Leptospira* spp. in the aqueous humor of horses with naturally acquired recurrent uveitis. *J. Clin. Microbiol.* **38,** 2731–2733.
8. Ausubel, F. M., Brent, R., Kingston, R. E., et al. (1992) Short Protocols in Molecular Biology, 2nd ed., John Wiley and Sons, New York, USA.
9. Murgia, R., Riquelme, N., Baranton, G., and Cinco, M. (1997) Oligonucleotides specific for pathogenic and saprophytic *Leptospira* occurring in water. *FEMS Microbiol. Lett.* **97,** 27–34.
10. Dohner D.E., Dehner M.S., and Gelb L.D. (1995) Inhibition of PCR by mineral oil exposed to UV irradiation for prolonged periods. *Biotechniques* **18,** 964–967.

14

Detection of Pathogenic Mycobacteria of Veterinary Importance

Robin A. Skuce, M. Siobhan Hughes, Malcolm J. Taylor, and Sydney D. Neill

1. Introduction

1.1. Importance of the Agents

There are a large variety of bacteria that are pathogenic for animals, including many opportunistic pathogens normally residing in the environment. Among these diverse veterinary pathogens, mycobacteria are highly significant, particularly for farmed animal species, as many are zoonotic, and their impact can have significant economic consequences. The genus *Mycobacterium* comprises more than 70 species *(1)*. Many of these are innocuous free-living saprophytes, but some are inherently pathogenic for animals (**Table 1**). *Mycobacterium bovis*, the causative agent of bovine tuberculosis, has an exceptionally broad host range that includes farmed and feral animals, wildlife, and also humans *(2)*. Bovine tuberculosis can be a serious barrier to the cattle trade within and between countries with significant agricultural economies. It is also an occupational zoonosis. Several countries in the developed world have been unable to eradicate bovine tuberculosis, despite implementing comprehensive and costly eradication schemes. The disease is epidemiologically complex, with interbovine transmission and wildlife reservoirs of infection suggested as the major obstacles to eradication *(3)*.

Johne's disease or paratuberculosis, a chronic hypertrophic enteritis observed in cattle and other ruminants, is caused by *M. avium* subspecies *paratuberculosis* and results in significant economic losses to agricultural industries worldwide. *M. avium* subspecies *paratuberculosis* has been implicated in Crohn's disease in humans and now is considered a potential zoonotic pathogen *(4)*. Significant

From: *Methods in Molecular Biology, vol. 216: PCR Detection of Microbial Pathogens: Methods and Protocols*
Edited by: K. Sachse and J. Frey © Humana Press Inc., Totowa, NJ

Table 1
Significant Mycobacterial Pathogens of Animals

Diseases in animals	Mycobacteria
Mammalian tuberculosis	*M. bovis* (*M. tuberculosis*, *M. microti*, *Micaprae*)
Avian tuberculosis	*M. avium* subsp. *avium* (*M. a. sylvaticum* in wood pigeons)
Lymphadenitis in pigs	*M. avium* subsp. *avium*
Lymphadenitis in deer	*M. avium* subsp. *avium*, *M. bovis*
Johne's disease, or paratuberculosis	*M. avium* subsp. *paratuberculosis*
Rat and cat leprosy	*M. lepraemurium*
Fish skin ulceration	*M. marinum*
Canine leproid granuloma syndrome	Mycobacterial species (*insertia sedis*)
Farcy	*M. farcinogenes*, *M. senegalense*

a Species in parentheses cause disease rarely in animals.

financial losses in agriculture can also be a consequence of other mycobacterial infections, such as avian tuberculosis in poultry, swine, cattle, and farmed deer. The aetiological agent, *M. avium* subspecies *avium* is again zoonotic and often found as a disseminated infection in AIDS patients (*5*). There are other mycobacteria that have veterinary significance, but possibly less so than those mentioned previously. These include: *M. lepraemurium*, the causative agent of a localized, leprosy-like skin infection in rats and cats (*6*) *M. marinum* in fish (*7*) and a *Mycobacterium* species (*insertia sedis*), which produces a pyogranulomatous disease in dogs (*8*).

1.2. Diagnostic Issues

Veterinary mycobacteriology laboratories have a key role to play in both diagnosis and surveillance of animal infections. Such laboratories often provide culture confirmation for field diagnosis of some significant mycobacterial pathogens, e.g., isolates from tuberculin reacting cattle identified during tuberculosis eradication programs. These specialist laboratories also identify those other mycobacteria of significance in animal diseases. Traditionally, culture on solid media and biochemical procedures have been used routinely for isolation and speciation of mycobacteria from clinical samples. There are disadvantages in using such procedures, as most of these mycobacteria are difficult and slow to culture and some at present are nonculturable. Primary isolation of members of the *M. tuberculosis* complex requires, on average, between 4–6 wk to recover on traditional solid culture media and even between 13–15 d in radiometric and automated culture systems (*9*). Similar difficulties can arise with identification, as some species are difficult to differentiate from closely related mycobacterial species or from species in genera such as *Corynebacterium, Nocardia*, and *Rhodococcus* .

Consequently in recent years, there has been considerable and justifiable interest in applying the modern technologies of molecular biology to detection, speciation, and even to the epidemiology of diseases caused by mycobacteria. Such technologies have potential to provide improvements in these areas, not least possibly obviating the process of time-consuming culture. However, there has been a tendency to underestimate the difficulties inherent in introducing and applying these technologies to such recalcitrant pathogens as mycobacteria. Their intracellular nature, the intractability of the mycobacterial cell wall and the potential presence of polymerase chain reaction (PCR) inhibitors in clinical specimens are limitations on PCR effectiveness. The paucibacillary nature of specimens in some scenarios can be particularly problematic. Additionally, the standard configuration of PCR does not currently distinguish between live and dead mycobacteria, a factor of importance, e.g., when considering its use for detecting the survival of *M. avium* subsp *paratuberculosis* post-pasteurization *(10)*. However, if molecular diagnostic methods are to be considered as acceptable replacements for traditional procedures, there must be sustained improvement in sensitivity, specificity, reproducibility, and convenience, together with possibly a capability for increased throughput capacity. Reduced costs and availability of appropriate resources and trained staff are also important factors for their successful adoption by routine laboratories. The more conventional technologies may be more appropriate in some countries and particularly so for some pathogens. For mycobacteria, in both developed and developing countries, culture is still considered to be the gold standard *(11)*, despite lacking specificity and being time-consuming.

There are several published protocols for PCR detection of clinically important mycobacteria. Many have focused on mycobacterial pathogens of humans *(12)*. However, among these, there appears to be no consensus regarding extraction or amplification methodologies. There are also difficulties in technology transfer and reproducibility between reputable laboratories in multicenter studies *(13)*. Potential problems with false negative reactions, possibly as a result of inhibitors, and also with false positive reactions, possibly due to contamination have been recorded. Stringent design and use of PCR facilities and inclusion of relevant controls should reduce and allow monitoring of such occurrences.

In veterinary use, the primary focus has been mainly on in-house amplification-based detection of *M. bovis (14–17)*. Other applications of nucleic acid-based diagnostics in veterinary mycobacteriology laboratories have tended to be the evaluation of more robust and sometimes prohibitively expensive commercial kits. Examples of these are the GenProbe® amplification-based *M. tuberculosis* Direct test and the AccuProbe® *M. avium* complex test kits. These are attractive because there are multiple 16S rRNA targets available, the amplification is isothermal, and the products degrade over time, thus reducing the potential for subsequent contamination *(18)*. Quality control, an important issue with in-house PCRs, is also likely to be more easily facilitated with a commercial kit. However, cost has been cited as a problem, reducing routine application of this test *(19)*. DNA amplification and detection technology is developing very rapidly. If early problems with sensitivity and specificity are overcome, there will be an increased demand for high-throughput molecular diagnostics in veterinary as well as human mycobacteriology. There is now a move towards real-time fluorescence detection of PCR products, which should facilitate development of the desired high-throughput molecular diagnostic assay *(20)* and provide a more practical approach to confirmation of the significant mycobacterial infections in animals.

1.3. PCR-Based Molecular Diagnostic Approach

The approach taken in this chapter to PCR detection is drawn from experiences of the mycobacterial reference and research laboratories of the Veterinary Sciences Division, Stormont, Belfast, which operate in support of Government strategies for controlling and ultimately eradicating tuberculosis from cattle in Northern Ireland. The laboratories have been successful in developing and applying PCR-based molecular diagnostics in several ways: for confirmation of *M. tuberculosis* complex (*M. tuberculosis, M. bovis, M. bovis* BCG, *M. microti, M. africanum*) using IS*6110 (21)*; for spoligotyping *(22)*; for identifying *M. avium* complex (*M. avium* subsp *avium, M. avium* subsp *paratuberculosis* and *M. avium* subsp *sylvaticum*) using IS*1245 (23)*, IS900 *(24)*, and IS*901/902 (24)* PCRs; and for identifying mycobacteria other than the *M. tuberculosis* complex (MOTTs) using a 16S rRNA PCR and sequence analysis *(25)*. Initially, these PCR analyses were performed on isolates grown on solid media and from radiometric culture from selected bovine lymph nodes. For *M. bovis* detection, specimens usually comprised fresh or frozen lymph tissue, lesions were sometimes calcified, and on occasions, archived formalin-fixed paraffin-embedded lymph node tissue was used. Specimens comprising a range of tissues taken from a variety of animals other than cattle have also been used with PCR methods (**Table 2**). These have been mostly from badgers and deer, but also from companion animals, such as cats, dogs, ferrets, and fish.

PCR-based detection of *M. bovis* in bovine tissues has been impeded by the small numbers of bacilli commonly associated with tuberculous tissues of cattle *(16)* and by difficulties in extracting mycobacterial DNA from a bacterial species with a resilient lipid-rich cell wall *(9)* embedded possibly within fibrous

Table 2
Detection of Mycobacteria in Different Hosts and Tissues
by DNA Sequence-Capture PCR

Species	Specimen site	Specimen type
Badger	Kidney	Decontaminated homogenate
Boar	Lung	Formalin-fixed paraffin-embedded
Bovine	Lymph node	Fresh
Companion animals	Muscle	Frozen
Deer	Nasal Mucus	Lypholized
Ferret		

Sequence capture procedure has enabled detection of mycobacteria in a wide range of host species and specimen types in this laboratory (as above). We intend to adapt specimen pretreatment further to apply sequence capture to other specimen types including milk. feces, blood, and urine.

tissues. Several mycobacterial DNA extraction methodologies for PCR analyses of veterinary specimens have been used previously, both by this laboratory *(26)* and by others *(14–16, 27)*, but PCR assays have been shown not to be as sensitive as culture methods. Sequence capture PCR was subsequently reported to be successful in the detection of *M. tuberculosis* in paucibacillary specimens from patients with tuberculous pleurisy *(28,29)*. This procedure involved enrichment of target DNA by hybridization to biotinylated oligonucleotides and subsequent capture of hybridized DNA onto streptavidin-coated magnetic beads (*see* **Fig. 1**). Magnetic bead attractors enabled potential inhibitors to be removed easily by washing the beads. The method was advantageous from a health and safety perspective in not requiring solvent extraction.

We have now adapted this approach for detection of *M. bovis* in animal specimens. For PCR-based confirmation from culture, DNA was released simply by heat treatment and a modified sequence capture protocol (*see* **Fig. 1**) was used to concentrate the target DNA. Pretreatment involving mechanical and enzymatic disruption *(28)* has been modified for different specimen types (*see* **Fig. 2**). This approach has flexibility with regard to specimen type and mycobacterial pathogen. In addition, the method was robust in allowing successful adoption and adaptation by our laboratory without recourse to the developers of the methodology.

The basic sequence capture and PCR protocols, introducing different capture oligonucleotides and PCR primer sets, have been modified to facilitate a range of different PCR amplifications. We have used this procedure in a number of ways: for definitive diagnosis and simultaneous strain typing of *M. tuberculosis* complex from radiometric cultures at early growth stages *(30)*; and directly in decontaminated tissue homogenates and in lesioned tissue from

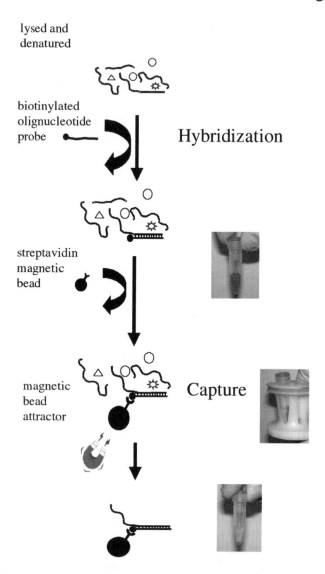

Fig. 1. Schematic representation of sequence capture. Denatured DNA from lysed specimens is enriched for target DNA by hybridization to biotinylated oligonucleotides. Hybridized DNA is subsequently captured onto streptavidin-coated magnetic beads (*see* **Subheading 3.2.**).

tuberculous cattle *(17)*. The sensitivity of sequence capture-spoligotyping PCR has been comparable to that of culture and has enabled detection of *M. bovis* from culture negative specimens. This protocol has been used also for mycobacterial species identification, including nonculturable mycobacteria

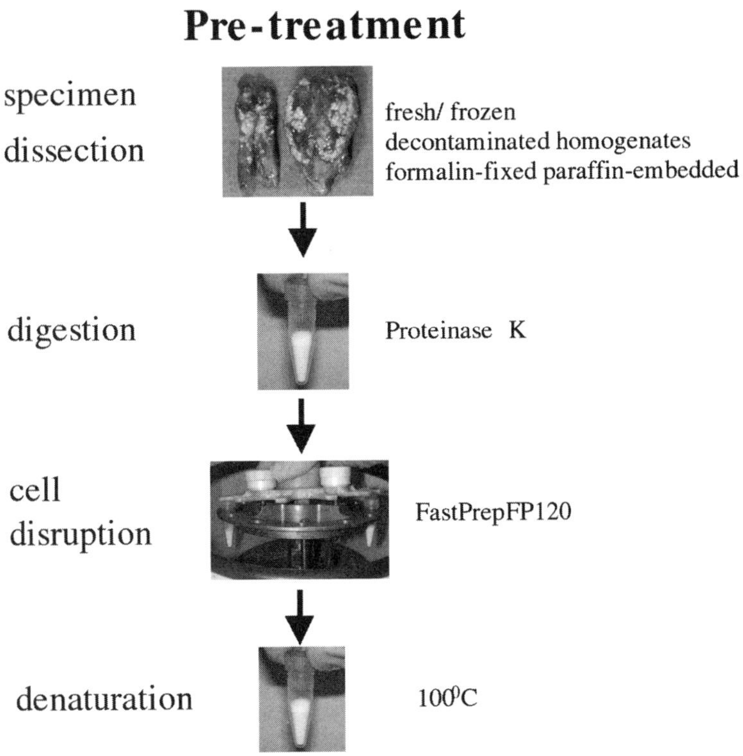

Pre-treatment

specimen

dissection

fresh/ frozen
decontaminated homogenates
formalin-fixed paraffin-embedded

digestion

Proteinase K

cell

disruption

FastPrepFP120

denaturation

100⁰C

Fig. 2. Schematic representation of the pretreatment of specimens prior to sequence capture. Tissue specimens, either fresh, frozen, decontaminated tissue homogenates, or formalin-fixed paraffin-embedded material are subjected to both enzymatic and mechanical disruption prior to lysis and denaturation by heat treatment (*see* **Subheading 3.1.**).

from paraffin-embedded and lyophylized tissue from cat *(31)*, ferret *(32)*, and dog specimens *(8)*.

In this chapter, we describe the basic specimen pretreatment, sequence capture, and PCR protocol, as applied to fresh or frozen bovine tissue specimens, but indicate the potential variations in methodology and application based on different specimens and PCR amplifications.

2. Materials

2.1. Pre-Treatment

1. Class I safety cabinet.
2. DNA Away™ (Molecular BioProducts, San Diego, CA, USA).
3. Disposable gloves.
4. Paper plates.

5. Disposable sterile scalpel.
6. TES buffer: 100 m*M* Tris-HCl, pH 7.4, 150 m*M* NaCl, and 50 m*M* EDTA.
7. Disposable specimen container: 20 mL (Polycon, Medical Wire and Equipment,Wiltshire, UK).
8. Proteinase K: 100 mg/mL (Sigma, Dorset, UK).
9. Sterile plugged Pasteur pipets.
10. Stomacher bag and Stomacher Lab-Blender 80 (Seward Medical, London, UK).
11. Zirconia beads: 0.1-mm diameter (BioSpec Products, Bartlesville, OK).
12. Screw-cap microcentrifuge tubes: 1.5 mL, sterile (Sarstedt, Leicester, UK).
13. FastPrep® FP120 Cell Disrupter (Qbiogene, Carlsbad, CA, USA).
14. Water bath set at 50°C.
15. P20, P100, P200, and P1000 pipets and appropriate sterile RNase-free, DNase-free aerosol-resistant pipet tips.
16. Heating block set at 100°C.
17. Micro Centaur Microcentrifuge (MSE, UK).

2.2. DNA Target Selection

Sequence capture probe solution containing 2.5 pmol of each capture probe in 200 µL 3.75 *M* NaCl, stored at −20°C (Gibco BRL, Life Technologies, Ltd., UK).

1. Cap DRa/b primers specific for DR region of *M. tuberculosis* complex *(22)*.
   Cap DRb 5'-Biotin AAAAACCGAGAGGGGACGGAAAC-3'
   Cap DRa 5'-Biotin AAAAAGGTTTTGGGTCTGACGAC-3'
2. Cap pA and Cap pH primers specific for mycobacterial 16S gene *(31)*, polyadenylated modifications of primers pA and pH *(33)*.
   Cap pA 5'-Biotin AAAAAAGAGTTTGATCCTGGCTCAG-3'
   Cap pH 5'-Biotin AAAAAAAGGAGGTGATCCAGCCGCA-3'

2.3. DNA Sequence Capture

1. Heating block set at 100°C.
2. Disposable container filled with ice.
3. Sequence capture probe: 2.5 pmol of each in 200 µL 3.75 *M* NaCl.
4. Thermomixer; Eppendorf Comfort mixer (Eppendorf-Netheler-Hinz, Cambridge, UK).
5. Prewashed streptavidin coated magnetic beads at 5 µg/µL, Dynabeads® M-280 Streptavidin (Dynal, Oslo, Norway).
6. Magnetic bead attractor (MBA): 6 × 1.5 mL (Stratagene, La Jolla, CA).
7. Wash buffer: 10 m*M* Tris-HCl, pH 8.0, 0.1 m*M* EDTA.
8. PCR quality water (Sigma).

2.4. PCR Amplification

1. P10 positive displacement pipet and disposable tips.
2. PCR buffer (10X): 500 m*M* KCl, 100 m*M* Tris-HCl, pH 9.0, 1% Triton® X-100, 15 m*M* MgCl$_2$ (Qiagen, Ltd., West Sussex, UK).

Table 3
Primers Used for Sequence Capture-PCR Amplifications

DNA target		Primer sequence
IS*6110* *(21)*	IS2 F	5'-CTC GTC CAG CGC CGC TTC GGA-3'
	IS1 R	5'-TCC TGC GAG CGT AGG CGT CGG-3'
Nested IS*6110* *(34)*	IS3 F	5'-TTC GGA CCA CCA GCA CCT AA-3'
	IS4 R	5'-TCG GTG ACA AAG GCC ACG TA-3'
16S Full Length *(33)*	pA F	5'-AGA GTT TGA TCC TGG CTC AG-3'
	pH R	5'-AAG GAG GTG ATC CAG CCG CA-3'
16S V2/V3 region[a] *(35)*	17 F	5'-CAT GCA AGT CGA ACG GAA AG-3'
	525 R	5'-TTT CAC GAA CAA CGCGAC AA-3' [b]
Nested 16S V2/V3 region[a]	53 F	5'-GAG TGG CGA ACG GGT GAG TAA-3'
	485 R	5'-TTA CGC CCA GTA ATT CCG GAC AA-3'
IS*900* *(24)*	P90 F	5'-GTT CGG GGC CGT CGC TTA GG-3'
	P91 R	5'-GAG GTC GAT CGC CCA CGT GA-3'
IS*902* *(24)*	P102 F	5'-CTG ATT GAG ATC TGA CGC-3'
	P103 R	5'-TTA GCA ATC CGG CCG CCC T-3'
IS*1245* *(23)*	P1 F	5'-GCC GCC GAA ACG ATC TAC-3'
	P2 R	5'-AGG TGG CGT CGA GGA AGA C-3'
DRa/b (Spoligotyping) *(22)*	DRb	5'-CCG AGA GGG GAC GGA AAC-3'
	DRa	5'-GGT TTT GGG TCT GAC GAC-3' [c]

[a] 16S V2/V3 region of nucleotide variation as previously described *(36)*.
[b] Mycobacterium genus-specific.
[c] Primer biotinylated at 5' end.

3. dNTP solution: 10 m*M* each of dATP, dCTP, dGTP, and dTTP, made from 100 m*M* stock solutions (Roche Diagnostics, Ltd., East Sussex, UK).
4. HotStar Taq DNA Polymerase (5U/μL) (Qiagen Ltd.).
5. Q-solution (Qiagen Ltd.).
6. 8-methoxypsoralen (Sigma).
7. PCR-quality water (Sigma).
8. Dimethyl sulfoxide (Sigma).
9. Primers 100 ng/μL stock solutions (**Table 3**)
   (For a typical 18-mer oligonucleotide, 100 ng/μL equates to 16.8 μM)

2.5. Analysis

2.5.1. Gel-Based Analysis

1. 1.2, 2.0 and 4.0% E-Gel (InVitrogen, Paisley,UK).
2. Molecular weight markers, 50-bp ladder (Sigma).
3. Digital camera, DC260 Zoom camera (Eastman Kodak Company, Rochester, NY, USA).
4. CorelDRAW Photo Paint (Corel Corporation, Ontario, Canada).

2.5.2. Sequence Analysis

1. PicoGreen® double stranded DNA quantitation kit (Molecular Probes, Eugene, OR, USA).
2. Commercial sequence analysis (MWG Biotech, Milton Keynes, UK).
3. DNASIS for Windows v2.1 (Hitachi Software Engineering, Ltd, Yokohama, Japan).

2.5.3. Membrane-Based Analysis

1. Spoligotyping Kit (Isogen Bioscience BV, Maarssen, Netherlands).
2. SSPE buffer (2X): 1X SSPE is 0.18 M HCl, 10 mM NaH$_2$PO$_4$, 1 mM EDTA. To be supplemented with 0.1 % sodium dodecyl sulfate (SDS).

3. Methods

The sequence-capture PCR methodology detailed here involves 3 stages; pretreatment, sequence capture, and PCR. Experimental controls for each stage are recommended strongly (*see* **Note 1**).

3.1. Pre-Treatment

3.1.1. Fresh–Frozen

1. Prior to commencing work, treat a Class 1 safety cabinet, in a Hazard Group 3 facility (*see* **Note 2a**) with DNA Away (*see* **Note 3a**). Spray and wipe surfaces prior to use, according to manufacturer's instructions (i.e. spray work area, wait 2 to 3 min then wipe dry with tissue.)
2. If specimen is frozen, thaw at room temperature in Class 1 safety cabinet or at 37°C in an incubator.
3. Use new gloves, scalpel, and plate for each specimen dissection (*see* **Note 3b**). Chop approx 500 mL of tissue (*see* **Note 4**) into small pieces (1-mm cubes) on the paper plate using only sterile disposable scalpels to cut and manipulate specimen (*see* **Note 3c**).
4. Transfer chopped material into 20-mL sterile specimen container.
5. Add 1 mL of TES buffer to specimen in container.
6. Add 20 µL of Proteinase K to specimen in container.
7. Mix well and incubate overnight in a 50°C waterbath. If possible, mix occasionally prior to overnight incubation or use a shaking water bath to provide continual gentle agitation.
8. Transfer specimen, if liquified, to a sterile 1.5-mL screw-cap microcentrifuge tube containing 500 mL of 0.1-mm diameter zirconia beads. The tube containing zirconia beads may be filled to capacity with specimen if large tissue homogenates are being used (*see* **Note 4**). Use a sterile Pasteur pipet to transfer the liquefied specimen (*see* **Note 2b**). If the specimen has not been fully liquefied, transfer digested material to stomacher bag, placed inside another stomacher bag (as a precautionary containment of specimen if first stomacher bag ruptures), homogenize for 2 min and transfer to screw-cap microcentrifuge tube as before.

Occasionally the specimen may not be fully liquefied still. If transfer necessitates pouring, care should be taken to avoid spillage of infectious material and for cross-contamination of specimens.

9. Place tube in the FastPrep FP120 Cell Disrupter, secure and shake vigorously at a speed setting 6.5 m/s for 45 s (*see* **Note 2c**).

3.1.2. Decontaminated Homogenates

Tissue homogenates, having been decontaminated in 5% (w/v) oxalic acid as described (*37*), were pretreated as described below.

1. DNA Away treat a Class 1 safety cabinet, as described in **Subheading 3.1.1., step 1**.
2. Thaw specimen if frozen as described in **Subheading 3.1.1., step 2**.
3. Change gloves after processing each specimen (*see* **Note 3b**).
4. Using a Pasteur pipet, transfer tissue homogenate to fill a sterile screw-cap microcentrifuge tube containing 500 µL of 0.1-mm diameter zirconia beads.
5. If the homogenate is very compact add some TES buffer to liquefy and assist homogenate transfer.
6. Centrifuge homogenate in a microcentrifuge at full speed for 10 min inside the safety cabinet (*see* **Note 2a**). Discard the supernatant into an empty specimen container for autoclaving and disposal.
7. Add 530 µL of TES pH 7.4 to the remaining pelleted material in the screw-cap microcentrifuge tube.
8. Place tube in the FastPrep FP120 Cell Disrupter, secure and shake vigorously as in **Subheading 3.1.1., step 9**.
9. Remove tube from FastPrep FP120 Cell Disrupter and add 20 µL of proteinase K to the tube and invert to mix.
10. Incubate overnight at 50°C in a heating block (*see* **Note 5**).
11. Shake vigorously as in **Subheading 3.1.1., step 9**.

3.1.3. Formalin-Fixed, Paraffin-Embedded Sections

Sequence capture has been effective in extracting DNA from formalin-fixed paraffin-embedded tissues. However, the potential adverse effect of fixation is not circumvented by sequence capture.

1. Either excise a small portion of tissue from a paraffin-embedded tissue block, using a disposable scalpel, ensuring that minimal wax is included or extract from 2 or 3 (5–10 µM) sections cut from a paraffin block using disposable microtome blade (single block use). If using excised tissue, each dissection or manipulation should be carried out using a new disposable scalpel and paper plate. Gloves should be changed after processing each specimen. Formalin-fixed material is considered noninfectious, and as the airflow in Class 1 safety cabinet may well disturb such light specimens, it is preferential that these procedures are carried out in another area of the laboratory (*see* **Note 6**).

2. Carefully transfer specimen to screw-cap microcentrifuge tube containing 500 µL 0.1-mm diameter zirconia beads.
3. Add 530 µL of TES pH 7.4 to the tube.
4. Heat-treat specimens in heating block at 100°C for 5 min.
5. Shake vigorously as in **Subheading 3.1.1., step 9**
6. Treat as in **Subheading 3.1.1., steps 6–9**

3.2. DNA Sequence Capture

3.2.1. DNA Target Selection

Biotinylated capture oligonucleotide probes are used for target selection; more than one set of probes can be used in the same sequence capture extraction.

1. Target DNA from *M. tuberculosis* complex strains may be selected using the DR region biotinylated capture oligonucleotides, Cap DRa and b.
2. Target DNA from all mycobacterial strains may be selected using the universal 16S rDNA biotinylated capture oligonucleotides, Cap pA and Cap pH (*see* **Note 7**).

3.2.2. DNA Sequence Capture

1. Using a P1000 pipet set at 550 µL, withdraw approx 200–300 µL of supernatant present in the screw-cap microcentrifuge tube after the vigorous shaking described in **Subheading 3.1.1., step 9.**
2. Carefully expel the fluid back into the same tube so as to create an indentation in the beads. We have found this trough facilitates the subsequent maximum transfer of supernatant out of the tube.
3. Transfer approx 550 µL of supernatant into a new sterile screw-cap microcentrifuge tube.
4. After transfer, if necessary, make vol of supernatant up to 550 µL with TES pH 7.4. Familiarization with this vol in the screw-cap tube aids this procedure.
5. Denature samples by incubation for 7 min at 100°C in heating block (*see* **Note 2d**).
6. Immediately immerse samples in an ice bath and incubate for 5 min (*see* **Note 3b**),
7. Add 200 µL of sequence capture probe solution to each sample, while still incubating on ice, and invert several times to mix.
8. Incubate samples in a thermomixer at 42°C for 3 h at 500 rpm (*see* **Note 5**).
9. Remove samples from thermomixer and reset to 21°C
10. Add 10 µL of prewashed streptavidin-coated magnetic beads to each tube, invert several times to ensure complete mixing and incubate at 21°C for 2 h in a thermomixer at 500 rpm. We have noted that capture works effectively at 36°C The latter temperature was chosen when a non-refrigerated thermomixer was used since the instrument was very slow to reach a temperature of 21°C
11. Periodically invert tubes to maintain suspension of beads during incubation.
12. Insert tubes into the MBA ensuring that the tubes are in close contact with the magnet and leave for 2 min.
13. Invert MBA and leave for a further 2 min to ensure that beads attached to tube lids are recovered.

14. With one hand, carefully loosen the caps on all tubes while using the other hand to secure the tubes tightly against the magnetic core so that the tubes do not rotate thus disturbing the magnetic attraction of beads.
15. Using a Pasteur pipet with the tip located at the bottom of the microcentrifuge tube, withdraw the fluid in one continuous action taking care not to disturb the pellet of magnetic beads. Carefully remove the supernatant including any residual zirconia beads or tissue debris from each tube, ensuring that the beads remain in contact with the MBA.
16. Discard the supernatant waste into a specimen container for autoclaving and disposal.
17. Wash beads twice with 750 µL of wash buffer, thoroughly resuspending the beads on each occasion by repeat pipeting and inverting the tubes several times.
18. Repeat wash procedure (**steps 12–15**).
19. After the final wash, remove the residual buffer with a micropipet tip, ensuring that no globules of fluid are remaining along the inside of the tubes.
20. Resuspend the magnetic bead pellet in 25 µL PCR quality water and store at –20°C if not proceeding with PCR on the same day.

3.3. PCR Amplification

3.3.1. PCR Master Mixture

1. Make up 45 µL of a PCR master mix, as follows: 5 µL 10X PCR buffer (containing 15 mM MgCl$_2$), 1 µL 10 mM dNTP (200 nM final concentration), 0.25 µL Hotstar Taq DNA Polymerase, 2 µL of each forward primer (673 nM final concentration for a typical 18-mer), 2 µL of each reverse primer (673 nM final concentration for a typical 18-mer), and PCR-quality water to make volume up to 45 µL. Variations from standard master mix are provided in **Table 4**.
2. Add 5 µL of resuspended bead solution (**Subheading 3.2.2., step 21**) to master mixture using a positive displacement pipet (*see* **Note 3d**).
3. For re-amplifications, 2 µL of primary PCR product is removed using positive displacement pipet and added to nested secondary PCR master mixture (*see* **Note 3d**).

3.4. Analysis

3.4.1. Gel-Based Analysis

1. Agarose gels may be prepared by standard procedure *(38)* or commercial gels may be used.
2. Pre-run appropriate E-Gel as instructioned by manufacturer at 20 mA for 60 s.
3. Remove disposable comb.
4. Add appropriate commercial standard molecular weight size marker in the first well (1 of 12).
5. Load 5 µL of each PCR product with 15 µL H$_2$O in an E-Gel well and resolve at 20 mA for 30 min.
6. View under UV light on a transilluminator at a wavelength of 366 nm.

Table 4
Recommended PCR Amplification and Gel Analysis Conditions

PCR target	Primers	Variations from **Subheading 3.3.1.**	Cycling conditions	Size of product	Analysis
IS6110	IS2 IS1		94°C for 15 min [94°C for 1 in, 68°C for 1 min, 72°C for 1 min] × 40 72°C for 7 min	123 bp	4% gel
Nested IS6110	IS3 IS4	use 2 μL of 1st amp products (reamplification)	94°C for 15 min [94°C for 40 s, 58°C for 1 min, 72°C for 1 min] × 40 72°C for 7 min	92 bp	4% gel
16S (full length)	PA pH	0.5 μL of 8-methoxypsoralen (*see* **Note 3D**) 10 μL Q solution	94°C for 15 min [94°C for 1 min, 68°C for 1 min, 72 °C for 1 min] × 40 72°C for 7 min	Size dependent on species	1.2% gel
16S (v2/v3 region)	17F 525R	0.5 μL of 8-methoxypsoralen (*see* **Note 3D**) 20 μL Q solution	94°C for 15 min [94°C for 1 min, 60°C for 1 min, 72°C for 1 min] × 40 72°C for 7 min	508 bp	2% gel
Nested 16S (v2.v3 region)	53F 485R	0.5 μL of 8-methoxypsoralen (*see* **Note 3D**) 10 μL Q solution use 2 μL of 1st amp products (reamplifiation)	94°C for 15 min [94°C for 1 min, 60°C for 1 min, 72°C for 1 min] × 40 72°C for 7 min	432 bp	2% gel
IS900	P90 P91		94°C for 15 min [94°C for 1 min, 55°C for 1 min, 72°C for 1 min] × 40 72°C for 7 min	400 bp	2% gel
IS902	P102 P103	Additional $MgCl_2$ to 3.7 mM final	94°C for 15 min [94°C for 1 min, 55°C for 1 min, 72°C for 1 min] × 40 72°C for 7 min	252 bp	4% gel
IS1245	1245 P1 1245 P2		94°C for 15 min [94°C for 1 min, 65°C for 1 min, 72°C for 1 min] × 4 72°C for 7 min	427 bp	1.2% gel
DRa/b	DRa DRb	4 μL of each primer (*see* **Note 3D**)	94°C for 15 min [94°C for 40 s, 55°C for 40 s, 72°C for 30 s] × 50 72°C for 7 min		Spoligotyping

7. PCR product size can be estimated by comparison with molecular weight markers.
8. The E-Gel is photographed using a digital camera, recorded as a digital image (e.g., *see* **Fig. 3**) and manipulated using appropriate image analysis software; we use Corel Photo Paint.

3.4.2. Sequence Analysis

1. 16S PCR products are quantified using PicoGreen double-stranded DNA quantitation kit.
2. Appropriate quantities of unpurified PCR products are sent for commercial double-strand sequence analysis.
3. Consensus sequence (*see* **Fig. 4**), using the double-strand sequence data, is facilitated by DNASIS software and compared to DNA sequence databases for sequence identification, e.g., GenBank® *(39)*, European Molecular Biology Laboratory (EMBL) *(40)*, and Ribosomal Databases Project II *(41)* databases.

3.4.3. Membrane-Based Analysis by Spoligoblot *(22)*

1. Add 20 μL of Dra/b amplified PCR products to 150 μL of 2X SSPE/0.1% SDS.
2. Heat denature for 10 min at 100°C
3. Transfer immediately to an ice bath.
4. Apply 170 μL to spoligo membrane and develop according to manufacturer's instructions (e.g., *see* **Fig. 5**).

4. Notes

1. Controls. Three sets of controls are used to monitor the pretreatment, sequence capture and PCR amplification stages of the methodology. Preferred pretreatment controls include previously culture and/or PCR positive and negative tissues. For sequence capture, an aliquot of appropriate reference culture is recommended as a positive control and TES buffer only as a negative control. Preferentially, more than one negative control should be used especially when processing several specimens (>10). The negative controls should be interspersed throughout processing the test specimens, and the positive control should be processed as the last specimen to minimize potential cross-contamination. Similarly for PCR amplification, positive (heat-inactivated culture) and negative (no template) controls should be incorporated as described for sequence capture controls.

2. Health and Safety.
   a. We routinely use this technique for the detection of mycobacterial species, which are Hazard Group 3 agents. For such biological agents, in the UK, Advisory Committee on Dangerous Pathogens (ACDP) guidelines *(42)* must be strictly adhered to, and all manipulations prior to heat inactivation at 100°C should be carried out in a safety cabinet.
   b. Pasteur pipets may become blocked if the specimen contains undigested tissue. Care should be taken to avoid this outcome. If blockage of the pipet occurs and attempts to release material by vigorous pipeting fail, as a last

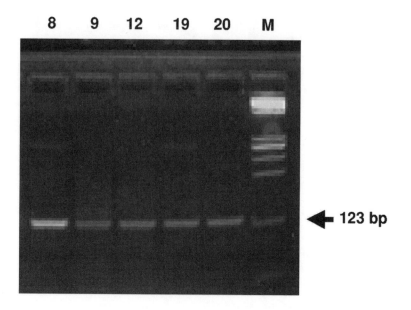

Fig. 3. PCR products, resulting from IS6110 IS1/IS2 amplification of NA extracted by sequence capture from lesioned tuberculous bovine lymph nodes (specimens 8, 0, 12, 19, 20) and molecular weight (M) were resolved in a 4% E-Gel *(20)*.

resort, very carefully break the glass pipet within the 20-mL specimen container. A balance must be drawn between sample loss and health and safety.

c. We have found that it is not feasible to operate the FastPrep FP120 Cell Disrupter within a safety cabinet because of size limitations of the cabinet. We have also noted that it is difficult to secure the tubes in the rotor. The manufacturer's instructions regarding securing tubes must be adhered to strictly, since tube release during operation may result in aerosol contamination of the equipment with a Category 3 pathogen. For this reason, we recommend that this equipment is not left unattended during operation and that the operator listens for evidence of tube release. The operator should respond instantly to any change in equipment tone and stop the instrument immediately.

d. We consider that it is a safer practice to use a heating block rather than a boiling water bath within a safety cabinet for heat treatment at 100°C.

3. Contamination. Measures taken to minimize the risk of cross-contamination during DNA extraction and PCR amplification include the following.

a. Treatment of appropriate work areas with DNA Away to eliminate contaminating DNA.

b. Single use of disposable consumables including gloves, scalpels, paper plates, appropriate sized microtiter plates (serving as tube racks for screw-cap microcentrifuge and PCR tubes), and disposable ice baths.

1   50
GCGGCGTGCTTAACACATGCAAGTCGAACGGAAAGGCCCCTTCGGGGGTA
51   100
CTCGAGTGGCGAACGGGTGAGTAACACGTGGGTAATCTGCCCTGCACTTC
101   150
GGGATAAGCCTGGGAAACTGGGTCTAATACCGGATATGACCACGAAGCGC
151   ←─┤ Helix 10
ATGCTTTGTGGTGGAAAGCTTTTGCGGTGTGGGATGGGCCCGCGGCCTAT
─────────────→   250
CAGCTTGTTGGTGGGGTGATGGCCTACCAAGGCGACGACGGGTAGCCGGC
251   300
CTGAGAGGGTGTCCGGCCACACTGGGACTGAGATACGGCCCAGACTCCTA
301   350
CGGGAGGCAGCAGTGGGGAATATTGCACAATGGGCGCAAGCCTGATGCAG
351   400
CGACGCCGCGTGGGGGATGACGGCCTTCGGGTTGTAAACCTCTTTCAGCA
401   ←─────
GGGACGAAGCGCAAGTGACGGTACCTGCAGAAGAAGCACCGGCCAACTAC
──────┤ Helix 18   ├─────────────→   500
GTGCCAGCAGCCGCGGTAATACGTAGGGTGCGAGCGTTGTCCGGAATTAC
501   519
TGGGCGTAAAGAGCTCGTA

Fig 4. Nucleotide sequence of the 16S rRNA gene amplicons derived from target DNA extracted by sequence capture from formalin-fixed paraffin-embedded canine specimens. The sequence starts at nucleotide position 2 and finishes at nucleotide position 544 of aligned myobacterial 16S rRNA sequences *(8)*.

c. Tissue dissection, which is an open process, should be carried out carefully and quickly to reduce exposure time to possible contaminants.

d. The standard precautions to prevent PCR contamination are employed. In addition, we use a positive displacement pipet for PCR mixture inoculation. 8-Methoxypsoralen in dimethyl sulfoxide (final concentration in PCR mix is 25 μg/mL) and a 4-min exposure to long-wavelength (366 nm) UV irradiation is used to eliminate *Taq* contaminants for 16S rRNA amplifications *(43)*. For spoligotyping from clinical specimens, only one round of amplification (50 cycles) is used to eliminate potential contamination from reamplification.

4. Quantity of tissue. Quantity of tissue is a limiting feature of this procedure since the FastPrep FP120 Cell Disrupter microcentrifuge tube capacity is restricted to 1.5 mL. This is not a problem when the foci of infection, e.g., lesions, are easily

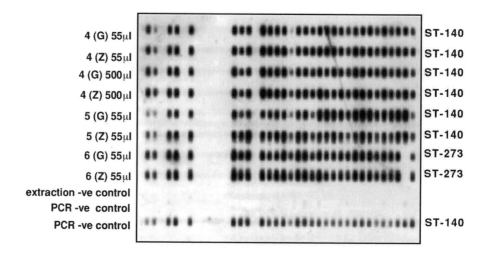

4 (G) 55 μl	ST-140
4 (Z) 55 μl	ST-140
4 (G) 500 μl	ST-140
4 (Z) 500 μl	ST-140
5 (G) 55 μl	ST-140
5 (Z) 55 μl	ST-140
6 (G) 55 μl	ST-273
6 (Z) 55 μl	ST-273
extraction -ve control	
PCR -ve control	
PCR -ve control	ST-140

**Tissue disruption with G (glass) or Z (zirconia) beads;
sequence capture on 55μl or 500 ml homogenates.**

Fig. 5. Spoligotype patterns derived from DNA extracted by sequence capture from lesioned tuberculous bovine lymph nodes (specimens 4, 5, 6) *(17)*.

identified. In such cases, 500 μL vol of tissue or less may be used. For tissue specimens where foci are not evident, homogenates of larger quantities of tissue are a preferable starting point.

5. Alternative equipment. The sequence capture procedure outlined here mentions specific pieces of equipment. Alternative equipment may be as effective or, in some instances, preferable with regard to efficacy and safety. Suggested substitutions include; heating blocks instead of water baths for tube incubations. A tube rotator, used within an incubator during target DNA selection with capture probes and sequence capture with magnetic beads, may improve extraction efficacy.

6. Formalin-fixed, paraffin-embedded specimens. Specimen excision directly from the paraffin block is preferred to sectioning although the integrity of the block for subsequent sectioning may be compromised. However, cutting sections may compromise the PCR with regard to cross-contamination; single use disposable blades are recommended for each block, but there is potential still for contamination in section collection. Sectioning is also disadvantaged in yielding samples with an excess of paraffin and less material. Sections may not be as representative as an excised fragment and may be difficult to work with in the airflow of the safety cabinet since they are easily airborne. As formalin-fixed material is considered to be noninfectious, the latter problem should be alleviated by working outside the safety cabinet.

7. Flexibility. Selection of DNA target with 16S rRNA capture primers has enabled amplification of other mycobacterial target sequences with some unrelated primer

sets. This has not been attempted with formalin-fixed paraffin-embedded tissues because of potential DNA degradation resulting from fixation.

References

1. Goodfellow, M., and Magee, J. G. (1997) Taxonomy of mycobacteria, in *Mycobacteria, vol. I. Basic Aspects* (Gangadharam, P. R. J., and Jenkins, P. A., eds.), Chapman and Hall Medical Microbiology Series, International Thomson Publishing, London, UK, pp. 1–71.
2. O'Reilly, L. M., and Daborn, C. J. (1995) The epidemiology of *Mycobacterium bovis* infections in animals and man: a review. *Tuberc. Lung Dis.* **76,** Suppl. I, 1–46.
3. Neill, S. D., Pollock, J. M., Bryson, D. B., and Hanna, J. (1994) Pathogenesis of *Mycobacterium bovis* infection in cattle. *Vet. Microbiol.* **40,** 41–52.
4. Caldow, C., and Dunn, G. J. (2001) Assessment of surveillance and control of Johne's disease in farm animals in GB. Available from URL:http//www.defra.gov.uk/animalh/diseases/default.htm.
5. Pavlik, I, Svastova, P., Bartl, J., Dvorska, L., and Rychlik, I. (2000) Relationship between IS*901* in the *Mycobacterium avium* complex strains isolated from birds, animals, humans, and the environment and virulence for poultry. *Clin. Diagn. Lab. Immunol.* **7,** 212–217.
6. Pattyn, S. R. (1984) *Mycobacterium lepraemurium,* in *The Mycobacteria-A Sourcebook, Ch. 54* (Kubica, G. P. and Wayne, L. G., eds.), Marcel Dekker, Inc., New York, USA, pp. 1277–1286.
7. Thoen, C. O. ,and Schliesser, T. A. (1984) Mycobacterial infections in cold-blooded animals, in *The Mycobacteria-A Sourcebook, Ch. 54* (Kubica, G. P. and Wayne, L. G., eds.), Marcel Dekker, New York, USA, pp. 1297–1311.
8. Hughes, M. S., James, G., Ball, N., et al. (2000) Identification by 16S rRNA gene analyses of a potential novel mycobacterial species as an etiological agent of canine leproid granuloma syndrome. *J. Clin. Microbiol.* **38,** 953–959.
9. Eisenach, K. D. (1998) Molecular diagnostics, in *Mycobacterium–Molecular Biology and Virulence* (Rutledge, C. and Dale, J., eds.), Blackwell Science, Oxford, UK, pp. 161–179.
10. Grant, I. R. (1997) Zoonotic potential of *Mycobacterium paratuberculosis,* in *Modern Perspectives on Zoonoses* (Holland, C. V., ed.), Royal Irish Academy, Dublin, Ireland, pp. 75–83.
11. Collins, C. H., Grange, J. M., and Yates, M. D. (1997) *Tuberculosis Bacteriology: Organization and Practice,* 2nd ed., Butterworth-Heinemann, Oxford, UK.
12 Pfyffer, G. E. (1999) Nucleic acid amplification for mycobacterial diagnosis. *J. Infect.* **39,** 21–26.
13 Nordhoek, G. T., van Embden, J. D. A., and Kolk, A. H. J. (1996) Reliability of nucleic acid amplification for detection of *Mycobacterium tuberculosis:* an international collaborative quality control study among 30 laboratories. *J. Clin. Microbiol.* **34,** 2522–2525.

14. Aranaz, A., Liébana, E., Mateos, A, Vidal, D., Domingo, M., and Dominguez, L. (1995) Direct detection of *M. bovis* from tissue samples. Improvement of a DNA extraction method for PCR amplification, in *Tuberculosis in Wildlife and Domestic Animals* (Griffin, F. and de Lisle, G., eds.), University of Otago Pres, Dunedin, New Zealand, pp. 60–63.

15. Liébana, E., Aranaz, A., Mateos, A., et al. (1995) Simple and rapid detection of *Mycobacterium tuberculosis* complex organisms in bovine tissue samples by PCR. *J. Clin. Microbiol.* **33,** 33–36.

16. Wards, B. J., Collins, D. M., and de Lisle, G. W. (1995) Detection of *Mycobacterium bovis* in tissues by polymerase chain reaction. *Vet. Microbiol.* **43,** 227–240.

17. Roring, S., Hughes, M. S., Skuce, R. A., and Neill, S.D. (2000) Simultaneous detection and strain differentiation of Mycobacterium bovis directly from bovine tissue specimens by spoligotyping. *Vet. Microbiol.* **74,** 227–236.

18. Bollo, E., Guarda, F., Capucchio, M. T., and Galietti, F. (1998) Direct detection of *Mycobacterium tuberculosis* complex and *M. avium* complex in tissue specimens from cattle through identification of specific rRNA sequences. *J. Vet. Med.* **B45,** 395–400.

19. Doern, G. V. (1996). Diagnostic mycobacteriology: where are we today? *J. Clin. Microbiol.* **34,** 1873–1876.

20. Taylor, M. J., Hughes, M. S., Skuce, R. A., and Neill, S. D. (2001) Detection of *Mycobacterium bovis* in bovine clinical specimens using real-time fluorescence and fluorescence resonance energy transfer probe rapid-cycle PCR. *J. Clin. Microbiol.* **39,** 1272–1278.

21. Eisenach, K. D., Cave, M. D., Bates, J. H., and Crawford, J. T. (1990) Polymerase chain reaction amplification of a repetitive DNA sequence specific for *Mycobacterium tuberculosis*. *J. Infect. Dis.* **161,** 977–981.

22. Kamerbeek, J., Schouls, L., Kolk, A., et al. (1997) Simultaneous detection and strain differentiation of *Mycobacterium tuberculosis* for diagnosis and epidemiology. *J. Clin. Microbiol.* **35,** 907–914.

23. Guerrero, C., Bernasconi, C., Burki, D., Bodmer, T., and Telenti, A. (1995) A novel insertion element from *Mycobacterium avium*, IS*1245*, is a specific target for analysis of strain relatedness. *J. Clin. Microbiol.* **33,** 304–307.

24. Moss, M. T., Sanderson, J. D., Tizard, M. L. V., et al. (1992) Polymerase chain reaction detection of *Mycobacterium paratuberculosis* and *Mycobacterium avium* subsp *silvaticum* in long term cultures from Crohn's disease and control tissues. *Gut* **33,** 1209–1213.

25. Hughes, M. S., Skuce, R. A., Beck, L-A., and Neill, S. D. (1993) Identification of mycobacteria from animals by restriction enzyme analysis and direct DNA cycle sequencing of polymerase chain reaction-amplified 16S rRNA gene sequences. *J. Clin. Microbiol.* **31,** 3216–3222.

26. Hughes, M. S., Ball, N. W., Beck, L–A., de Lisle, G. W., Skuce, R. A., and Neill, S. D. (1997) Determination of the etiology of presumptive feline leprosy by 16S rRNA gene analysis. *J. Clin. Microbiol.* **35,** 2464–2471.

27. Vitale, F., Capra, G., Maxia, L., Reale, S., Vesco, G., and Caracappa, S. (1998) Detection of *Mycobacterium tuberculosis* complex in cattle by PCR using milk, lymph node aspirates and nasal swabs. *J. Clin. Microbiol.* **36,** 1050–1055.

28. Mangiapan, G., Vokurka, M., Schouls, L., et al. (1996) Sequence capture-PCR improves detection of mycobacterial DNA in clinical specimens. *J. Clin. Microbiol.* **34,** 1209–1215.

29. Brugière, O., Vokurka, M., Lecossier, D., et al. (1997) Diagnosis of smear-negative pulmonary tuberculosis using sequence capture polymerase chain reaction. *Am. J. Resp. Crit. Care Med.* **155,** 1478–1481.

30. Roring, S., Hughes, M. S., Beck, L. -A., Skuce, R. A., and Neill, S. D. (1998) Rapid diagnosis and strain differentiation of Mycobacterium bovis in radiometric culture by spoligotyping. *Vet. Microbiol.* **61,** 71–80.

31. Hughes, M. S., Ball, N. W., Love, D. N., et al. (1999) Disseminated *Mycobacterium genavense* infection in a FIV-positive cat. *J. Feline Med. Surg.* 1, 23–29

32. Lucas, J., Lucas, A., Furber, H., et al. (2000) *Mycobacterium genavense* infection in two aged ferrets with conjunctival lesions. *Aust. Vet. J.* **78,** 12–16.

33. Edwards, U., Rogall, T., Blöcker, H., Emde, M., and Böttger, E. C. (1989) Isolation and direct complete nucleotide determination of entire genes. Characterization of a gene coding for 16S ribosomal RNA. *Nucleic Acids Res.* **17,** 7843–7853.

34. Taylor, G. M., Crossey, M., Saldanha, J., and Waldron, T. (1996) DNA from *Mycobacterium tuberculosis* identified in mediaeval human skeletal remains using polymerase chain reaction. *J. Archaeol. Sci.* **23,** 789–798.

35. Hashimoto, A., Koga, H., Kohno, S., Miyazaki, Y., Kaku, M., and Hara, K. (1996) Rapid detection and identification of mycobacteria by combined method of polymerase chain reaction and hybridization protection assay. *J. Infect.* **33,** 71–77.

36. Kempsell, K. E. Ye, Ji., Estrada, I. C., Colston, M. J., and. Cox, R. A. (1992) The nucleotide sequence of the promoter, 16S rRNA and spacer region of the ribosomal RNA operon of *Mycobacterium tuberculosis* and comparison with *Mycobacterium leprae* precursor rRNA. *J. Gen. Microbiol.* **138,** 1717–1727.

37. Strong, B. E., and Kubica, G. P. (1987) Demonstration and isolation of mycobacteria, in *Isolation and Identification of Mycobacterium tuberculosis., A Guide for the Level II Laboratory*, CDC Lab Manual, U.S. Dept of Health and Human Services/Public Health Service CDC, Atlanta, GA, USA, pp. 43–88.

38. Sambrook, J., and Russell, D. W. (2001) *Molecular Cloning. A Laboratory Manual,* 3rd ed., CSH Laboratory Press, Cold Spring harbor, NY, USA, Ch. 5.

39. GenBank database. Internet address: URL: (http://www.ncbi.nlm.nih.gov/Genbank/GenbankSearch.html)

40. EMBL database. Internet address: URL: (http://www.ebi.ac.uk/embl.index.html)

41. Ribosomal Databases Project II. Internet address: URL:(http://rdp.cme.msu.edu/html)

42. ACDP guidelines. Internet address: URL: (http://www.hse.gov.uk/hthdir/noframes/agent1.pdf)

43. Meier, A., Persing, D. H., Finken, M., and Böttger, E. C. (1993) Elimination of contaminating DNA within polymerase chain reaction reagents: implications for a general approach to detection of uncultured pathogens. *J. Clin. Microbiol.* **31,** 646–652.

15

Multiplex PCR of Avian Pathogenic Mycoplasmas

Mazhar I. Khan

1. Introduction

More than 20 mycoplasma species have been isolated and characterized from avian sources (*1*). Only four avian mycoplasmas species are known to cause economic losses in commercial poultry production. *Mycoplasma gallisepticum* (Mg) infection commonly causes chronic respiratory disease (CRD) in chickens and infectious sinusitis in turkeys (*2*), *M. synoviae* (Ms) infection most frequently occurs as a subclinical upper respiratory infection and synovitis in chickens and turkeys (*3*), *M. iowae* (Mi) causes decrease in hatchability and high embryo mortality in turkeys (*4*), and *M. meleagridis* (Mm) is the cause of an egg-transmitted disease of turkeys in which the primary lesion is an airsacculitis in the progeny, which leads to lower hatchability and skeletal abnormalities in young turkeys (*5*).

Identification of any of these avian mycoplasma organisms is of great importance to the poultry industry, where prompt diagnosis is of paramount importance. Both serologic and isolation procedures have been used for diagnosis of avian mycoplasmas. However, interspecies cross-reactions and nonspecific reactions (*6*) often hamper serologic tests, while isolation of mycoplasmas is difficult and time-consuming. Molecular methods, such as DNA probes (*7–9*) and polymerase chain reaction (PCR) (*10–14*), have been developed as alternatives to conventional serologic and culture methods to detect specific types of mycoplasmal microorganisms in clinical samples. On the other hand, multiple infections of avian pathogenic mycoplasmas are also not uncommon in chicken and turkey flocks. Especially commercial breeder chicken flocks infected with both Mg and Ms (*15–17*) need to be differentiated and diagnosed with culture and serology, as well as PCR amplification tests (*6*).

From: *Methods in Molecular Biology, vol. 216: PCR Detection of Microbial Pathogens: Methods and Protocols*
Edited by: K. Sachse and J. Frey © Humana Press Inc., Totowa, NJ

Simultaneous detection of multiple bacterial *(18–20)* or viral *(21,22)* infections and defective genes *(23)* have been described using multiplex PCR amplification techniques. This approach can be highly specific, sensitive, and cost-effective, making it an attractive alternative to conventional culture and specific PCR methods for individual avian mycoplasmas in clinical samples suspected of multiple infections. This chapter describes a multiplex PCR protocol for pathogenic avian mycoplasmas that was developed and optimized in our laboratory *(24)*.

2. Materials

1. Proteinase K (2 mg/mL stock solution).
2. Phenol-chloroform-isoamyl alcohol: 25:24:1(v/v/v), equilibrated with 50 m*M* Tris-HCl, pH 8.0.
3. Sodium dodecyl sulfate (SDS): 10% (w/v) stock solution.
4. Sodium acetate: 3 *M*, pH 5.2.
5. Ethanol: 100% (v/v) ice-cold, and 70% (v/v).
6. TE buffer: 10 m*M* Tris-HCl pH 7.6, 1 m*M* EDTA.
7. PCR amplification buffer (10X): 500 m*M* KCl, 100 m*M* Tris-HCl, pH 8.3, 15 m*M* MgCl$_2$, 0.1% (w/v) gelatin (Applied Biosystems, Foster City, CA, USA).
8. MgCl$_2$: 25 m*M* solution.
9. dNTP mixture (consists of dATP, dCTP, dGTP, and dTTP): 2.5 m*M* each (Applied Biosystems).
10. AmpliTaq® Gold DNA polymerase: 5 U/µL (Applied Biosystems).
11. Oligonucleotide primers for PCR: working solution Mg 11.14 nmol/mL each primer; Ms 14.17 nmol each primer; Mm 30.66 nmol/mL each primer; Mi 420 nmol/mL each primer in sterile water (for sequences, *see* **Table 1**; for storage conditions, *see* **Note 1**).
12. Positive control DNA: 100 ng/µL of each avian pathogenic mycoplasma species, i.e., Mg, Mm, Mi, Ms.
13. Microfuge tubes: 1.8 and 0.5 mL.
14. Automated thermal cycler: Model 480 GeneAmp® PCR System (Applied Biosystem), or another comparable model.
15. Horizontal gel electrophoresis apparatus.
16. Agarose: 1.5% (w/v), ultrapure grade (Bethesda Reseach Laboratories, Bethesda, MD, USA).
17. Electrophoresis buffer: Tris-borate buffer (0.045 *M* Tris-borate, 0.001 *M* EDTA, pH 8.0).
18. Ethidium bromide: 10 mg/mL stock solution, keep in dark bottle at room temperature (*see* **Note 2**).
19. Tracking dye (6X stock solution): 0.25% bromophenol blue and 40% (w/v) sucrose in water, store at 4°C.
20. Polaroid films and camera.

Table 1
PCR Primers for Avian Pathogenic Mycoplasmas

Primer	Specificity	Sequence (5'–3')	Reference
MG 1	*M. gallisepticum*	GGATCCCATCTCGACCAGGAGAAAA	*11*
MG 2		CTTTCAATCAGTGAGTAACTGATGA	
MS 1	*M. synoviae*	GAAGCAAATAGTGATATCA	*10*
MS 2		GTCGTCTCGAAGTTAACAA	
MM 1	*M. meleagridis*	GGATCCTAATATTAATTTAAACAAATTAATGA	*14*
MM 2		GAATTCTTCTTTATTATTCAAAAGTAAAGTAC	
MI 1	*M. iowae*	GAATTCTGAATCTTCATTTCTTAAA	*13*
MI 2		CAGATTCTTTAATAACTTATGTATC	

3. Methods
3.1. Sample Collection

1. Tracheal samples from suspected live birds should be obtained on sterile swabs.
2. Place the swab samples in transport medium (sterile water) at 4°C.
3. Keep the transport medium at 4°C.

3.2. Sample Preparation

Sample preparation is crucial for insuring the quality and reproducibility of PCR, particularly when working with clinical samples (*see* **Note 3**).

1. Squeeze swabs samples in the transport media and discards the swab.
2. Transfer 1.5 mL of transport medium to the microfuge tube.
3. Microfuge the sample for 15 min at 14,000g at 4°C.
4. Decant supernatant.
5. Resuspend the pellet in 400 µL of TE buffer.

3.3. DNA Isolation

1. Add 40 µL of 10% SDS to lyse the cells in 400 µL of TE buffer.
2. Add 10 µL of proteinase K (final concentration will be 20 µg/mL).
3. Incubate at 37°C for 1 h.
4. Extract the samples in microfuge tube with an equal vol of phenol-chloroform-isoamyl alcohol (25:24:1) equilibrated with 50 mM Tris-HCl, pH 8.0.
5. Vortex mix the mixture for 30 s.
6. Microfuge the sample for 2 min at 14,000g at room temperature.
7. Transfer the upper layer into new microfuge tube.
8. Repeat **steps 4–6**.
9. Precipitate DNA by addition of 1/10 volume of 3 M sodium acetate and 2–2.5 volumes of ice-cold 100% ethanol.

10. Incubate for 20 min at $-70°C$.
11. Centrifuge the sample for 15 min at $14,000g$.
12. Decant supernatant.
13. Wash the pellet with 70% ethanol.
14. Air-dry the pellet under the hood and resuspend in 20 µL of TE buffer.
15. Use 10 µL of sample DNA extract for PCR test (*see* **Subheading 3.4., steps 10–12,** and **Subheading 3.5.**).

3.4. Preparation of the Reaction Mixture for Amplification (see Note 4)

1. The amplification reaction is carried out in 100-µL vol. (50-µL reactions are also possible. In that case, the amounts of all reagents have to be halved.)
2. Add 10 µL of PCR amplification buffer into a 500-µL microfuge tube.
3. Add 2 µL (200 µ*M*) of each dNTP.
4. Add 2 µL (final concentration 2.5 m*M*) of $MgCl_2$.
5. Add 1 µL (2.5 U) of AmpliTaq Gold DNA polymerase.
6. Add 3 µL of each primer (total of 24 µL) as listed in **Table 1**. The final concentrations will be 334 pmol for Mg, 425 pmol for Ms, 920 pmol for Mm, and 420 pmol for Mi.
7. Add 10 µL of the DNA extract from swab samples prepared according to **Subheading 3.3.** (For positive controls, use 100 ng of DNA of each mycoplasma species dissolved in 10 µL of water.)
8. Add sterile distilled water to make up the total vol to 100 µL.
9. (Optional) If using a thermal cycler requiring oil overlays, add 50 µL of mineral oil.

3.5. Amplification

Run the PCR according to the following temperature–time profile: initial denaturation at $94°C$ for 5 min, then 35 cycles of denaturation at $94°C$ for 1 min, annealing at $50°C$ for 1 min, extension at $72°C$ for 2 min, final extension at $72°C$ for 10 min, and then set the final temperature to $4°C$ until further use.

3.6. Detection of Amplified Products by Agarose Gel Electrophoresis

1. Prepare 1.5% agarose gel in electrophoresis buffer.
2. The setup of the apparatus, preparation and running of the gels should be done according to the procedure described in **ref. 25**.
3. Inject an appropriate vol of amplified PCR product with tracking dye into gel slot, e.g., 10–15 µL of product plus 3 µL of tracking dye.
4. Run the electrophoresis.
5. Stain the gel with ethidium bromide, then expose to UV light to visualize bands and be photographed. An example is shown in **Fig. 1**.

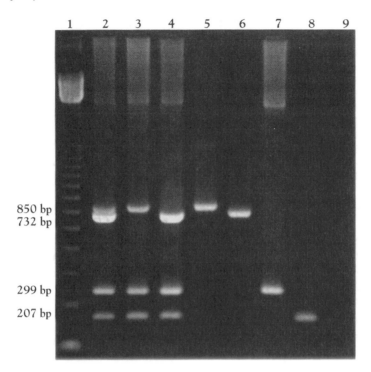

Fig. 1 Agarose gel electrophoresis of multiplex PCR-amplified products from puri-fied DNA of known avian mycoplasmas. Lane 1, 123 bp marker; lane 2, Mm (RY39), Mg (S6), Mi (695), Ms (WVU1853); lane 3, Mm (RY39), Mi (695), Ms (WUV 1853); lane 4, Mg (S6), Mi (695), Ms (WVU1853); lane 5, Mm (RY39); lane 6, Mg (S6); lane 7, Mi (695); lane 8, Ms (WUV1853); lane 9, Negative control (PCR buffer). Reprinted with permission from **ref. 24.**

4. Notes

1. Primers should be aliquoted in small quantities, and one aliquot of PCR primers should be used at a time and placed in 4°C until used. All other aliquots should be stored −20°C.
2. Ethidium bromide is a powerful mutagen and is moderately toxic. Gloves should be worn when working with solutions that contain this dye. After its use these solutions should be decontaminated by one of the methods described in **ref. 25.**
3. Tracheal swab samples can be used directly without incubating overnight in Frey's mycoplasma medium *(26)*. In the event of negative flocks, the overnight incubation in Frey's medium will enhance the sensitivity of the multiplex PCR.
4. Pre-PCR processing of mycoplasma samples and mixing of reagents should be performed in a room designated for that purpose. The thermal cycler should be kept in another room to prevent carryover contamination.

5. Periodically, multiplex PCR amplification should be performed on known positive and nonrelated mycoplasmas, such as Mg, Ms, Mm, Mi, as well as *M. gallinarum* and *M. gallinacium*, to check the validity of the assay's specificity and sensitivity.

Acknowledgments

The author thanks Ms. Hang Wang and Dr. Amin A. Fadl for development and optimization of multiplex PCR for avian pathogenic mycoplasmasas. This work was supported in part by funds provided by the U.S. Department of Agriculture's Hatch grant.

References

1. Kleven, S. H. (1997) Mycoplasmosis, in *Diseases of Poultry* (Calnek, B. W., Barnes, H. J., Beard, C. W., McDougald, L. R., and Saif, Y. M. eds.), Iowa State University Press, Ames, IA, USA, pp. 191–193.
2. Ley, D. H., and Yoder, H. W. (1997) *Mycoplasma gallisepticum* infection, in *Diseases of Poultry* (Calnek, B. W., Barnes, H. J., Beard, C. W, McDougald, L. R., and Saif, Y. M. eds.), Iowa State University Press, Ames, IA, USA, pp. 194–207.
3. Kleven, S. H. (1997) *Mycoplasma synoviae* infection, in *Diseases of Poultry* (Calnek, B. W., Barnes, H. J., Beard, C. W., McDougald, L. R., and Saif, Y. M. eds.), Iowa State University Press, Ames, IA, USA, pp 220–227.
4. Kleven, S. H. (1997) *Mycoplasma iowae* infection, in *Diseases of Poultry* (Calnek, B. W., Barnes, H. J., Beard, C. W., McDougald, L. R., and Saif, Y. M. eds.), Iowa State University Press, Ames, IA, USA, pp. 228–231.
5. Yamamoto, R. (1997) *Mycoplasma meleagridis* infection in, *Diseases of Poultry* (Calneck, B. W., Barnes, H. J., Beard, C. W., McDougald, L. R., and Saif, Y. M. eds.), Iowa State University Press, Ames, IA, USA, pp. 208–219.
6. U. S. Dept. of Agriculture (1985) National poultry improvement plan and auxiliary provisions. USDA, APHIS. VS, Bulletin 91–40.
7. Dohms, J. E., Hnatow, L. L., Whetzel, P., Morgan, R., and Keeler, Jr., C. L. (1993) Identification of the putative cytadhesin gene of *Mycoplasma gallisepticum* and Its Use as a DNA Probe. *Avian Dis.* **37,** 380–388.
8. Khan, M. I., B. C. Krikpatrick, B. C and Yamamoto, R. (1989) *Mycoplasma gallisepticum* species and strain specific recombinant DNA probes. *Avian Pathol.* **18,** 135–146.
9. Razin, S. (1994) DNA probes and PCR in diagnosis of mycoplasma infections. *Mol. Cell. Probes* **8,** 497–511.
10. Lauerman, L. H., Hoerr, F. J., Sharpton, A. R., Shah, S. M and Van Santen, V. L. (1993) Development and application of polymerase chain reaction assay for Mycoplasma synoviae. *Avian Dis.* **37,** 829–834.
11. Nascimento, E. R., Yamamoto, R., Herrick, K. R., and Tait, R. C. (1991) Polymerase chain reaction for detection of *Mycoplasma gallisepticum. Avian Dis.* **35,** 62–69.

12. Zhao, S., and Yamamoto, R. (1993) Amplification of *Mycoplasma iowae* using polymerase chain reaction. *Avian Dis.* **37**, 212–217.
13. Zhao, S., and Yamamoto, R. (1993) Detection of *Mycoplasma synoviae* by polymerase chain reaction. *Avian Pathol.* **22**, 533–542.
14. Zhao, S., and Yamamoto, R. (1993) Detection of *Mycoplasma meleagridis* by polymerase chain reaction. *Vet. Microbiol.* **36**, 91–97. 1993.
15. Bradbury J. M., and McClenaghan M. (1982) Detection of mixed mycoplasma species. *J. Clin. Microbiol.* **16**, 314–318.
16. Rott M., Pfützner, H., Gigas, H., and Rott, G.(1989) Diagnostic experiences in the routine restrained inspection of turkey stock for mycoplasma infections. *Arch. Exper. Vet. Med.* **43**, 743–746.
17. Sahu, S. P., and Olson, N. O. (1981) Characterization of an isolate of *Mycoplasma* WVU 907 which possesses common antigens to *Mycoplasma gallisepticum*. *Avian Dis.* **25**, 943–953.
18. Way, J. S., Josephson, K. L., Pillai, S. D., Abbaszadegan, M., Gerba, C. P, and Pepper, I. L. (1993) Specific detection of Salmonella spp. by multiplex polymerase chain reaction. *Appl. Environ. Microbiol.* **59**, 1473–1479.
19. Kulski, J. K., Khinsoe, C., Pryce, T., and Christiansen, K. (1995) Use of multiplex PCR to detect and identify *Mycobacterium avium* and *M. intercellulare* in blood culture fluids of AIDS patients. *J. Clin Microbiol.* **33**, 668–74.
20. Lawrence, L. M., and Gilmour, A. (1994) Incidence of *Listeria spp.* And *Listeria monocytogenes* in a poultry-processing environment and in poultry products and their confirmation by multiplex PCR. *Appl. Environ. Microbiol.* **60**, 4600–4604.
21. Karlsen, F., Kalantari, M., Jenkins, A., et al. (1996) Use of multiplex PCR primer sets for optimal detection of human papillomavirus. *J. Clin. Microbiol.* **34**, 2095–2100.
22. Wang, X., and Khan, M. I. (1999) A multiplex PCR for Massachusetts and Arkansas serotypes of infectious bronchitis virus. *Mol. Cell. Probes* **13**, 1–7.
23. Chamberlain, J. S., Gibbs, R. A., Ranier, J. E., Nguyen, P. N., and Caskey, C. T. (1981) Deletion screening of the duchenne muscular dystrophy locus via multiplex DNA amplification. *Nucleic Acid Res.* **16**, 11141–56.
24. Wang, Han., Fadl, A. A, and Khan, M. I. (1996) Multiplex PCR for avian pathogenic mycoplasmas. *Mol. Cell Probes* **11**, 211–216.
25. Sambrook, J., Fritsch, E. T., and Maniatis, T. (1989) *Molecular Cloning: A Laboratory Manual.* pp. #1.25, 1.85, 6.3, 9.47, A.1, E.3, E.5. CSH Laboratory Press, Cold Spring Harbor, NY, USA.
26. Frey, M. L., Hanson, R. P., and Anderson, D. P. (1968) A medium for the isolation of avian mycoplasmas. *Am. J. Vet. Res.* **29**, 2163–2171.

16

Detection and Differentiation
of Ruminant Mycoplasmas

Helmut Hotzel, Joachim Frey,
John Bashiruddin, and Konrad Sachse

1. Introduction

1.1. Importance of the Agents

More than 20 different species of mollicutes, most of them belonging to the genus *Mycoplasma*, have been identified from ruminant hosts to date. While a considerable part of this group is conceived to be of minor epidemiological relevance, it contains some important pathogenic agents that have specific host ranges. Due to several peculiar properties of mycoplasmas, which include the absence of a cell wall and the capability of surface antigen variation, diseases caused by mycoplasmas are difficult to control in the conventional fashion by chemotherapy or immunoprophylaxis. Another general feature of mycoplasma infections is their protracted and occasionally chronic course. Five mycoplasmas of economic and welfare importance are mentioned here and serve as examples for the variety of clinical and diagnostic circumstances that can be resolved using polymerase chain reaction (PCR).

Mycoplasma (M.) bovis, one of the etiological agents of bovine mycoplasmosis occurring most frequently in Europe and North America, was associated with mastitis in cows, arthritis and pneumonia in calves and young cattle, as well as genital disorders in bulls and cows *(1)*. While mastitis outbreaks mainly occur in the larger dairy herds, pneumonia and arthritis in calves represent typical mycoplasma diseases in small farms.

M. agalactiae causes contagious agalactia, a severe infectious disease of sheep and goats expressed clinically as mastitis, arthritis, or keratoconjunctivi-

From: *Methods in Molecular Biology, vol. 216: PCR Detection of Microbial Pathogens: Methods and Protocols*
Edited by: K. Sachse and J. Frey © Humana Press Inc., Totowa, NJ

tis *(2)*. Particularly the countries of the Mediterranean and the Middle East are affected, but the agent seems to be much more widespread.

M. mycoides subsp. *mycoides* small colony type (*Mmm*SC) is the causative agent of contagious bovine pleuropneumonia (CBPP), a contagious disease of cattle causing severe losses in livestock production. The disease is especially widespread in Africa, where mortality rates of 30–80% were reported in affected herds *(3)*. Due to its high sanitary, economic, and socio-economic impact, CBPP is a disease of List A of the Office International des Epizooties and, consequently, belongs to the animal diseases which must be eradicated.

A phylogenetically closely related pathogen, known as Mycoplasma *mycoides* subsp. *mycoides* large colony type (*Mmm*LC) or as *Mycoplasma mycoides* subsp. *capri* has been isolated mainly from goats, more rarely from sheep and cattle, and was reported to be responsible for cases of mastitis, keratoconjunctivitis, arthritis, pulmonary disease, and septicemia *(4)*.

M. conjunctivae was demonstrated to be the etiological agent of infectious keratoconjunctivitis in domestic sheep and goats, as well as in European alpine ibex and chamois *(5)*. The disease is characterized by inflammation of the conjunctiva and cornea.

1.2. Diagnostic Methods

There are several specific reasons why a change from conventional detection methods to DNA-based techniques is bound to bring about significant improvements in the diagnosis of mycoplasma-associated diseases.

First of all, the slow growth of mycoplasmas represents a great obstacle. Identification by culture usually takes 5–10 d, which may be late for effective control measures in infected herds to be taken. In contrast, a PCR assay delivering results in 1 d can make all the difference, for instance, in animal trade. In herd diagnosis, antigen enzyme-linked immunosorbent assay (ELISA) may be an economical and relatively fast alternative, but specificity is often limited due to cross-reactions with related mycoplasmas, and its sensitivity is inferior to that of PCR by two or more orders of magnitude, thus rendering it unsuitable for the identification of clinically inapparent shedders. Similarly, antibody detection ELISAs have their limitations in mycoplasma diagnosis, since antibody titers emerge only 10–14 d after infection at the earliest, and sensitivity is often insufficient to identify chronically infected carriers.

Meanwhile, PCR-based detection assays for all important ruminant mycoplasmas have been published. Being the most extensively studied genomic region of many mycoplasma species, 16S ribosomal (r) RNA genes emerged as a favorite target of amplification assays *(6)*. As a result of evolutionary processes, they harbor both conserved and variable sequences, thus allowing the selection of primers in a wide range of specificity, from class- to species-specific, in many cases.

General PCR assays for mycoplasmas using these target sequences were proposed by several authors *(7,8)*. Species-specific PCR detection systems based on 16S rRNA gene amplification were described for *M. bovis* and *M. agalactiae (9)*, *M. capricolum* subsp. *capripneumoniae (10)*, *M. bovirhinis*, *M. alkalescens*, *M. bovigenitalium (11)*, *M. conjunctivae (5)*, and *Mmm*SC *(12)*. The adjacent 16S–23S intergenic spacer region, known as an interesting marker for phylogenetic studies *(13,14)*, was also shown to be suitable as a PCR target, not least because of its size variation within the *Mollicutes (15)*.

There are, however, limitations to the use of rRNA gene targets because of intraspecies sequence heterogeneity between the various rRNA operons (there are up to 3 operons in certain species) or lack of interspecies sequence variation within certain groups (*see* **Chapter 1**).

Genes for housekeeping enzymes, such as the DNA repair gene *uvr*C, which encodes deoxyribodipyrimidine photolyase *(16)*, or oligopeptide permease genes *(17,18)* proved to be robust targets for species-specific detection of *M. agalactiae* and *M. bovis*, respectively. A gene encoding a major lipoprotein of several mycoplasmas belonging to the *M. mycoides* cluster was shown to be suitable for identification of *Mmm*SC and *Mmm*LC *(19,20)*. Furthermore, insertion sequence IS*1634* was reported to have been found exclusively in *Mmm*SC and was also proposed as a target for identification of this pathogen by PCR *(21)*.

This chapter describes a set of PCR amplifications for the identification of pathogenic ruminant mycoplasmas. Several of these target sequences also served for direct detection of the agents in clinical samples, milk, and semen. Therefore, we have included the pre-PCR treatment methods of these biological samples, e.g., antigen capture from milk and various DNA extraction procedures, as well as the corresponding PCR conditions for direct detection.

2. Materials
2.1. DNA Extraction

1. Water. Deionized water must be used for all buffers and dilutions.
2. Lysis buffer: 100 mM Tris-base, pH 8.5, 0.05% (v/v) Tween® 20.
3. Proteinase K: 10 mg/mL in water.
4. Carbonate-bicarbonate buffer (CBB): 50 mM Na$_2$CO$_3$, 50 mM NaHCO$_3$, pH 9.6. Adjust pH by adding NaHCO$_3$ solution.
5. Monoclonal antibody (MAb) 4F6 recognizing a 32-kDa protein of *M. bovis*: 40 µg IgG per mL CBB (available from K.S. upon request).
6. Phosphate-buffered saline (PBS): 10 mM Na$_2$HPO$_4$, 10 mM NaH$_2$PO$_4$, 145 mM NaCl, pH 7.0. Adjust pH by adding NaH$_2$PO$_4$.
7. Tris-EDTA (TE) buffer: 10 mM Tris-HCl, pH 8.0, 1 mM EDTA (ethylene diamine tetraacetic acid).

8. Sodium dodecyl sulfate (SDS) solution: 10 mg/mL in TE.
9. Phenol: saturated solution in TE buffer. If two separate phases are visible, use the lower phase only.
10. Chloroform-isoamyl alcohol: 24:1 (v/v).
11. Isopropanol, analytical or molecular biology-grade.
12. Commercially available DNA extraction kit for PCR template preparation (*see* **Note 1**). In our hands, the following products worked well: High Pure PCR Template Preparation Kit (Roche Diagnostics, Mannheim, Germany), QIAamp® DNA Mini Kit (QIAGEN, Hilden, Germany), E.Z.N.A. Tissue DNA Kit II (PEQLAB, Erlangen, Germany).

2.2. PCR

1. *Taq* DNA polymerase. We use MasterTaq (5 U/µL) from Eppendorf (Hamburg, Germany).
2. 10X reaction buffer for *Taq* DNA polymerase: provided by the manufacturer of the enzyme, contains 1.5 mM MgCl$_2$.
3. dNTP mix: dATP plus dGTP plus dCTP plus dTTP, 2 mM each. Store in aliquots at –20°C.
4. Primer oligonucleotides according to **Table 1**.

2.3. Electrophoresis and Visualization

1. Agarose, molecular biology-grade: 1% gels for PCR products of 300–1000 bp, 2% gels for products below 300 bp.
2. Tris-borate electrophoresis buffer (TBE): 0.09 M Tris-borate, 0.002 M EDTA, pH 8.0. For 1 L of 10X TBE, mix 108 g Tris-base, 55 g boric acid, and 80 mL of 0.25 M EDTA, make up with water. Dilute 1:10 before use.
3. Gel loading buffer (GLB): 20% (v/v) glycerol, 0.2 M EDTA, 0.01% (w/v) bromophenol blue, 0.2% (w/v) Ficoll® 400.
4. Ethidium bromide stock solution: 1% (10 mg/mL) solution in water. Caution: Ethidium bromide is presumed to be mutagenic. Avoid direct contact with skin. Wear gloves when handling it.
5. DNA size marker: We mostly use the 100-bp DNA ladder (Gibco/Life Technologies, Eggenstein, Germany). For large fragments, *Hin*dIII-digested λ DNA (Roche Diagnostics) may be used.

2.4. General Equipment

1. Thermal cycler. We use the T3 Thermal cycler (Biometra, Goettingen, Germany).
2. Vortex shaker, e.g., MS1 Minishaker (IKA Works, Wilmington, DE, USA).
3. Benchtop centrifuge with Eppendorf rotor, e.g., Model 5402 (Eppendorf) and/or a mini centrifuge, e.g., Capsule HF-120 (Tomy Seiko, Tokyo, Japan).
4. Heating block, for incubation of Eppendorf tubes, adjustable temperature range 30–100°C.
5. Apparatus for horizontal gel electrophoresis.

Table 1
Primers for Detection of Mycoplasmas

Denomination	Sequence (5'-3')	Reference
Myc23F1729	CTAAGGTDAGCGAGWDAACTATAG*	*(22)*
Myc23R1837	CCCCYCWTSYTTYACTGMGGC*	
P1	TAT ATG GAG TAA AAA GAC	*(23)*
P2	AAT GCA TCA TAA ATA ATT G	
PpMB920-1	GGCTCTCATTAAGAATGTC	*(17)*
PpMB920-2	TTTTAGCTCTTTTTGAACAAAT	
PpSM5-1	CCAGCTCACCCTTATACATGAGCGC	*(18)*
PpSM5-2	TGACTCACCAATTAGACCGACTATTTCACC	
MBOUVRC2-L	TTACGCAAGAGAATGCTTCA	*(16)*
MBOUVRC2-R	TAGGAAAGCACCCTATTGAT	
MAGAUVRC1-L	CTCAAAAATACATCAACAAGC	*(16)*
MAGAUVRC1-R	CTTCAACTGATGCATCATAA	
SC3NEST1-L	ACAAAAGAAGATATGGTGTTGG	*(19)*
SC3NEST1-R	ATCAGGTTTATCCATTGGTTGG	
SC3VII	ATTAGGATTAGCTGGTGGAGGAAC	*(19)*
SC3IV-S	TCTGGGTTATTCGAACCATTAT	
MMMLC2-L	CAATCCAGATCATAAAAAACCT	*(20)*
MMMLC1-R	CTCCTCATATTCCCCTAGAA	
MOLIGEN1-L	ACTCCTACGGGAGGCAGCA	*(5)*
16SUNI-R	GTGTGACGGGCGGTGTGTAC	
Mcor1	CAGCGTGCAGGATGAAATCCCTC	*(5)*
McoF1	GTATCTTTAGAGTCCTCGTCTTTCAC	

*Degenerate nucleotides: D=A,G,T; W=A,T; Y=C,T; S=G,C; M=A,C.

6. UV transilluminator, 254 and/or 312 nm.
7. Video documentation or photographic equipment.
8. Set of pipets covering the whole vol range from 0.1–1000 μL. We use the Eppendorf Research series (Eppendorf).
9. 8-channel pipet (optional).
10. Aerosol-resistant pipet tips (filter tips).
11. 0.2-, 0.5-, 1.5-, and 2.0-mL Plastic tubes; sterile, DNase- and RNase-free (Eppendorf).
12. 96-Well microtiter plates, round bottom (e.g., Fisher Scientific, Nidderau, Germany).

3. Methods

3.1. DNA Extraction from Different Sample Matrixes

3.1.1. Broth Culture

The simplest method to release DNA suitable for PCR from broth cultures of mycoplasmas is 5-min of boiling. After removal of cellular debris by centrifugation at 12,000g for 1 min, the supernatant can be used directly as template. Failure to amplify a specific target could be due to the presence of PCR inhibitors. In these instances, a commercial DNA extraction kit should be tried (*see* **Note 1**).

3.1.2. Swabs from Nasal Mucus, Conjunctival, Pleural, Synovial, or Bronchial Lavage Fluid

1. Pipet 500 µL of lysis buffer into a 2-mL Safe-Lock tube containing the cotton swab.
2. Vortex mix thoroughly for 1 min.
3. Centrifuge at 12,000g for 30 s.
4. Put the swab into a 1-mL pipet tip whose lower half was cut off and place it all into a fresh tube.
5. Centrifuge at 12,000g for 1 min to force the remaining liquid out of the cotton.
6. Add the liquid to that in the first tube from **step 3**.
   If you have mucus or fluid samples (no swabs), start with **step 7**.
7. Centrifuge the liquid or mucus at 12,000g for 15 min.
8. Discard the supernatant and resuspend the pellet in 50 µL of lysis buffer.
9. Add 20 µL of proteinase K and incubate at 60°C for 2 h.
10. Inactivate the proteinase K by heating at 97°C for 15 min.
11. Centrifuge at 12,000g for 5 min to remove debris.
12. Use 5 µL of the supernatant for PCR.

3.1.3. Milk Samples for the Detection of M. bovis

The following protocol includes pre-PCR enrichment by antigen capture and subsequent DNA extraction and is recommended for milk samples containing *M. bovis* (*see* **Note 2**).

3.1.3.1. COATING OF MICROTITER PLATES

1. Pipet 100 µL of the solution containing MAb 4F6 in CBB into the cavities of a 96-well microtiter plate.
2. Incubate overnight at 4° C.
3. Rinse 3× with 200 µL of PBS per well and empty the wells.
4. The plates can now be sealed with parafilm and stored at −20° C until use.

3.1.3.2. SAMPLE PROCESSING

1. Introduce 200 µL of milk samples into the wells of a MAb-coated microtiter plate.

2. Seal with parafilm and incubate at 37°C overnight.
3. Remove the liquid from the wells using an 8-channel pipet.
4. Add 200 µL of 1% SDS and 20 µL proteinase K to the wells and incubate at 37°C for 1 h.
5. Transfer the content of each well to a 1.5-mL tube.
6. Wash the wells with 100 µL of TE buffer.
7. Transfer the washing liquid to the respective 1.5-mL tube to unite the vol.
8. Add 300 µL of phenol to each tube for DNA extraction and vigorously vortex mix the mixture for 1 min.
9. Centrifuge at 12,000*g* for 5 min.
10. Transfer the (upper) aqueous phase into fresh tubes.
11. Add 300 µL of chloroform-isoamyl alcohol to each tube.
12. Vortex mix at highest intensity for 1 min.
13. Centrifuge at 12,000*g* for 5 min.
14. Transfer the (upper) aqueous phase into a fresh tubes.
15. Add 200 µL of isopropanol for DNA precipitation.
16. Mix reagents and incubate at room temperature for 10 min.
17. Centrifuge at 12,000*g* for 10 min. Discard supernatant.
18. Allow pellet (with DNA) to air-dry for 30 min.
19. Dissolve pellet in 20 µL water. Use 1 µL in an amplification reaction.

3.1.4. Tissue from Lung and Other Organs

The following procedure is the simplest method for DNA extraction from tissue specimens. Alternatively, commercial DNA extraction kits can be used (*see* **Note 1**).

1. Boil 100 mg of homogenized tissue in 200 µL of water in a plastic tube for 10 min. Subsequently, allow the tube to cool to room temperature.
2. Optionally, proteinase digestion can be carried out to increase the final yield of DNA: add 200 µL SDS solution and 20 µL of proteinase K to the tube and incubate at 55°C for 1 h.
3. Add 200 µL of phenol.
4. Vortex mix vigorously for 1 min.
5. Centrifuge at 12,000*g* for 5 min.
6. Transfer the (upper) aqueous phase into a fresh tube.
7. Add 200 µL of chloroform-isoamyl alcohol.
8. Vortex mix at highest intensity for 1 min.
9. Pipet the (upper) aqueous phase into a fresh tube.
10. Precipitate DNA by adding 120 µL of isopropanol. Thoroughly mix reagents and incubate at room temperature for 10 min.
11. Collect DNA by centrifugation at 12,000*g* for 10 min. Discard supernatant.
12. Allow DNA pellet to air-dry for 30 min.
13. Redissolve pellet in 20 µL of water. Use 1 µL for an amplification reaction.

3.1.5. Semen (see **Note 1**)

1. Dilute 50 μL of semen in a plastic tube with 150 μL of SDS solution and homogenize by intensive vortex mixing.
2. Digest proteins by adding 20 μL of proteinase K solution and vortex mix for 1 min.
3. Incubate at 55°C for 1 h.
4. Continue extraction procedure as described for tissue (*see* **Subheading 3.1.4.**) beginning with **step 3**.

3.2. DNA Amplification

3.2.1. Preparation of Reaction Mixtures and Controls

Generally, reaction mixtures should be made in final vol of 50 μL. Each tube may contain: 1 U *Taq* DNA polymerase, e.g., 0.2 μL of a solution containing 5U/μL, 5 μL of reaction buffer (10X), between 1.5–3.0 mM of MgCl$_2$ either contained in the reaction buffer or from addition of 3–6 μL of 25 mM stock solution (the amount may need to be optimized for best results), 200 μM of each dNTP obtained by the addition of 1 μL of 50 mM stock solution, 20 pmol of each primer diluted from a 100 μM stock solution, 1–5 μL of sample with between 100 fg and 100 ng of template DNA, and PCR-grade water to adjust the vol to 50 μL.

In each series, include amplification controls that contain the following templates: DNA of a reference strain of the mycoplasma species expected in the sample (positive control), DNA of a related mycoplasma species that could be present in the sample, but should not be amplified with the present primers (specificity control), and water (negative control) instead of sample extract.

3.2.2. General Detection of Mycoplasmas

For general detection of mycoplasmas, primer pair Myc23F1729/ Myc23R1837 is recommended *(22)*. The target sequence in the central domain of the 23S rRNA gene represents a genomic region that is highly conserved among mycoplasmas, but distinct from bacteria outside the class *Mollicutes*. The following temperature–time program should be used for amplification: initial denaturation at 95°C for 60 s, 40 cycles of denaturation at 95°C for 15 s, primer annealing at 48°C for 30 s, and primer extension at 72°C for 30 s, and then final extension at 72°C for 60 s. The size of the amplicon is 102–110 bp.

A PCR system for the *M. mycoides* cluster based on the CAP-21 genomic target region can be used for group-specific, as well as species-specific detection (*see* **Note 3**). Primer pair P1/P2 was designed for amplification of all members of the cluster with 30 cycles of 94 °C for 30 s, 46 °C for 60 s, 72 °C for 90 s *(23)*. A resulting fragment size of 253–265 bp indicates a positive result.

Table 2
Temperature-Time Profiles of *M. bovis* Detection Assays

	*opp*D/F System		*uvr*C System
Primers	First round PpMB920-1/2	Second round PpSM5-1/2	MBOUVRC2-L/R
Initial denaturation	96°C for 60 s	94°C for 60 s	
Denaturation	96°C for 15 s	94°C for 45 s	94°C for 30 s
Primer annealing	48°C for 60 s	54°C for 60 s	52°C for 30 s
Primer extension	72°C for 150 s	72°C for 120 s	72°C for 60 s
Final extension	72°C for 180 s	72°C for 180 s	
Number of cycles	35	30	35
Amplicon size	1911 bp	409 bp	1626 bp

3.2.3. Detection of M. bovis

Here we propose two different detection systems based on *opp*D/F or *uvr*C target sequences, respectively (*see* **Notes 4** and **5**). Experimental parameters and amplicon sizes are given in **Table 2**. The sensitivity of both detection systems was established during extensive application over several yr (*see* **Note 5**). Primer pairs MBOUVRC2-L/R for the *uvr*C-based PCR and PpMB920-1/2 for the *opp*D/F-based PCR can both be used in single-step PCR assays, where detection limits will be around 50 colony forming units (cfu). However, the sensitivity of detection can be increased considerably by running a nested PCR using the *opp*D/F system. After the first round using primer pair PpMB920-1/2, the products (5 µL of a 1:100 dilution) are subjected to a second amplification with primers PpSM5-1/2. The detection limit of the nested PCR protocol with bulk tank milk samples was found to be 0.85 cfu equivalents/mL *(18)*. **Figure 1** shows the results of the nested *opp*D/F system being used for examination of DNA extracts from tissue samples of an infected calf.

3.2.4. Detection of M. agalactiae

Analysis of *uvr*C genes revealed that there were sufficient differences between *M. agalactiae* and *M. bovis* to design primers MAGAUVRC1-L and MAGAUVRC1-R that are specific for *M. agalactiae*, thus suggesting its utilization as a target in a species-specific PCR assay *(16)* (*see* **Note 6**).The following temperature–time profile should be used in 35 cycles: denaturation at 94 °C for 30 s, annealing at 50 °C for 30 s, extension at 72 °C for 60 s. The size of the amplicon is 1624 bp.

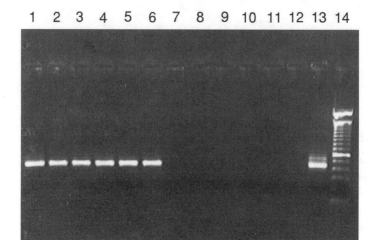

Fig. 1 Detection of *M. bovis* from tissue of different lung lobes and fluid samples of an experimentally infected calf using the nested *opp*D/F assay. For DNA extraction, the High Pure PCR Template Preparation Kit was used. The eluted DNA fraction was processed as described in **Subheading 3.1.3.2., steps 15–19**. Lane 1, accessory lobe; lane 2, left diaphragmatic lobe; lane 3, right diaphragmatic lobe; lane 4, right apical cranial lobe; lane 5, left apical cranial lobe; lane 6, endo/pericardium; lane 7, cerebrospinal fluid; lane 8, left carpal joint fluid; lane 9, right carpal joint fluid; lane 10, left tarsal joint fluid; lane 11, right tarsal joint fluid; lane 12, reagent control; lane 13, reference strain PG 45 of *M. bovis*; lane 14, DNA marker, 100-bp ladder.

3.2.5. Detection of M. mycoides *subsp.* mycoides SC (MmmSC)

The following nested PCR is particularly suitable for the detection of *Mmm*SC from bronchial lavage fluid samples prepared as described under **Subheading 3.1.2.** The first amplification reaction is done using 5 µL of DNA extract and primer pair SC3NEST1-L/SC3NEST1-R with 35 cycles of 94°C for 30 s, 52°C for 30 s, 72°C for 30 s. The resulting fragment is 716 bp long. The second round uses 1 µL of the product of the first reaction as a template and primer pair SC3VII/SC3IV-S in 35 cycles with the same temperature–time program. For species identification from culture, a single-step amplification with either of the two primer pairs is sufficient. Important general advice is given in **Note 7**.

3.2.6. Detection of M. mycoides *subsp.* mycoides LC (MmmLC) and M. mycoides *subsp.* capri

A detection system for these caprine mycoplasmas was developed on the basis of their *lpp*A genes, which encode a 62-kDa surface lipoprotein *(20)*.

Table 3
Temperature-Time Profiles of *M. conjunctivae* Detection Assays

	PCR 1[a] MOLIGEN1-L/16SUNI	PCR 2[b] McoF1/McoR1
Primers		
Denaturation	94°C for 30 s	94°C for30 s
Primer annealing	51°C for 30 s	54°C for 30 s
Primer extension	72°C for 60 s	72°C for 60 s
Number of cycles	35	35
Amplicon size	1063 bp	748 bp

[a]PCR 1, followed by PCR 2, may be used as a nested procedure.
[b]PCR 2 is specific for *M. conjunctivae*.

Using primer pair MMMLC2-L/MMMLC1-R, a 1049-bp fragment is amplified (*see* **Note 8**).The recommended cycling program is: denaturation 94°C for 30 s, annealing 49°C for 30 s, extension 72°C for 60 s.

3.2.7. Detection of M. conjunctivae

Species-specific identification of *M. conjunctivae* can be accomplished using a 16S rDNA-based assay *(5)*, which can be run as a one-round or nested PCR. In the former, primer pair McoF1/McoR1 is used to produce an amplicon of 748 bp. In the nested assay, the outer primers MOLIGEN1-L/16SUNI are used in the first round of amplification, and McoF1/McoR1 represent the inner primer pair. The detection limit of the nested PCR was determined to be at 20 cfu per swab *(5)*.

PCR amplification programs and amplicon sizes are given in **Table 3**.

3.3. Electrophoresis and Visualization

1. Prepare 1 or 2% (w/v) solution of agarose in TBE. Store gel in Erlenmeyer flasks at room temperature.
2. Liquefy gel by microwave heating (approx 30 s at 600 W) prior to use.
3. Pour gel on a horizontal surface using an appropriately sized frame.
4. Fill electrophoretic tank with TBE.
5. Run the gel at a voltage corresponding to 5 V/cm of electrode distance for approx 30 min.
6. Load each well with 10 μL of PCR product mixed with 5 μL GLB.
7. Stain DNA bands by immersing the gel in ethidium bromide solution containing 5 μL stock solution in 200 mL of water. (Alternatively, ethidium bromide-containing agarose gels can be used. Add 5 μL of ethidium bromide solution to 100 mL of melted agarose in TBE buffer.)
8. Visualize bands under UV light using a transilluminator.

4. Notes

1. The use of DNA preparation kits can be recommended for samples of organ tissue, broth culture, and with some qualification, also for semen and feces. In the latter instances, the kit should be tested with a series of spiked samples containing defined numbers of mycoplasma cells in order to examine its suitability.

    Most commercial kits are easy to work with. They contain a special buffer reagent for lysis of the bacterial and tissue cells, the effectiveness of which is decisive for the kit's performance. An optional RNase digestion is intended to remove cellular RNA. The lysate is then centrifuged through a mini-column, where the released DNA is selectively bound to a solid phase (modified silica, hydroxyl apatite, or filter membrane). After washing, the DNA can be eluted with an elution buffer or water. DNA prepared in this manner is usually of high purity and largely free of PCR inhibitors.

    It should be noted, however, that the yield of extracted DNA is limited by the binding capacity of the mini-column. If maximum recovery of mycoplasmal DNA is important, e.g., in quantitative assays or for preparation of reference DNA, an alternative extraction protocol should be followed.

2. Extraction of bacterial DNA from milk samples seems to be particularly difficult in the case of mycoplasmas. This may be due to adhesive properties and/or the small size of mycoplasma cells. In a previous study, several extraction procedures were compared, which included cold and hot phenol extraction, Tween 20 treatment, protease digestion, commercial DNA extraction kits, and, finally, a combination of enzymatic treatment and selective membrane binding of DNA *(17)*. Using the latter, 50–500 cfu of *M. bovis*/mL milk were detectable. The present method of pre-PCR enrichment by antigen capture allowed an improvement of the detection limit up to 20 cfu/mL milk when amplification products were visualized on agarose gels, and 2 cfu/mL when visualized on Southern blots *(24)*.

3. Several sets of primers have been designed that detect all members of the *M. mycoides* cluster, i.e., *M. mycoides* subsp. *mycoides* SC, *M. mycoides* subsp. *mycoides* LC, *M. mycoides* subsp. *capri*, *M. capricolum* subsp. *capricolum*, *M. capricolum* subsp. *capripneumoniae*, *Mycoplasma* bovine group 7. Those described by Bashiruddin et al. *(25)*, Dedieu et al. *(26)*, Hotzel et al. *(23)*, and Rodriguez et al. *(27)* are derived from the same set of sequence data from the CAP-21 genomic region, and one or more of them could be used additionally for the confirmation of results of further differentiation.

4. The target region of the primers PpMB920-1/2 and PpSM5-1/2, used for the nested PCR system, is the operon of oligopeptide permease *(opp)* genes encoding ATP-binding proteins, which are members of the so-called ABC-transporter family. The *opp*D and *opp*F genes of *M. bovis* were identified on the basis of sequence homology to the analogous genes of *M. hominis*.

5. *M. bovis*-specific primer systems MBOUVRC2-L/R and PpMB920-1/2 were tested together with other assays in a ring trial in COST Action 826. Both produced identical results with field isolates from different origins and surpassed other PCR test systems. All field isolates of *M. bovis* tested so far were found to

produce a positive signal, whereas cross-amplification of DNA from other mycoplasma species was not observed. Primer pair MBOUVRC2-L/R was derived from *uvr*C, the gene encoding deoxyribodipyrimidine photolyase, which is involved in DNA repair. It is of particular value when differentiation between the closely related species *M. bovis* and *M. agalactiae* has to be conducted.

6. Primer system MAGAUVRC1-L/R was also tested in the same ring trial and demonstrated to be specific for *M. agalactiae* only.

7. In view of the high importance of CBPP caused by *M. mycoides* subsp. *mycoides* SC, we strongly recommend the additional use of an alternative PCR method to confirm the diagnostic results as well as the species identity **(refs. *21,23,25,26,* and *28)*.**

8. In a previous study, the characteristic product of 1049 bp was obtained with many field strains of different geographic origin *(20)*. The system was also shown to be specific, as no cross-amplification of other mycoplasmas, notably the closely related members of the *M. mycoides* cluster, was observed. However, several atypical strains with deletions or mutations in the target sequence were identified *(20)*. Furthermore, some strains that were serologically typed as *M. mycoides* subsp. *mycoides* SC failed to produce an amplicon with these primers *(29)*. In these cases, sequencing the 16S rRNA gene region may be an alternative approach to identify the isolates. However, a recent study showed that several strains serologically typed as M. *mycoides* subsp. *mycoides* LC had 16S rRNA sequences that were identical or similar to those of other mycoplasma species and were negative with the *lpp*A-based PCR identification method *(30)*.

References

1. Pfützner, H., and Sachse K. (1996) *Mycoplasma bovis* as an agent of mastitis, pneumonia, arthritis and genital disorders in cattle. *Rev. Sci. Tech.* **15,** 1477–1494.

2. Bergonier, D., Berthelot, X., and Poumarat, F. (1997) Contagious agalactia of small ruminants: current knowledge concerning epidemiology, diagnosis and control. *Rev. Sci. Tech. Off. Int. Epiz.* **16,** 848–873.

3. Nicholas, R. A., and Bashiruddin, J. B. (1995) *Mycoplasma mycoides* subspecies *mycoides* (small colony variant): the agent of contagious bovine pleuropneumonia and member of the "Mycoplasma mycoides cluster". *J. Comp. Pathol.* **113,** 1–27.

4. Rodriguez, J. L., Poveda, J. B., Oros, J., Herraez, P., Sierra, M. A., and Fernandez, A. (1995) High mortality in goats associated with the isolation of a strain of *Mycoplasma mycoides* subsp. *mycoides* (large colony type). *Zentralbl. Veterinärmed. B* **42,** 587–593.

5. Giacometti, M., Nicolet, J., Johansson, K. -E., Naglic, T., Degiorgis, M. P., and Frey, J. (1999) Detection and identification of *Mycoplasma conjunctivae* in infectious keratoconjunctivitis by PCR based on the 16S rRNA gene. *Zentralbl. Veterinärmed. B* **46,** 173–180.

6. Johansson, K. -E., Heldtander, M. U., and Pettersson, B. (1998) Characterization of mycoplasmas by PCR and sequence analysis with universal 16S rDNA primers

in *Mycoplasma Protocols*, (Miles, R. J., and Nicholas, R. A. J., eds.), *Methods in Molecular Biology, vol. 104,* Humana Press, Totowa, NJ, USA, pp. 145–165,

7. Wirth, M., Berthold, E., Grashoff, M., Pfützner, H., Schubert, U., and Hauser, H. (1994) Detection of mycoplasma contaminations by the polymerase chain reaction. *Cytotechnology* **16**, 67–77.

8. Hotzel, H. and Sachse, K. (1998) Improvement and acceleration of the diagnosis of contagious bovine pleuropneumonia by direct detection of the pathogen using the polymerase chain reaction. *Berl. Münch. Tierärztl. Wochenschr.* **111**, 268–272.

9. Chávez González, Y. R., Ros Bascunana C., Bölske, G., Mattsson, J. G., Fernández Molina, C., and Johansson, K. -E. (1995) In vitro amplification of the 16S rRNA genes from *Mycoplasma bovis* and *Mycoplasma agalactiae* by PCR. *Vet. Microbiol.* **47**, 183–190.

10. Bölske, G., Mattsson, J. G., Ros Bascunana, C., Bergström, K., Wesonga, H., and Johansson, K. -E. (1996) Diagnosis of contagious caprine pleuropneumonia by detection and identification of *Mycoplasma capricolum* subsp. *capripneumoniae* by PCR and restriction enzyme analysis. *J. Clin. Microbiol.* **34**, 785–791.

11. Kobayashi, H., Hirose, K., Worarach, A., et al. (1998) In vitro amplification of the 16S rRNA genes from *Mycoplasma bovirhinis, Mycoplasma alkalescens* and *Mycoplasma bovigenitalium* by PCR. *J. Vet. Med. Sci.* **60**, 1299–1303.

12. Persson, A., Pettersson. B., Bölske, G., and Johansson, K.-E. (1999) Diagnosis of contagious bovine pleuropneumonia by PCR-laser-induced fluorescence and PCR-restriction endonuclease analysis based on the 16S rRNA genes of *Mycoplasma mycoides* subsp. *mycoides* SC. *J. Clin. Microbiol.* **37**, 3815–3821.

13. Harasawa, R. (1999) Genetic relationships among mycoplasmas based on the 16S–23S rRNA spacer sequence. *Microbiol. Immunol.* **43**, 127–132.

14. Harasawa, R., Hotzel, H., and Sachse, K. (2000) Comparison of 16S–23S rRNA intergenic spacer regions among strains of the *Mycoplasma mycoides* cluster and reassessment of the taxonomic position of *Mycoplasma* bovine group 7. *Int. J. Syst. Evol. Microbiol.* **50**, 1325–1329.

15. Scheinert, P., Krausse, R., Ullmann, U., Söller, R., and Krupp, G. (1996) Molecular differentiation of bacteria by PCR amplification of the 16S–23S rRNA spacer. *J. Microbiol. Meth.* **26**, 103–117.

16. Subramaniam, S., Bergonier, D., Poumarat, F., et al. (1998) Species identification of *Mycoplasma bovis* and *Mycoplasma agalactiae* based on the *uvr*C genes by PCR. *Mol. Cell. Probes.* **12**, 161–169.

17. Hotzel, H., Sachse, K., and Pfützner, H. (1996) Rapid detection of *Mycoplasma bovis* in milk samples and nasal swabs using the polymerase chain reaction. *J. Appl. Bacteriol.* **80**, 505–510.

18. Pinnow, C. C., Butler, J. A., Sachse, K., Hotzel, H., Timms, L. L., and Rosenbusch, R. F. (2001) Detection of *Mycoplasma bovis* in preservative-treated field milk samples. *J. Dairy Sci.* **84**, 1640–1645.

19. Miserez, R., Pilloud, P., Cheng, X., Nicolet, J., Griot, C., and Frey, J. (1997) Development of a sensitive nested PCR method for the specific detection of *Mycoplasma mycoides* subsp. *mycoides* SC. *Mol. Cell. Probes* **11**, 103–111.

20. Monnerat, M. -P., Thiaucourt, F., Poveda, J. B., Nicolet, J., and Frey, J. (1999) Genetic and serological analysis of lipoprotein LppA in *Mycoplasma mycoides* subsp. *mycoides* LC and *Mycoplasma mycoides* subsp. *capri. Clin. Diagn. Lab. Immunol.* **6,** 224–230.

21. Vilei, E.M., Nicolet, J. and Frey, J. (1999) IS*1634*, a novel insertion element creating long, variable-length direct repeats, which is specific for *Mycoplasma mycoides* subsp. *mycoides* SC small-colony type. *J. Bacteriol.* **181,** 1319–1323.

22. Sachse, K., Diller, R., and Hotzel. H. (2002) Analysis of a signature region in the 23S ribosomal RNA gene of *Mycoplasma* and *Acholeplasma* spp., a potential target region for identification and differentiation at class, genus, cluster, and species levels. *Syst. Appl. Microbiol.,* in press.

23. Hotzel, H., Sachse, K., and Pfützner, H. (1996) A PCR scheme for differentiation of organisms belonging to the *Mycoplasma mycoides* cluster. *Vet. Microbiol.* **49,** 31–43.

24. Hotzel, H., Heller, M., and Sachse, K. (1999) Enhancement of *Mycoplasma bovis* detection in milk samples by antigen capture prior to PCR. *Mol. Cell. Probes* **13,** 175–178.

25. Bashiruddin, J. B., Taylor, T. K., and Gould, A. G. (1994) A PCR-based test for the specific identification of *Mycoplasma mycoides* subspecies *mycoides* SC. *J. Vet. Diagn. Invest.* **6,** 428–434.

26. Dedieu, L., Mady, V. and Lefèvre, P. C. (1994) Development of a selective polymerase chain reaction assay for the detection of *Mycoplasma mycoides* subsp. *mycoides* SC (contagious bovine pleuropneumonia agent). *Vet. Microbiol.* **42,** 327–339.

27. Rodriguez, J. L., Ermel, R. W., Kenny, T. P., Brooks, D. L., and DaMassa, A. J. (1997) Polymerase chain reaction and endonuclease digestion of selected members of the "*Mycoplasma mycoides* cluster" and *Mycoplasma putrefaciens. J. Vet. Diagn. Invest.* **9,** 186–190.

28. Nicholas, R., Bashiruddin, J., Ayling, R., and Miles, R. (2000) Contagious bovine pleuropneumonia—a review of recent developments. *Vet. Bull.* **70,** 827–838.

29. Naglic, T., Hotzel, H., Ball, H. J., Seol, B., and Busch, K. (2001) Studies on the aetiology of caprine mycoplasmosis in Croatia, in *Mycoplasmas of Ruminants: Pathogenicity, Diagnostics, Epidemiology and Molecular Genetics* (Poveda, J. B., Fernandez, A., Frey, J., and Johansson, K.-E., eds.) EUR 19693–COST Action 826, vol. 5, EC, Brussels, Belgium, pp. 137–140.

30. Frey, J. (unpublished results).

17

Detection of *Mycoplasma hyopneumoniae* from Clinical Samples and Air

Marylène Kobisch and Joachim Frey

1. Introduction

Mycoplasma hyopneumoniae is the etiological agent of enzootic pneumonia, a worldwide disease that causes economic losses in swine production *(1)*. Generally, transmission of *M. hyopneumoniae* occurs by direct contact or aerosol in chronically infected herds when young susceptible pigs are in contact with older pigs. Piglets can be infected by gilts, low parity sows, but also by older sows *(2)*. According to these authors, the percentage of sows carrying *M. hyopneumoniae* decreased with age. However, carrier sows represent the most likely source of transmission, of *M. hyopneumoniae* to their piglets. Another risk of *M. hyopneumoniae* contamination is due to airborne transmission, which can occur over short distances inside the herd *(3)* and between herds *(4)*. In these circumstances, specific and sensitive tests are necessary to control enzootic pneumonia.

M. hyopneumoniae is one of the mycoplasmas most difficult to isolate, cultivate, and identify. Thus, isolation of this species is generally not performed by diagnostic laboratories. In routine bacteriological diagnosis, detection of *M. hyopneumoniae* infection is carried out by an immunofluorescence test or by serology *(5,6)*. More recently, several polymerase chain reaction (PCR) or nested PCR assays have been developed to detect *M. hyopneumoniae* in the respiratory tract of pigs at necropsy or from live pigs in field conditions. Mattsson et al. *(7)* and Sorensen et al. *(5)* have described a PCR test to detect *M. hyopneumoniae* in nasal cavities. Baumeister et al. *(8)* and Blanchard et al. *(9)* have developed PCR methods for the detection of *M. hyopneumoniae* in bronchoalveolar fluids or tracheobronchiolar washings. However, in order to increase the sensitivity of the method, Verdin et al. *(10)* introduced a nested

From: *Methods in Molecular Biology, vol. 216: PCR Detection of Microbial Pathogens: Methods and Protocols*
Edited by: K. Sachse and J. Frey © Humana Press Inc., Totowa, NJ

PCR assay to detect *M. hyopneumoniae* in tracheobronchiolar washings from pigs and from lung samples. Using this method, they showed that samples of pigs at 2 mo of age, in the post-weaning period, gave positive results with the nested PCR, while only a few sera of the same animals were seropositive as measured by enzyme-linked immunosorbent assay (ELISA). Moreover, these authors showed that the nested PCR test was able to detect *M. hyopneumoniae* at necropsy in lungs of pigs without signs of pneumonia. Here, we propose nested PCR methods as very sensitive tools to detect *M. hyopneumoniae* in diseased pigs as well as in healthy carrier animals which play an important role in transmission of *M. hyopneumoniae*. Furthermore we propose a method for air sampling combined with an alternative nested PCR assay, which allows the detection of *M. hyopneumoniae* in the air of pig housings or in expectoration of coughing pigs.

The nested PCR method "ABC" using the primer pairs Hp1/Hp3 and Hp4/Hp6 is based on a putative ABC transporter gene, which is specific to *M. hyopneumoniae (11)*. The PCR product is detected by Southern blot hybridization, with a radioactively labeled oligonucleotide matching the PCR product in order to enhance the sensitivity of detection to a very low number of *M. hyopneumoniae* organisms per sample and provide an additional verification of the amplified product. The second nested PCR method REPhyo using the primer pairs MHP950-1L/MHP950/1R and MHP950-2L/MHP950/2R is based on a repeated DNA fragment which is specific to the species *M. hyopneumoniae (3)* and is detected directly by agarose gel electrophoresis and photography of ethidium bromide-stained gels.

2. Materials

1. *M. hyopneumoniae* reference strain NCTC 10110 cells at 10^8 cells/mL in phosphate-buffered saline (PBS).
2. Heating block for Eppendorf® tubes at 57°, 60°, and at 95°C.
3. Vortex blender.
4. 20-mL syringes.
5. Catheter to introduce and aspirate the tracheobronchiolar fluids (1-mm diameter and 1.5-m long).
6. Plastic tubes to collect tracheobronchiolar fluids.
7. Polyethersulfone membrane filters, Ø = 47-mm, pore size = 0.2 µm, (Supor 200; Gelman Sciences, Ann Arbor, MI, USA).
8. Filter holder device for Ø = 47-mm filters.
9. Air pump with tubing connected to filter holder (*see* **Fig. 1**).
10. Centrifuge with rotor (12,000–13,000*g*) for Eppendorf tubes.
11. Safety cabinet for PCR preparations (Clean Spot; Coy Laboratories, Grass Lake, MI, USA).
12. Cotton swabs.
13. PBS buffer: 50 m*M* Na-phosphate, pH 7.5, 150 m*M* NaCl.

Fig. 1. Sampling of air from expectoration of a coughing pig. The device with the filter is held at a distance of 10–20 cm in front of the mouth of a possibly coughing pig, and air is pumped during 1–2 min (courtesy of Dr. Katharina Stärk, IVI, Mittelhäusen).

14. TE buffer: 10 m*M* Tris-HCl, 1 m*M* EDTA, pH 8.5.
15. PCIA: phenol:chloroform:isoamylalcohol (49.5:49.5:1; v/v/v).
16. 3M Sodium-acetate, pH 4.5.
17. 80% Ethanol, prechilled in freezer.
18. Instagene matrix (Bio-Rad, Hercules, CA, USA).
19. Oligonucleotide primers, each 10 µM (*see* **Table 1**).
20. 10 X PCR buffer: 10 m*M* Tris-HCl, pH 8.3, 1.5 m*M* MgCl$_2$, 50 m*M* KCl, 0.005% Tween® 20-detergent.
21. Lysis buffer: 100 m*M* Tris-HCl, 0.05% Tween 20 detergent, 0.2 mg/mL proteinase K, pH 8.5.
22. 100 m*M* dATP, 100 m*M* dCTP, 100 m*M* dGTP, 100 m*M* dTTP.
23. Oligonucleotides, 10 µM (*see* **Table 1**).
24. *Taq* DNA polymerase, 5 U/µL.
25. Thermal cycler and corresponding thin-walled tubes for PCR applications.
26. 10X Kinase buffer: 0.5 M Tris-HCl, 0.1 M MgCl$_2$, 50 m*M* dithithreitol, 1 m*M* spermidine HCl, 1 m*M* EDTA, pH7.6.
27. Bacteriophage T4-polynucleotide kinase (10 U/µL).
28. [γ-32P]ATP (1000 Ci/mmol, 10 µCi/µl).

Table 1
Oligonucleotide Primers[a]

Name	Utilization	Sequence 5'–3' direction	Annealing temperature	Fragment size
Hp1	1st step L- primer	TTCAAATTATAACCTCGGTC	57°C	
Hp3	1st step R- primer	AGCAAATTTAGTCTCTCTGC	57°C	
Hp4	2nd step L- primer	CGCTTTAGTACCGATATGGG	58°C	702 bp
Hp6	2nd step R- primer	GCCATTCGCTTATATGGTGA	58°C	
Hp42	hybridization	ACTGCCCCAAATGGAACAGG		
MHP950-1L	1st step L- primer	AGGAACACCATCGCGATTTTTA	52°C	
MHP950-1R	1st step R- primer	ATAAAAATGGCATTCCTTTTCA	52°C	
MHP950-2L	2nd step L- primer	CCCTTTGTCTTAATTTTTGAA	52°C	807 bp
MHP950-2R	2nd step R- primer	GCCGATTCTAGTACCCTAATCC	52°C	

[a] Based on the nucleotide sequence of GenBank Accession no. AF004388

29. 10X SSC buffer, 1X SSC is: 15 mM Na-citrate, 150 mM NaCl.
30. 0.5 M EDTA.
31. Agarose gel equipment including agarose, TBE running buffer (90 mM Tris-base, 90 mM boric acid, 2 mM EDTA, pH 8.3), gel loading buffer, ethidium bromide, DNA size marker, as well as photographic equipment for documentation. For more details *see* **refs.** *12* and *13* and other chapters of this volume.
32. Equipment for Southern blot hybridization.
33. Autoradiography film for detection of radioactively labeled DNA probes.

3. Methods

The methods described below are aimed at *(i)* the detection of *M. hyopneumoniae* in tracheobronchiolar washings; *(ii)* the detection of *M. hyopneumoniae* in lung tissue samples; and *(iii)* the detection of *M. hyopneumoniae* from air samples from pig housing or expectoration of coughing pigs.

3.1. Detection of M. hyopneumoniae in Tracheobronchiolar Washings

3.1.1. General Remarks

Detection of *M. hyopneumoniae* in tracheobronchiolar washings, done on non-anesthetized pigs, requires some experience in handling with pigs. Lung lavages obtained can be kept frozen until analysis by PCR in the laboratory. After extraction of DNA from the samples *(14)*, detection of *M. hyopneumoniae* DNA is done preferentially by nested PCR "ABC" with the primer pairs Hp1/Hp3 and Hp4/Hp6, followed by a detection step involving a radioactively labeled oligonucleotide (*see* **Subheadings 3.4.1., 3.4.2.,** and **3.4.4.**) in order to warrant highest possible sensitivity (*see* **Notes 1** and **2**). It is recommended to prepare 2 positive control samples by artificially contaminating 1 mL of tracheobronchiolar lavage fluids from healthy pigs with 1000 and with 100 colony forming units (cfu) of *M. hyopneumoniae* NCTC 10110. In addition, at least 2 negative control samples from healthy pigs should be included.

3.1.2. Tracheobronchiolar Lavage Fluids

1. Open the mouth of nonanaesthetized pigs.
2. Introduce the catheter into the trachea of pigs.
3. Introduce 10–20 mL (according to the age of the pig) of PBS buffer.
4. Aspirate the fluid immediately (approx 1–2 mL).
5. Keep the lavage fluid until further processing at –20°C.

3.1.3. Extraction of DNA from Tracheobronchiolar Lavage Fluids

1. Centrifuge 1-mL samples from tracheobronchiolar lavage fluid at 12,000g for 30 min.
2. Resuspend the pellets in 1 mL PBS.
3. Centrifuge at 12,000g for 30 min.
4. Resuspend pellets in 0.5 mL lysis buffer.
5. Incubate 30 min at 60°C in heating block.
6. Incubate 10 min at 95°C in heating block.
7. Cool on ice.
8. Extract 2× with 0.5 mL PCIA.
9. Separate aqueous phase carefully.
10. Add 50 µL of 3M sodium-acetate, pH 4.5.
11. Add 0.4 mL isopropanol.
12. Cool on ice for 10 min.
13. Centrifuge at 12,000g for 10 min.
14. Remove liquid.
15. Wash pellet with 80% ethanol and dry it.
16. Resuspend the final DNA pellet in 100 µL of sterile deionized water.
17. Use 5 µL of this preparation as a template for the nested PCR method, ABC, with primers Hp1/Hp3 and Hp4/Hp6 and Southern blot hybridization (*see* **Subheadings 3.4.1., 3.4.2., 3.4.4.,** and **Note 2**).

3.2. Detection of M. hyopneumoniae *in Lung Tissue*

For the detection of *M. hyopneumoniae* in lung tissue, DNA of the samples is recovered using Instagene matrix prior to the use as a template for the nested PCR method REPhyo, using primers MHP950-1L/MHP950-1R and MHP950-2L/MHP950-2R which are based on repetitive elements specific to *M. hyopneumoniae (3)* (*see* **Note 1**).

3.2.1. Preparation of Template from Lung Tissue (see **Note 1**)

M. hyopneumoniae can be detected directly from pathological lung lesions by PCR. The preparation of the template DNA is done using the following method: it is recommended to prepare 2 positive control samples by artificial contamination of a 0.25-g lung sample with 1000 and with 100 cfu of *M. hyopneumoniae* NCTC 10110. In addition, 2 negative control samples from healthy pigs should be included.

1. Excise 0.25 g lung tissue from the edge of a lesion.
2. Place in a sterile tube and homogenize manually using a sterile pestle.
3. Add 0.5 mL PBS to facilitate homogenization.
4. Centrifuge for 5 s at 10,000*g* to remove large debris.
5. Retain 200 µL of the supernatant, discard the rest.
6. Centrifuge the supernatant for 5 min at 13,000*g*.
7. Keep the pellet and discard supernatant fluid.
8. Add 200 µL Instagene matrix and resuspend the pellet.
9. Heat to 56°C for 30 min.
10. Vortex mix briefly.
11. Heat to 95–97°C (or boil) for 8 min.
12. Vortex mix briefly.
13. Centrifuge for 5 min at 13,000*g*.
14. Keep the supernatant and transfer to separate tube.
15. Use 5 µL of the supernatant as a template for PCR REPhyo with primers MHP950-1L/ MHP950-1R and MHP950-2L/ MHP950-2R (*see* **Subheadings 3.4.1., 3.4.2.,** and **3.4.3.**).

3.3. Detection of M. hyopneumoniae *from Air*

Detection of *M. hyopneumoniae* from the air requires the analysis of a large air vol by pumping air through a micro-mesh membrane filter. The DNA is then extracted by dissolving the filter in organic solvents and extracting the DNA by aqueous buffers therefrom. The extracted DNA is used as a template for the nested PCR method REPhyo using primers MHP950-1L/MHP950-1R and MHP950-2L/MHP950-2R.

1. Assemble the polyethersulfone membrane filter in a filter holder and connect to air pump.
2. Pump air through filter at a rate of 20 L/min. To control the air in pig housings, pump for 100 min (2000 L), and to control individual pigs, hold the filter close to the nose of the pig and pump for 1–2 min (20–40 L) (*see* **Fig. 1**).
3. Remove the filter from the filter holder and place in a petri dish to dry for 1–2 h.
4. Fold filter and place it in a 10-mL reaction tube, which is chloroform-resistant (glass tube, or Falcon plastic tube); (Becton Dickinson, Lincoln Park, NJ, USA.).

5. Dissolve filter in 5 mL chloroform.
6. Add 5 mL TE buffer and shake vigorously to extract DNA.
7. Separate organic from aqueous phase by centrifugation at 10,000g for 10 min.
8. Recover aqueous phase and transfer it to a separate tube.
9. Extract aqueous phase with 5 mL of phenol-chloroform-isoamyl alcohol (49.5: 49.5: 1; v/v/v).
10. Recover aqueous phase and add 8 mL chilled ethanol and 400 µL 3 M sodium-acetate, pH 5.5.
11. Mix well and cool at −20°C for 10 min.
12. Recover precipitated DNA by centrifugation at 10,000g for 20 min.
13. Dry the pellet.
14. Resuspend pellet in 50 µL TE buffer
15. Use 5 µL of the supernatant as a template for PCR REPhyo with primers MHP950-1L/MHP950-1R and MHP950-2L/MHP950-2R (*see* **Subheadings 3.4.1., 3.4.2.,** and **3.4.3.**).

3.4. PCR Amplification

3.4.1. PCR Premixture (for 20 Reactions)

Label tube with corresponding primer pair (e.g., Hp1/ Hp3 or MHP950-1L/ MHP950-1R). Add: 843 µL H$_2$O bidistilled, 100 µL 10X PCR buffer, 1.7 µL dATP 100 mM, 1.7 µL dCTP 100 mM, 1.7 µL dGTP 100 mM, 1.7 µL dTTP 100 mM, 25 µL 10 µM Primer-L (*see* **Table 1**), and 25 µL 10 µM Primer-R (*see* **Table 1**).

3.4.2. Nested PCR

Two highly sensitive 2-step nested PCR methods are used. Method ABC uses the oligonucleotide primer pairs Hp1/Hp3 and Hp4/Hp6, followed by detection of the PCR product by Southern blot hybridization with a radioactively labeled oligonucleotide. Method REPhyo uses primer pairs MHP950-1L/MHP950-1R and MHP950-2L/MHP950-2R and uses direct detection of the PCR products by agarose gel electrophoresis. Particular care has to be taken for both methods in order to avoid any cross-contamination by spillover and by aerosols originating from samples containing PCR amplicons, especially from the first amplification step. Care must also be taken when opening PCR tubes (avoid aerosol formation). The preparation of the second step, involving the pipeting of the PCR amplicon of the first step, should be done, if ever possible, in a separate safety cabinet for PCR preparations, which should be irradiated by UV light after each use. It is strongly recommended to include both positive and negative control samples prepared analogously to the clinical samples.

First round of amplification:

1. Use thin-walled thermal cycler tubes precooled at 4°C.
2. Add for each first-round reaction: 45.0 μL PCR premixture with corresponding primers, 0.25 μL *Taq* DNA polymerase, and 5.0 μL template.
3. Preheat thermal cycler to 95°C.
4. Place the cooled tubes directly into preheated thermal cycler.
5. Run 35 cycles with: 30 s at 94°C, 30 s at annealing temperature (according to **Table 1**), and 45 s at 72°C.

Second round of amplification:

6. Prepare the second round in a precooled thin-walled tube by mixing: 45.0 μL PCR premixture with corresponding primers, 0.25 μL *Taq* DNA polymerase, and 1 μL amplicon of the first round.
7. Preheat thermal cycler to 95°C.
8. Place the cooled tubes directly into the preheated thermal cycler.
9. Run 35 cycles with: 30 s at 94°C, 30 s at annealing temperature (according to **Table 1**), and 30 s at 72°C.
10. Take the tubes to a different room for analysis.
11. Proceed to either direct visualization of the PCR products on agarose gel electrophoresis (*see* **Subheading 3.4.3.**) or Southern blot hybridization with a radioactively labeled probe (*see* **Subheading 3.4.4.**).

3.4.3. Direct Visualization of PCR Products on Agarose Gels

1. Analyze 10 μL of each PCR amplicon on a 0.7% agarose gel.
2. Stain the gel with ethidium bromide and photograph fluorescence by UV light.
3. Analyze the correct sizes of the bands (*see* **Table 1**) and verify all negative and positive controls.

3.4.4. Analysis of PCR Products by Southern Blot Hybridization

In order to insure a very high sensitivity of the nested PCR method with primer pairs Hp1/Hp3 and Hp4/Hp6, which is used preferentially for the amplification of samples from tracheobronchiolar washings, the products of the second PCR amplification are submitted to agarose gel electrophoresis, followed by transfer on a nylon membrane and hybridization with the radioactively labeled primer Hp43.

3.4.4.1. LABELING OF THE OLIGONUCLEOTIDE PRIMER

1. Prepare kinase reaction for labeling primer Hp43: 2 μL (20 pmol) Oligonucleotide primer Hp43 (10 μM), 5 μL 10X Kinase buffer, 5 μL (total ATP minimal 1 μM) [γ-^{32}P]ATP (1000 Ci/mmol, 10 μCi/μL), 2 μL bacteriophage T4-polynucleotide kinase (10 U/μL), and 46 μL H$_2$O.
2. Incubate 30 min at 37°C.

3. Add 2 µL of 0.5 M EDTA.
4. Extract once with 50 µL PCIA.
5. Purify labeled DNA from unincorporated ATP by chromatography over a Sephadex® G25 column.
6. Precipitate labeled oligonucleotide with 80% cooled ethanol.
7. Resuspend in 50 µL H_2O.

3.4.4.2. SOUTHERN BLOT HYBRIDIZATION

1. Analyze 10 µL of PCR amplicon from the nested PCR ABC on a 0.7% agarose gel.
2. Transfer DNA onto a positively charged nylon membrane using alkaline transfer with 0.4 M NaOH.
3. Wash the membrane in 10 mL hybridization buffer.
4. Incubate the membrane in 5 mL hybridization buffer plus 10 µL labeled Hp43 primer at 42°C for 16 h.
5. Wash membrane 4× in 2X SSC buffer supplemented with 0.5% SDS at 60°C.
6. Wash membrane 2× in 1X SSC buffer supplemented with 0.5% SDS at 60°C.
7. Dry membranes and expose 3–10 h for autoradiography
8. Analyze hybridization results for presence of the specific band (*see* **Table 1**).

4. Notes

1. The sensitivity of the method for detection of *M. hyopneumoniae* in tracheobronachiolar washings using the nested PCR method "ABC" with primers Hp1/Hp3 and Hp4/Hp6, followed by Southern blot hybridization with the radioactively labeled probe Hp42, was reported to reach as few as 1 cfu/PCR *(10)*. This corresponds to at least 20 cfu/mL tracheobronchiolar washing, assuming that no loss occurred during extraction of DNA. When tracheobronchiolar lavage fluid, artificially contaminated with *M. hyopnemoniae* strain NCTC 10110, were analyzed using the same extraction method, but using the nested PCR REPhyo with the primer pairs MHP950-1L/MHP950-1R and MHP950-2L/MHP950-2R and direct detection of PCR products (*see* **Subheading 3.3.4.**), the detection limit ranged between 100–200 cfu/mL tracheobronchiolar lavage or 100–500 cfu/g of lung tissue. The two nested PCR methods can, therefore, substitute for each other in the different applications. It has to be taken into account that the method with direct detection of the PCR products by agarose gel electrophoresis has a somewhat lower sensitivity, but is significantly easier to perform.
2. In order to detect possible failure of the PCR amplification due to PCR inhibitors or loss of DNA during the preparative purification steps, an internal control system can be used as described previously *(15)*.

Acknowledgments

We are grateful to Pia Wyssenbach for editorial help with the manuscript and to Yvonne Schlatter for valuable technical help. This work was supported by the Swiss Federal Veterinary Office.

References

1. Ross, R. F. (1999) Mycoplasmal diseases, in *Diseases of swine*, (Straw, B. E., D'Allaire, S., Mengeling, W. L., and Taylor, D. J., eds.), Iowa State University Press, IA, USA, pp. 495–510.
2. Calsamiglia, M., and C. Pijoan. (2000) Colonisation state and colostral immunity to *Mycoplasma hyopneumoniae* of different parity sows. *Vet. Rec.* **146,** 530–532.
3. Stark, K. D. C., Nicolet, J., and Frey, J. (1998) Detection of *Mycoplasma hyopneumoniae* by air sampling with a nested PCR assay. *Appl. Environ. Microbiol.* **64,** 543–548.
4. Goodwin, R. F. (1985) Apparent reinfection of enzootic-pneumonia-free pig herds: search for possible causes. *Vet. Rec.* **116,** 690–694.
5. Sorensen, V., Ahrens, P., Barfod, K., et al. (1997) *Mycoplasma hyopneumoniae* infection in pigs: Duration of the disease and evaluation of four diagnostic assays. *Vet. Microbiol.* **54,** 23–34.
6. Wallgren, P., Bolske,G., Gustafsson, S., Mattsson, S., and Fossum, C. (1998) Humoral immune responses to *Mycoplasma hyopneumoniae* in sows and offspring following an outbreak of mycoplasmosis. *Vet. Microbiol.* **60,** 193–205.
7. Mattsson, J. G., Bergstrom, K., Wallgren, P., and Johansson, K. E. (1995) Detection of *Mycoplasma hyopneumoniae* in nose swabs from pigs by in vitro amplification of the 16S rRNA gene. *J. Clin. Microbiol.* **33,** 893–897.
8. Baumeister, A. K., Runge, M., Ganter, M., Feenstra, A. A., Delbeck, F., and Kirchhoff, H. (1998) Detection of *Mycoplasma hyopneumoniae* in bronchoalveolar lavage fluids of pigs by PCR. *J. Clin. Microbiol.* **36,** 1984–1988.
9. Blanchard, B., Kobisch, M., Bove, J. M., and Saillard, C. (1996) Polymerase chain reaction for *Mycoplasma hyopneumoniae* detection in tracheobronchiolar washings from pigs. *Mol. Cell. Probes* **10,** 15–22.
10. Verdin, E., Saillard, C., Labbe, A., Bove, J. M., and Kobisch, M. (2000) A nested PCR assay for the detection of *Mycoplasma hyopneumoniae* in tracheobronchiolar washings from pigs. *Vet. Microbiol.* **76,** 31–40.
11. Blanchard, B., Saillard, C., Kobisch, M., and Bove, J. M. (1996) Analysis of putative ABC transporter genes in *Mycoplasma hyopneumoniae*. *Microbiology-UK* **142,** 1855–1862.
12. Ausubel, F. M., Brent, R., Kingston, R. E., et al. (1999) *Current Protocols in Molecular Biology*. John Wiley & Sons, New York, NY, USA.
13. Sambrook, J., Fritsch, E. F., and Maniatis, T. (1989) *Molecular Cloning. A Laboratory Manual*. CSH Laboratory Press, Cold Spring Harbor, NY, USA.
14. Marmur, J. (1961) A procedure for the isolation of deoxyribonucleic acid from micro-organisms. *J. Mol. Biol.* **3,** 208–218.
15. Verdin, E., Kobisch, M., Bove, J. M., Garnier, M., and Saillard, C. (2000) Use of an internal control in a nested-PCR assay for *Mycoplasma hyopneumoniae* detection and quantification in tracheobronchiolar washings from pigs. *Mol. Cell. Probes* **14,** 365–372.

18

PCR-Detection of *Hemophilus paragallinarum*, *Hemophilus somnus*, *Mannheimia (Pasteurella) hemolytica*, *Mannheimia* spp., *Pasteurella trehalosi*, and *Pasteurella multocida*

Henrik Christensen, Magne Bisgaard, Jesper Larsen, and John Elmerdahl Olsen

1. Introduction

1.1. Diseases of Veterinary Importance Caused by Hemophilus paragallinarum, Hemophilus somnus, Mannheimia hemolytica, Mannheimia spp., Pasteurella trehalosi and Pasteurella multocida

Most members of the bacterial family *Pasteurellaceae* are usually regarded as opportunistic secondary invaders, which under normal conditions might inhabit the mucosal membranes of the upper respiratory and lower genital tracts of mammals and birds *(1)*. Out of the almost 100 species or species-like taxa that might be isolated from mammals, reptiles, and birds, only *Pasteurella multocida*, *Actinobacillus pleuropneumoniae*, and *Hemophilus paragallinarum* are regarded as major pathogens *(1,2)* while *A. suis*, *H. somnus*, and *Mannheimia hemolytica* are considered as potential animal pathogens. The pathogenic potential is incompletely known for most taxa, probably due to limitations in their classification and identification, and inappropriate detection methods.

H. paragallinarum is responsible for infectious coryza in chickens characterized by relatively mild respiratory signs, non-thriving chickens, and significant reductions in egg production *(3)*. *H. somnus* may affect the respiratory and genital tracts of cattle and sheep, resulting in respiratory disease, reproductive problems, myocarditis, otitis, conjunctivitis, mastitis, arthritis, septicemia, and

From: *Methods in Molecular Biology, vol. 216: PCR Detection of Microbial Pathogens: Methods and Protocols*
Edited by: K. Sachse and J. Frey © Humana Press Inc., Totowa, NJ

thrombotic meningoencephalomyelitis *(4,5)*. Members of *M. hemolytica* are mainly associated with pneumonic pasteurellosis in feedlot cattle *(6)*. In sheep, pasteurellosis is mainly associated with *M. hemolytica*, other species of *Mannheimia*, or *P. trehalosi (7)*. Different lineages of *P. multocida* may be responsible for various diseases in both birds and mammals with hemorrhagic septicemia (HS) in cattle and buffalo, caused by certain members of serotypes B:2 and E:2 *(8)*, while progressive atrophic rhinitis (PAR) in pigs is caused by representatives of capsular types D or A *(9, 10)*. PAR has also been reported in goats and rabbits caused by *P. multocida (11,12)*. The toxin responsible for *P. multocida*, which caused PAR in pigs and other animals, is encoded by the *toxA* gene *(13)*. Members of *P. multocida* belonging to capsule types A and D may also cause pneumonia in pigs *(9)*. Fowl cholera is caused by certain pathogenic lineages of *P. multocida*, most frequently of capsular type A *(14)*.

1.2. Benefits of PCR Compared to Detection Based on Biochemical, Physiological and Serological Tests

Complex phenotypic and genotypic relationships in addition to complex growth requirements for *Pasteurellaceae* and lack of evaluated commercially available diagnostic kit-systems and diagnostic tables have made diagnostic improvements difficult within this family *(2)*. For the same reasons, and as these bacteria have been found to die in transport or be overgrown by other bacteria before they can be identified by culture-dependent methods, polymerase-chain reaction (PCR) tests may improve their identification *(3,4,9,14,16)*. For *H. paragallinarum*, the PCR test additionally was found specific for biochemical and serological variants *(3,17)*. For *H. somnus*, which shows extensive variability in morphology, biochemistry, and serology *(5,18)*, the PCR test was also found to be specific and to detect "*Hemophilus agni*" and "*Histophilus ovis*" *(19)*. Difficulties of isolating *H. somnus* from animals treated with antibiotics have been reported, and PCR might improve the diagnostic situation in such cases *(19, 20)*. For identification of *P. multocida*, PCR may further be preferred when isolates are nonpathogenic to mice *(16)*. The use of PCR also eliminates ethical problems related to the use of laboratory animals for isolation of these organisms. Finally, the development of a highly specific multiplex PCR assay for *P. multocida (21)* also adds important information on vaccine prophylaxis, because capsular types can be recognized easily without the use of the more laborious indirect hemagglutination test *(22)*.

PCR tests have been developed for the most important members of *Pasteurellaceae* in relation to their potential to cause disease, such as *A. pleuropneumoniae, H. paragallinarum, H. somnus, M. hemolytica, Mannheimia* spp., *P. trehalosi* , and *P. multocida*.

In this chapter, PCR detection of *H. paragallinarum, H. somnus, M. hemolytica, Mannheimia* spp., *P. trehalosi,* and *P. multocida* is described, while PCR tests for the detection of *A. pleuropneumoniae* are treated in **Chapter 5**. It might be relevant for diagnostic veterinary laboratories to set up PCR methods for detection of these different pathogens belonging to the family. In addition, a comparison of procedures will allow multiplex protocols to be set up. To date, a multiplex PCR test has only been reported for *P. multocida (21, 23)*.

1.3. Choice of Target DNA Sequence

The relations between target DNA sequences and specificity of the PCR tests are listed in **Table 1**. For all species except certain lineages of *P. multocida* causing arthropic rhinitis, the genetic background for virulence is insufficiently known. Not all lineages of the different species cause disease, and species-specific detection methods give no information about such differences. Some tests are based on the detection of genes encoding virulence factors in order to predict the disease potential of an organism, however, the interpretation of such tests is difficult since the existence of other virulence factors cannot be ruled out. Horizontal DNA transfer and recombination have been important evolutionary mechanisms in diversification of virulence factors. These tests lack species specificity for *Mannheimia* and *P. trehalosi*. Potentially virulent genotypes of *M. hemolytica, Mannheimia* spp., and *P. trehalosi* might be traced but without species-specific detection.

Although PCR methods available at present allow species-specific detection of *H. paragallinarum* and *H. somnus*, potentially virulent genotypes might not be identified. *Pasteurella multocida* might be detected at the species level, certain virulent PAR genotypes might be identified, as well as genotypes with the potential to cause HS. Moreover, typing at the capsular level is possible, which also might help to identify virulent isolates as it is known that certain diseases are associated with certain capsular types (*see* **Subheading 1.1.** and **Notes 1–5**).

1.4. Choice of Sample Material and Outline of Protocols

PCR may be performed with DNA extracted directly from swabs or DNA extracted after bacterial isolation on plates. (Bacterial cultivation in broth is also possible, but has not been included in the published protocols.) Both direct extraction from swabs and pre-isolation on plates allow pre- and post-mortem identification. However, to save time, PCR performed directly on DNA extracts from swabs should be preferred, as described for *H. paragallinarum, H. somnus,* and *P. multocida (17,32,33)*. Detection from swabs after pre-enrichment has also been reported for *P. multocida. (14,27)*. In the present chapter, protocols for all pathogens mentioned above are summarized and set

Table 1
Specificity of Published PCR Tests Toward the Treated Pathogens

Species	Target Gene	Species-specific	Reference
H. paragallinarum	Unknown	Yes	*17*
H. somnus	rrs (16S rRNA)	Yes	*19*
M. hemolytica, M. granulomatis, M. glucosida, M. spp. and P. trehalosi	lkt (leukotoxin)	No	*24,25*
P. multocida	Unknown	Yes	*23*
P. multocida	rrl	Yes	*26*
P. multicida	psl P6-like protein	Yes, but verification by hybridization required.	*27*
P. multocida	Unknown	Yes, but only serotypes B:2, B:5, and B:2.5 causing HS.	*23*
P. multocida	16S to 23S rRNA spacer	Yes, but only serotype B:2 causing HS.	*28*
P. multocida	cap (capsule)	Yes, but five capsular types.	*21*
P. multocida	toxA (toxin)	Yes, but only isolates causing PAR.	*10, 29–31*

up against each other in order to allow simultaneous detection of several pathogens by similar protocols or the use of multiplex formats.

1.5. Controls

Positive control strains must be included in each assay. They are of special importance for verification of the efficiency of DNA extraction and the performance of amplification. Negative controls without template DNA should also be included to detect contamination by carryover of DNA.

2. Materials

2.1. Collection of Samples and Bacterial Reference Strains

2.1.1. Material and Media for Collection of Suspect Material

1. Sterile cotton swabs.
2. Blood agar plates.
3. Brain Heart Infusion Broth (BHI) (Difco, Detroit, MI, USA).
4. Chocolate agar for H. paragallinarum.

2.1.2. Bacterial Reference Strains for Use as Positive Controls

1. *H. paragallinarum* (NCTC 11296^T).
2. *H. somnus* (CCUG 18779, CCUG 12839).
3. *M. hemolytica* (NCTC 9380^T, ATCC 43270 [*lktA* positive]).
4. *P. trehalosi* (NCTC 10370^T, ATCC 29703 [*lktA* positive]).
5. *P. multocida* (NCTC10322^T, NCTC 12178 [*toxA* positive], ATCC 6530 [HS positive]).

   Strains might be obtained through CCUG: (www.ccug.gu.se), NCTC: (http://www.phls.co.uk/services/nctc/index.htm), and ATCC (http://www.atcc.org).
6. The reference strains for the *P. multocida* capsular types might not be available from commercial culture collections. It is recommended to request these strains from national veterinary reference laboratories. The reference strains A 1113, B 925, D 42, E 978, and F 4679 were cited to have been obtained from The Royal Veterinary College, London *(34)* and the strains P 1201 (A), P 932 (B), P 934 (D), P 1234 (E) and P 4218 (F) from National Animal Disease Center, Ames, IA, USA *(21)*.

2.2. Extraction of DNA

1. Lysozyme.
2. Phosphate-buffered saline: 0.01 *M* in phosphate, 0.3 *M* in sodium, pH 7.0.
3. Potassium acetate: 3 *M*.
4. Proteinase K.
5. 10:10 TE buffer: 10 m*M* Tris, 10 m*M* EDTA, pH 7.6.
6. 50:50 TE buffer: 50 m*M* Tris, 50 m*M* EDTA, pH 8.

2.3. PCR

1. PCR cycler with heated lid cover for 0.2- or 0.5-mL tubes.
2. PCR tubes: 0.2- or 0.5-mL, tested for use with the cycler.
3. *Taq* DNA polymerase with corresponding buffer. Main suppliers are Life Technologies (Rockville, MD, USA; www.lifetech.com,) Amersham Pharmacia Biotech (Piscataway, NJ, USA; www.apbiotech.com), and Applied Biosystems (Foster City, CA, USA; www.appliedbiosystems.com).
4. MgCl$_2$ (PCR grade): 25 m*M* stock solution.
5. Double-distilled water purified through a Millipore (Bedford, MA, USA) water purification system or an equivalent device (referred to as MilliQ®-water). MilliQ-water should be stored at 4 °C after purification. During prolonged storage, MilliQ-water should be sterile-filtered (0.45 µm) to exclude any microbial contamination.
6. dNTP: 100 m*M* solutions of dATP, dCTP, dGTP and dTTP (PCR-grade). 100 µL stock solution is made from 20 µL of each dNTP and 20 µL of MilliQ-water and stored at –20 °C.
7. Oligonucleotide primers (*see* **Table 2**): 50X dilutions are prepared from the stock solution of synthetic oligonucleotides, so that 1 µL of the dilution added to a 50-µL PCR will yield a concentration of 0.5 µ*M* in the amplification reac-

Table 2
Oligonucleotide Primers for PCR Amplification

Target organism	Target gene	Forward (5'-3')	Reverse (5'-3')	Reference
H. paragallinarum	Unknown	N1- TGAGGGTAGTCTTGCACGCGAAT	R1-CAAGGTATCGATCGTCTCTCTACT	17
H. somnus	rrs	HS-453-GAAGGCGATTTAGTTTAAGAG	HS-860-TTCGGGCACCAAGTRTTCA	19
Mannheimia spp.	lktA	TGTGGATGCGTTTGAAGAAGG	ACTTGCTTTGAGGTGATCCG	24
P. trehalosi	lktA coding region	AMU-GGGCAACCGTGAAGAAAAAATAG	C3575-CGCCATTTTGACCGATGATTTC	25
P. multocida	rrl	PM23F1-GGCTGGGAAGCCAAATCAAAG	PM23R2-CGAGGGACTACAATTACTGTAA	26
P. multocida	Unknown	KMT1SP6-GCTGTAAACGAACTCGCCAC	KMT1T7-ATCCGCTATTTACCCAGTGG	23
P. multocida	psl	TCTGGATCCATGAAAAAACTAACTAAAGTA (including BamHI-site)	AAGGATCCTTAGTATGCTAACACAGCACGACG (including BamHI-site)	27
P. multocida	Unknown but in serotypes B:2/B:5/ B:2.5 causing HS	KTSP61-ATCCGCTAACACACTCTC	KTT72-AGGCTCGTTTGGATTATGAAG	23
P. multocida	16S-23 rRNA spacer of serotypes B:2 causing HS	IPFWD-CGAAAGAAACCCAAGGCGAA	IPREV-ACAATGAATAACCGTGAGAC	28
P. multocida	Capsular type			21
A		CAPA-FWD-TGCCAAAATCGCAGTCAG	CAPA-REV-TTGCCATCATTGTCAGTG	
B		CAPB-FWD-CATTTATCCAAGCTCCACC	CAPB-REV-GCCCGAGAGTTTCAATCC	
D		CAPD-FWD-TTACAAAGAAAGACTAGGAGCCC	CAPD-REV-CATCTACCCACTCAACCATATCAG	
E		CAPE-FWD-TCCGCAGAAAATTATTGACTC	CAPE-REV-GCTTGCTGCTTGATTTTGTC	
F		CAPF-FWD-AATCGGAGAACGCAGAAATCAG	CAPF-REV-TTCCGCCGTCAATTACTCTG	
P. multocida	toxA	Oligo 1- TACTCAATTAGAAAAAGCGCTTTATCTTCC	Oligo 4- TCTACTACAGTTGCTGGTATTTTTAAATAT	10
P. multocida	toxA	CTTAGATGAGCGACAAGG	GAATGCCACACCTCTATAG	31
P. multocida	toxA	Set 1: GGTCAGATGATGCTAGATACTCC Set 3: CAAGTCTTAACTCCTCCACAAGG	Set 1: CCAAACAGGGTTATATTCTGGAC Set 3: GGGCTTACTGAATCACAAGAGCC	30
P. multocida	toxA	PDNT-1 AAGCTTTCAAGCTTTGAAA	PDNT-2 AAGCTTTCTGAAAAGCACCATTAAT	29

tion. The 50X dilution has to be 25 μ*M*, and for 100 μL of 50X dilution, this requires 2.5 nmol stock solution of primer. A vol equivalent to 2.5 nmol of oligonucleotide (10–20 μL) is taken from the stock solution and diluted with MilliQ-water to 100 μL. Diluted primer solutions are stored at –20 °C.

2.4. Agarose Gel Electrophoresis

1. Horizontal slab-gel apparatus and power supply.
2. Tris-acetate (TAE) buffer for preparation of agarose gel and gel tank: 50X stock solution is prepared from 242 g Tris base, 57.1 mL glacial acetic acid, 100 mL of 0.5 *M* EDTA, and MilliQ-water to 1 L with subsequent adjustment of pH to 8.0. The working solution is diluted 50-fold *(35)*.
3. Agarose (electrophoresis grade): the gel is made 1% (w/v) in TAE-buffer. If a procedure to cast the gel is required, *see* **ref.** *35*.
4. Gel-loading buffer: 0.4 mL of 0.5 M EDTA, pH 8.0, 0.1 g bromophenol blue, 35.3 mL 85% glycerol, and MilliQ-water to 100 mL.
5. DNA size marker: e.g., 100 bp ladder, no. XIV (Roche Diagnostics, Mannheim, Germany)
6. Ethidium bromide solution: 10 mg/mL *(35)*.

3. Methods

3.1. Collection of Samples

1. *H. paragallinarum:* Chickens suspected of being infected with *H. paragallinarum* are swabbed from affected tissues, i.e., sinus *(17)*.
2. *H. somnus*: Cattle or sheep suspected of being infected with *H. somnus* are swabbed from cross-sections of pneumotic lungs, bronchial surfaces, or other affected tissues *(33)*.
3. *P. multocida*:
   a. Cattle suspected for HS or other infections caused by *P. multocida* are sampled for blood or swabbed in the heart within few h after death. Swabs of other affected tissues may also be used. If the carcass has undergone considerable decomposition, the bone marrow may be examined. To obtain pure culture, mouse inoculation might be performed *(8)*.
   b. Pigs and other animals suffering from PAR are swabbed from the affected nasoturbinalia, and swabs are streaked on blood agar plates or incubated further in BHI *(15)*.
   c. Chickens and other birds suspected of being infected with *P. multocida* are swabbed from the affected tissues *(27)*. If the carcass is decomposed, the bone marrow may be used for detection. Chickens suspected of being intestinal carriers of *P. multocida* are sampled for intestinal material, or mucosal surfaces are swabbed *(32)*.
4. Pre-enrichment: overnight culture at 37 °C in BHI is used for detection of *P. multocida* by the *psl*-PCR *(27)* and for detection of *P. multocida* in intestinal material *(32)*.

3.2. Extraction of DNA

3.2.1. DNA Extraction by Boiling for Bacterial Cultures

1. Harvest 1 mL of cultured bacteria by centrifugation and resuspend pellet, or if bacteria were grown on plates, bacterial colonies, in 1 mL MilliQ-water.
2. Boil the suspension in water bath or heating block for 10 min.
3. Centrifuge at 10,000g for 5 min and collect the supernatant with DNA. Storage of extract is not recommended.

3.2.2. Extended DNA Extraction of Pure Cultured Bacteria (36)

1. Harvest 1 mL cultured bacteria by centrifugation and resuspend pellet, or if bacteria were grown on plates, bacterial colonies, in 1 mL of 50:50 TE buffer.
2. Centrifuge at 10,000g for 5 min and resuspend pellet in 0.5-mL of 50:50 TE.
3. Lyse bacteria by addition of 50 μL of 10 mg/mL lysozyme with incubation at 37°C for 30 min, and proceed with the addition and mixing of 50 μL of 10 mg/mL of proteinase K in 50:50 TE and 20 μL of sodium dodecyl sulfate (SDS). Incubate for 2 h at 56 °C.
4. Precipitate cell debris and protein with 297 μL of 3 M potassium-acetate.
5. Centrifuge for 10 min at 16,000g and add 0.54 vol of isopropanol to the supernatant by gentle mixing.
6. Centrifuge and wash the pellet 2× in ice-cold ethanol by centrifugation.
7. Vacuum dry the pellet and resuspend in MilliQ-water and store at –20 °C.

3.2.3. DNA Extraction from Swabs (see Notes 6–8)

1. *H. paragallinarum (17)*
   a. Soak *H. paragallinarum* suspected swabs in 1 mL of phosphate-buffered saline.
   b. Spin down blood cells (1000g, 3 min).
   c. Centrifuge the supernatant at 10,000g for 15 min.
   d. Resuspend the pellet in 20 μL of PCR buffer with 0.5% (v/v) Nonidet® P-40, 0.5% (v/v) Tween® 20, and 20 μg/mL of proteinase K, and incubate at 56°C for 1 h.
   e. Inactivate the proteinase at 98°C for 10 min and store extracts on ice.

2. *H. somnus (33)*
   a. Place *H. somnus* suspected swabs in a solution of 1 mL 0.1% Triton® X-100.
   b. Mix 0.5 mL of this solution with 0.5 mL chloroform/phenol.
   c. Centrifuge at 10,000g for 5 min and mix the supernatant with 0.4 mL chloroform.
   d. Centrifuge at 10,000g for 5 min and precipitate DNA from the supernatant with 0.75 mL of 96% ethanol and 5 μL of 5 M NaCl at –20 °C for 30 min.
   e. Wash the DNA pellet with 70 % ethanol and dissolve it in 25 μL of 10:10 Tris-EDTA buffer.

3. *P. multocida*: extract *P. multocida*-suspected swabs using the same protocol as described above for *H. paragallinarum*.

3.3. PCR Protocols

3.3.1. Preparation of PCR solutions

All solutions should be stored on crushed ice when thawed. All work is performed in a DNA-free flow hood, and all manipulations are done with micropipettes only used for PCR.

The standard 50-μL reaction mixture contains: 1 U *Taq* DNA polymerase, 1.5 m*M* MgCl$_2$, 200 μ*M* dNTP, 0.5μ*M* of each primer (*see* **Table 3**). The composition of the mixture may be altered for specific applications, e.g., to increase the primer concentration, the vol of added primer stock solution can be increased and the amount of MilliQ-water reduced accordingly.

For a series of 10 amplification reactions, the premixture (or master mixture) should be prepared for 11 tubes (excess of 10%) because of the usual pipeting errors. Each tube will receive 5 μL of 10X PCR-buffer, 0.2 μL of enzyme (if stock solution is 5000 U/mL), 0.5 μL of premixed dNTPs, 1 μL of each primer dilution, and 3 μL of MgCl$_2$ leaving the vol of MilliQ-water at 36.8 μL if 2.5 μL of DNA extract is added. For 10 reactions, the following vol should be mixed: 405 μL of MilliQ-water is added to a fresh 1.5 mL Eppendorf® tube followed by: 55 μL of 10X PCR-buffer, 33 μL of MgCl$_2$, 5.5 μL of dNTPs, and 11 μL of each of the diluted primers.

Vortex mix the tube briefly to avoid foaming and carefully pipet 2.2 μL of enzyme into the solution. Invert the tube 5× to mix the reagents. Pipet DNA extract from bacteria or tissue onto the bottom of each of 8 PCR tubes. Add an identical vol of MilliQ-water to the tube containing the negative control and add DNA from a bacterial reference strain to the positive control. When adding the premixture leave the tubes on ice. Do not to pipet the solution too vigorously and do not vortex mix after the enzyme has been added.

3.3.2. PCR Amplification

Immediately insert tubes kept on ice into the PCR cycler and run the specified program according to **Table 4**. When the program is completed, take out the tubes immediately and electrophorese the products on agarose gel. Samples can be stored frozen at –20°C.

3.3.3. Agarose Gel Electrophoresis (see **Note 9**)

This step should be performed in a fume hood with gloves to avoid contact to the toxic ethidium bromide and a Perspex shield for protection from UV radiation. The ethidium bromide bath should be changed every wk, and the waste including gels should be inactivated *(35)*.

1. Mix 5–10 μL of PCR product with 5 μL loading dye and load onto the gel.

Table 3
PCR Protocols for the Pathogens Treated (Reaction Mixture)

Target Organism	Amount of DNA or cell extract	50 mL reaction vol[a]			
		Taq U	MgCl$_2$ mM	dNTP µM	Primer µM
H. paragallinarum (17)	10–200 ng of extracted DNA or 10–20 µL extract of one colony extracted by boiling	1.25[c]	2.0	200	0.4
H. somnus (19,33)	2 µL extract from swab or colony lysed by boiling	0.5	1.5	100	0.2
Mannheimia/lkt (24)	0.5 µL of intact cells	1.5	1.5	150	1.9
P. trehalosi/lkt (25)	10 ng	2	1.5	200	1
P. multocida rrl (26)	1 µL cells boiled extracted	1.25[c]	3	200	0.5
P. multocida psl (27)	5 µL extract from swab	1.25[b]	1.0	200[b]	0.5
P. multocida serotype B:2 causing HS (28)	2.5 µL cells boiled extracted	5	1.5	200[b]	0.25
P. multocida/P.multocida serotopes B:2/B:5/B:2.5 causing HS, multiplex-format and capsular types A, B, D, E, and F multiplexformat (16, 21, 23, 32)	10 µL of extracted cells lysed by boiling or one colony suspended directly in the PCR mixture	1.5	2.0	200	1.6 PM 3.2 HS capsular types multiplex
P. multocida/toxA (31)	0.5 µL of extract of cells lysed by boiling 10 min (Amigot used 4µL extracted DNA)	1[b]	1.5[b]	400	0.2
P. multocida/toxA (10)	10 µL extract of cells lysed by boiling	1	1.52	200	1
P. multocida/toxA (30)	25 µL DNA extract from swab	10	2.0	200	0.3
P. multocida/toxA (29)	100 ng extracted DNA or 5 µL of boiled lysate	2	1.5	200	0.4

[a] Vol adjusted to 50 mL in all protocols.
[b] Standard conditions assumed.
[c] Hot start (addition of enzyme after initial 98 °C for 2 min).

Table 4
PCR Conditions (Temperature–Time Profiles)

Test	Cycles	Initial denaturation (°C/s)	Denaturation (°C/s)	Annealing (°C/s)	Extension (°C/s)	Final extension (min at 72 °C)	Amplicon size (bp)
P. multocida serotype B:5 causing HS (*28*)	35	No information	94/30	60/30	72/ 30	10[a]	334
Mannheimia/lkt (*24*)	30	94/60	94/60	55/60	72/60	10[a]	1146
P. trehalosi/lkt (*25*)	9 +30	95/270	95/30 95/60	54/60 54/35	72/120 72/120	7	563
H. somnus (*19*)	35	94/180	94/60	55/60	72/60	10[a]	400
P. multocida/rrl (*26*)	30	98/150	94/60	69/60	72/60	1	1432[c]
P. multocida/psl (*27*)	35	94/120	94/60	50/30	72/180	7	471
H. paragallinarum (*17*)	25	98/150	94/60	65/60	72/120	10	500 approx
P. multocida/P. multocida serotypes B:2/B:5/B:2.5 causing HS/multiplex and capsular type A, B, D, E, and F multiplex (*16,21,23,32*)	30	95/300	95/60	55/60	72/60	9	460 (*P.multocida*) 590 (*P.multocida* HS) 1044 (*P.multocida* A) 760 (*P.multocida* B) 657 (*P.multocida* D) 511 (*P.multocida* E) 851 (*P.multocida* F)
P. multocida/toxA (*10*)	25	94/120[b]	94/60	55/120	72/120	10[a]	1230
P. multocida/toxA (*31*)	40	94/120[b]	94/30	55/30	72/30	10[a]	846
P. multocida/toxA (*30*)	32	94/120[b]	94/35	65/60	72/150	20	Set 1. 338 Set 2. 217
P. multocida/toxA (*29*)	35	94/120	94/30	61/60	72/120	5	1501

[a]10 assumed.
[b] 94/120 assumed.
[c] Double products might be observed probably due to intervening sequences.

2. Apply fragment length size markers on left and right lanes of the gel and run electrophoresis for 1 to 2 h at 100 V.
3. Stain the gel in a solution of 1 μg/mL of ethidium bromide in MilliQ- water (e.g., 20 μL of stock solution of ethidium bromide in 200 mL of MilliQ-water) for 10–30 min.
4. Destain the gel in another MilliQ-water bath without ethidium bromide for 20 min and visualize the DNA fragments by UV light and photograph the gel.

4. Notes

1. PCR confirmed the diagnosis of *H. paragallinarum* of both challenged and naturally infected chickens and detected more positive diseased animals than culture *(37)*.
2. A genomic fingerprinting method based on repetitive-sequence-based-PCR (REP-PCR) and enterobacterial repetitive intergenic consensus PCR (ERIC-PCR) was specific for *H. somnus* when compared to *A. seminis*, *A. pleuropneumoniae*, *M. hemolytica*, *P. trehalosi*, and *P. multocida (38)*. This method might be used as an alternative to the 16S rRNA gene-based method of Angen et al. *(19)* when simultaneous genotyping of *H. somnus* isolates is required.
3. Leukotoxin was shown to play a significant role in pathogenesis *(39,40)* and is responsible for the β-hemolytic phenotype on blood agar *(41,42)*. The leukotoxin structural gene, *lktA*, has been found in all serotypes of *M. hemolytica*, *P. trehalosi*, *M. varigena*, and in all biotypes of *M. glucosida (43–46)*, even though the latter species is normally not associated with disease *(47)*. Correlation between PCR-positive isolates for the *lkt* gene and β-hemolytic activity was found *(24)*, and isolates with β-hemolytic activity had a greater potential to cause disease *(48)*. Green et al. *(25)* were able to differentiate between cytotoxic and noncytotoxic *P. trehalosi* strains using a PCR targeting the coding region of *lktA*. PCR methods for detection of *lktA* lack species and sometimes even genus specificity. This is probably due to intragenic and assortive (whole gene) recombination between *Mannheimia* spp. and *P. trehalosi*, which has played an important role in the evolution of *lktA (45)*. For example, the PCR method of Fisher et al. *(24)* yielded bands of the expected size in representatives of *P. trehalosi*, *M. hemolytica*, *M. glucosida*, *M. granulomatis*, and *M. varigena (24,46)*. Thus, conventional typing seems to be needed for the interpretation of positive results due to differences in virulence and species specificity. As an alternative to the *lktA*-based PCR, *M. (P.) granulomatis* was detected by random-amplified polymerase DNA (RAPD) *(49)*, however the specificity was only evaluated against a few other species of *Pasteurellaceae*.
4. Capsular typing and detection of *P. multocida* causing PAR or HS:
   a. The detection of capsular types and species-specific detection can be performed in a multiplex format using the seven primer-sets, CAPA-FWD/REV, CAPB-FWD/REV, CAPD-FWD/REV, CAPE-FWD/REV, CAPF-FWD/REV, KMT1SP6/KMT1T7, and KTSP61/KTT72 (**Table 2**). Isolates of *P. avium* biovar 2 and *P. canis* biovar 2 were also detected by the *P. multocida*

specific PCR *(23,26)*, however this is probably related to misclassification, because *P. avium* biovar 2 and *P. canis* biovar 2 and *P. multocida* have the same 16S rRNA gene sequences *(50,51)*. Recently the multiplex PCR test *(21)* was updated. Problems were reported in relation to the separation of capsular types A and F, and the HS-specific assay was in rare cases also found to detect capsular type D strains *(52)*.

  b. Good correspondence has been reported between detection by *tox*A-PCR and in vitro and in vivo detection methods of the toxic protein of *P.multocida* causing PAR *(10,30,31,53)*. Four different protocols have been described for detection of the *toxA* gene in *P. multocida* causing PAR. Comparative studies have not been performed, and no protocol can be recommended over the others. With two isolates from pig tonsils, amplification of a 1.5-kb fragment was observed by the PCR of Lichstensteiger et al. *(31)* as opposed to the expected 0.8-kb fragment expected. The fragment was found unrelated to the *toxA* gene by DNA sequence comparison *(16)*. In the study of Amigot et al. *(15)*, the presence of faint bands did probably not indicate the presence of the *toxA* gene these bands were found in strains negative for the toxic protein by enzyme-linked immunosorbent assay (ELISA) and cell culture. It is necessary to identify false-positive isolates by comparison with the positive control PCR. For detection of the *tox*A gene in strains of *P. multocida* the assay of Kamp et al. *(30)* was found best for large-scale analysis of swabs including a multiscreen method, whereas that of Lichtensteiger et al. *(31)* was found better for small-scale studies *(54)*. For use of the multiscreen method, see Kamp et al. *(30)*. Two primer-sets are given with the protocol of Kamp et al. *(30)*, and they recommended that both be used to reduce the detection of false positives. The *toxA* test of Lichtensteiger et al. *(31)* worked best with extracted DNA of cultured bacteria *(55)*.

  c. With the two protocols for HS causing isolates of *P. multocida* good correlation to an ELISA test, it was found, however, that the protocols will only identify isolates of serotype B:2 *(28)* and B:2, B:5, and B:2,5 *(23)*, respectively.

5. A specific PCR test for detection of *H. parasuis*, which is the causative agent of porcine polyserositis and arthritis, was recently described based on species-specific 16S rRNA gene amplification *(56)*. An assay specific for *A. suis* based on the amplification of fragments of either the *apxI* or *apxII* genes was described by van Ostaaijen et al. *(57)*, however the test was not evaluated for other representatives of *Pasteurellaceae* than *A. pleuropneumoniae*.

6. Swabs should be spiked with positive control strains before extraction, to control the efficiency of swab extraction. If the outcome is negative, the DNA extraction step must be controlled. This can be done by measuring the DNA concentration in swab extracts by use of Picogreen (Molecular Probes, Eugene, OR, USA; http://www.probes.com) or Hoechst33258 *(58)*. The amount of DNA added to each PCR tube should be at least 150 ng. If the present extraction protocol proves inefficient, an alternative extraction procedure should be tried (*see* **Note 4**).

7. To reduce labor during extraction of DNA from swabs, extraction by boiling might be tried, but it is important to control the efficiency of extraction by spiking swabs with positive controls (*see* **Note 1**). The samples are boiled for 15 min and centrifuged at 14,000*g* for 10 min. The supernatant containing DNA is transferred into a sterile tube *(27)*. For detection of *P. multocida* in intestinal material, it has been recommend to wash the mucosal swabs or intestinal material by centrifugation and further extract DNA from the bacterial pellet by boiling *(32)*.

8. GuSCN extraction was used as an alternative method for extraction of DNA from *P. multocida*-suspected swabs from nose and tonsils of pigs. The swabs are incubated overnight in lysis buffer including GuSCN, Triton X-100, and diatomite suspension. The diatomite–DNA complex is washed 2× with buffer, dried, and the DNA is eluted from diatomite with PCR buffer (*see* **ref.** *30* for further details).

9. For all protocols reported, amplicons were visualized by agarose-gel electrophoresis with ethidium bromide staining. The only exceptions include *psl* amplification of *P. multocida (27)*, where secondary hybridization was required to verify the fragment because the *psl* gene has also been found in *H. influenzae (59)*. It is recommended to consult the paper of Kasten et al. *(27)* to set up this hybridization. For precise fragment length determination, an automatic sequencer might be used. This will require labeling of primers or dNTPs with specific fluorochromes *(60)*.

Acknowledgment

The project was funded by the Danish Agricultural and Veterinary Research Council through grant nos. 9600803 and 9702797.

References

1. Bisgaard, M. (1993) Ecology and significance of *Pasteurellaceae* in animals. *Zbl. Bakt.* **279,** 7–26.
2. Bisgaard, M. (1995) Taxonomy of the family *Pasteurellaceae* Pohl 1981, in *Haemophilus, Actinobacillus*, and *Pasteurella* (Donachie, W., Lainson, F. A., and Hodgson, J. C., eds.), Plenum Press, New York, USA, pp. 1–7.
3. Blackall, P. J. (1999) Infectious coryza: overview of the disease and new diagnostic options. *Clin. Microbiol. Rev.* **12,** 627–632.
4. Harris, F. W. and Janzen, E. D. (1989). The *Haemophilus somnus* disease complex (Hemophilosis): a review. *Can. Vet. J.* **30,** 816–822.
5. Ward, A. C. S., Jaworski, M. D., Eddow, J. M., and Corbeil, L. B. (1995) A comparative study of bovine and ovine *Haemophilus somnus* isolates. *Can. J. Vet. Res.* **59,** 173–178.
6. Frank, G. H. (1989) Pasteurellosis of cattle, in *Pasteurella and Pasteurellosis* (Adlam, C. and Rutter, J. M., eds.), Academic Press, London, UK, pp. 197–222.
7. Gilmour, N. J. L. and Gilmour, J. S. (1989) Pasteurellosis of sheep, in *Pasteurella and Pasteurellosis*, (Adlam, C. and Rutter, J. M. eds.), Academic Press, London, UK, pp. 223–262.

8. Carter, G. R. and De Alwis, M. C. L. (1989) Haemorrhagic septicaemia, in *Pasteurella and Pasteurellosis* (Adlam, C. and Rutter, J. M. eds.), Academic Press, London, UK, pp. 131–160.

9. Chanter, N. and Rutter, J. M. (1989) Pasteurellosis in pigs and the determinants of virulence of toxigenic *Pasteurella multocida*, in *Pasteurella and Pasteurellosis* (Adlam, C., and Rutter, J. M. eds.), Academic Press, London, UK, pp. 161–195.

10. Nagai, S., Someno, S. and Yagihashi, T. (1994) Differentiation of toxigenic from nontoxigenic isolates of *Pasteurella multocida* by PCR. *J. Clin. Microbiol.* **32,** 1004–1010.

11. Baalsrud, K. J. (1987) Atrophic rhinitis in goats in Norway. *Vet. Rec.* **121,** 350–353.

12. DiGiacomo, R. F., Deeb, B. J., Giddens, W. E. Jr., Bernard, B.L., and Chengappa, M.M. (1989) Atrophic rhinitis in New Zealand white rabbits infected with *Pasteurella multocida*. *Am. J. Vet. Res.* **50,** 1460–1465.

13. Petersen, S. K. (1990) The complete nucleotide sequence of the *Pasteurella multocida* toxin gene and evidence for a transcriptional repressor, TxaR. *Mol. Microbiol.* **4,** 821–830.

14. Christensen, J. P. and Bisgaard, M. (1997) Avian pasteurellosis: taxonomy of the organisms involved and aspects of pathogenesis. *Avian Pathol.* **26,** 461–483.

15. Amigot, J. A., Torremorell, M. and Pijoan, C. (1998) Evaluation of techniques for the detection of toxigenic *Pasteurella multocida* strains from pigs. *J. Vet. Diagn. Invest.* **10,** 169–173.

16. Townsend, K. M., Hanh, T. X., O'Boyle, D., et al. (2000) PCR detection and analysis of *Pasteurella multocida* from the tonsils of slaughtered pigs in Vietnam. *Vet. Microbiol.* **72,** 69–78.

17. Chen, X., Miflin, J. K., Zhang, P. and Blackall, P. J. (1996) Development and application of DNA probes and PCR tests for *Haemophilus paragallinarum*. *Avian Dis.* **40,** 398–407.

18. Canto, G. J. and Biberstein, E. L. (1982) Serological diversity in *Haemophilus somnus*. *J. Clin. Microbiol.* **15,** 1009–1015.

19. Angen, Ø., Ahrens, P., and Tegtmeier, C. (1998) Development of a PCR test for identification of *Haemophilus somnus* in pure and mixed cultures. *Vet. Microbiol.* **63,** 39–48.

20. Tegtmeier, C., Uttenthal, Aa., Friis, N. F., Jensen, N. E., and Jensen H. E. (1999) Pathological and microbiological studies on pneumononic lungs from Danish calves. *J. Vet. Med. B* **46,** 693–700.

21. Townsend, K. M., Boyce, J. D., Chung, J. Y., Frost, A. J., and Adler, B. (2001) Genetic organization of *Pasteurella multocida cap* loci and development of a multiplex capsular PCR typing system. *J. Clin. Microbiol.* **39,** 924–929.

22. Rimler, R. B. and Rhoades, K. R. (1989) *Pasteurella multocida*, in *Pasteurella and Pasteurellosis* (Adlam, C. and Rutter, J. M., eds.), Academic Press, London, UK, pp. 37–73.

23. Townsend, K. M., Frost, A. J., Le, C. W., Papadimitriou, J. M., and Dawkins, H. J. S. (1998) Development of PCR assays for species- and type-specific identification of *Pasteurella multocida* isolates. *J. Clin. Microbiol.* **36,** 1096–1100.

24. Fisher, M. A., Weiser, G. C., Hunter, D. L., Ward, A. C. S. (1999) Use of a polymerase chain reaction method to detect the leukotoxin gene *lktA* in biogroup and biovariant isolates of *Pasteurella hemolytica* and *P. trehalosi. Am. J. Vet. Res.* **60,** 1402–1406.

25. Green, A. L., DuTeau, N. M., Miller, M. W., Triantis, J. M., and Salman, M. D. (1999) Polymerase chain reaction techniques for differentiaitng cytotoxic and noncytotoxic Pasteurella terhalosi form Rocky Mountain bighorn sheep. *Am. J. Vet. Res.* **60,** 583–588.

26. Miflin, J. K. and Blackall, P. J. (2001) Development of a 23S rRNA based PCR Assay for the identification of *Pasteurella multocida. Lett. Appl. Microbiol.* **33,** 216–221.

27. Kasten, R. W., Carpenter, T. E., Snipes, K. P. and Hirsh, D. C. (1997) Detection of *Pasteurella multicida*-specific DNA in turkey flocks by use of the polymerase chain reaction. *Avian Dis.* **41,** 676–682.

28. Brickell, S. K., Thomas, L. M., Long, K. A., Panaccio, M. and Widders, P. R. (1998) Development of a PCR test based on a gene region associated with the pathogenicity of Pasteurella multocida serotype B:2, the causal agent of Haemorrhagic septicaemia in Asia. *Vet. Microbiol.* **59,** 295–307.

29. Hotzel, H., Erler, W. und Schimmel D. (1997) Nachweis des Dermonekrotoxin-Gens in *Pasteurella-multocida*-Stämmen mittels Polymerase-Kettenreaktion (PCR). *Berl. Münch. Tierärztl. Wschr.* **110,** 139–142.

30. Kamp, E. M., Bokken, G. C. A. M., Vermeulen, T. M. M., et al. (1996) A specific and sensitive PCR assay suitable for large-scale detection of toxigenic *Pasteurella multocida* in nasal and tonsillar swabs specimens of pigs. *J. Vet. Diagn. Invest.* **8,** 304–309.

31. Lichtensteiger, C. A., Steenbergen, S. M., Lee, R. M., Polson, D. D. and Vimr, E. R. (1996) Direct PCR analysis for toxigenic *Pasteurella multocida. J. Clin. Microbiol.* **34,** 3035–3039.

32. Lee, C. W., Wilkie, I. W., Townsend, K. M., and Frost, A. J. (2000) The demonstration of *Pasteurella multocida* in the alimentary tract of chickens after experimental oral infection. *Vet. Microbiol.* **72,** 47–55.

33. Tegtmeier, C., Angen, Ø. and Ahrens, P. (2000) Comparison of bacterial cultivation, PCR, in situ hybridization and immunohistochemistry as tools for diagnosis of *Haemophilus somnus* pneumonia in cattle. *Vet. Microbiol.* **76,** 385–394.

34. Dziva, F., Christensen, H., Olsen, J. E. and Mohan, K. (2002) Random amplification of polymorphic DNA and phenotypic typing of Zimbabwean isolates of *Pasteurella multocida. Vet. Microbiol.*, in press.

35. Sambrook, J., Fritsch, E. F. and Maniatis, T. (1989) *Molecular Cloning: A Laboratory Manual,* 2nd ed. CSH Laboratory Press, Cold Spring Harbor, NY, USA.

36. Leisner, J. J., Pot, B., Christensen, H., et al. (1999) Identification of lactic acid bacteria from Chili Bo, a Malaysian food ingredient. *Appl. Environ. Microbiol.* **65,** 599–605.

37. Chen, X., Chen, Q., Zhang, P., Feng, W. and Blackall, P. J. (1998) Evaluation of a PCR test for the detection of *Haemophilus paragallinarum* in China. *Avian Pathol.* **27,** 296–300.

38. Appuhamy, S., Parton, R., Coote, J. G. and Gibbs, H. A. (1997) Genomic finger-printing of *Haemophilus somnus* by a combination of PCR methods. *J. Clin. Microbiol.* **35,** 288–291.

39. Petras, S. F., Chidambaram, M., Illyes, E. F., Froshauer, S., Weinstock, G. M. and Reese, C. (1995) Antigenic and virulence properties of Pasteurella hemolytica leukotoxin mutants. *Infect. Immunol.* **63,** 1033–1039

40. Highlander, S. K., Fedorova, N. D., Dusek, D. M., Panciera, R., Alvarez, L. E. and Rinehart, C. (2000) Inactivation of *Pasteurella (Mannheimia) hemolytica* leukotoxin causes partial attenuation of virulence in a calf challenge model. *Infect. Immunol.* **68,** 3916–3922.

41. Murphy, G. L., Whitworth, L. C., Clinkenbeard, K. D. and Clinkenbeard, P.A. (1995) Hemolytic activity of the *Pasteurella hemolytica* leukotoxin. *Infect. Immunol.* **63,** 3209–3212.

42. Fedorova, N. D., Highlander, S. K. (1997) Generation of targeted nonpolar gene insertions and operon fusions in Pasteurella hemolytica and creation of a strain that produces and secretes inactive leukotoxin. *Infect. Immunol.* **65,** 2593–2598.

43. Burrows, L.L., Olah-Winfield, E. and Lo, R. Y. (1993) Molecular analysis of the leukotoxin determinants from *Pasteurella hemolytica* serotypes 1 to 16. *Infect. Immunol.* **61,** 5001–5007.

44. Chang, Y. F., Ma, D. P., Shi, J. and Chengappa, M. M. (1993) Molecular character-ization of a leukotoxin gene from a *Pasteurella hemolytica*-like organism, encoding a new member of the RTX toxin family. *Infect. Immunol.* **61,** 2089–2095.

45. Davies, R. L., Whittam, T. S., and Selander, R. K. (2001) Sequence diversity and molecular evolution of the leukotoxin (*lktA*) gene in bovine and ovine strains of *Mannheimia (Pasteurella) hemolytica. J. Bacteriol.* **183,** 1394–1404

46. Larsen, J. (unpublished).

47. Biberstein, E. L. and Gills, M. G. (1962) The relationship of the antigenic types of the A and T types of *Pasteurella hemolytica. J. Comp. Pathol.* **72,** 316–320.

48. Jaworski, M. D., Hunter, D. L. and Ward, A. C. S. (1998) Biovariants of isolates of *Pasteurella* from domestic and wild ruminants. *J. Vet. Diagn. Invest.* **10,** 49–55.

49. Veit, H. P., Wise, D. J., Carter, G. R. and Chengappa, M. M. (1998) Toxin pro-duction by Pasteurella granulomatis. *Ann. N. Y. Acad. Sci.* **849,** 479–484.

50. Dewhirst, F. E., Paster, B. J., Olsen, I., and Fraser, G. J. (1993) Phylogeny of the *Pasteurellaceae* as determined by comparison of 16S ribosomal ribonucleic acid sequences. *Zbl. Bakt.***279,** 35–44.

51. Christensen, H. (unpublished).

52. Townsend, K., Gunawardana, G. A., and Briggs, R. E. (2002) PCR Identification and Typing of Pasteurella Multocida: An Update. Poster 13. International Pasteurellaceae Society Conference. Banff, AB, Canada. May 5–10, 2002.

53. Jong, M. F. De, Kamp, E., van der Schoot, A., and Banniseth, T. von (1996) Elimi-nation of AR toxinogenic Pasteurella from infected sow herds by a combination of ART vaccination and testing sows with a PCR and ELISA test. Proceedings of the 14th IPVS Congress, Bologna, Italy, 7–10 July 1996.

54. Hunt, M. L., Adler, B. and Townsend, K. M. (2000) The molecular biology of *Pasteurella multocida. Vet. Microbiol.* **72,** 3–25.

55. Dziva, F. (unpublished).

56. Angen, Ø., Ahrens, P. and Leser, T. D. (2001) Development of a PCR test for detection of *Haemophilus parasuis.* Abstract Z–43. ASM 101 st General Meeting Orlando Florida. ASM, Washington DC, USA.

57. Van Ostaaijen, J., Frey, J., Rosendal, S., and MacInnes, J. I. (1997). *Actinobacillus suis* strains isolated from healthy and diseased swine are clonal and carry *apxICABD* var.suis toxin genes. *J. Clin. Microbiol.* **35,** 1131–1137.

58. Paul, J. H., and Myers, B. (1982) Fluorometric determination of DNA in aquatic microorganisms by use of Hoechst 33258. *Appl. Environ. Microbiol.* **43,** 1393–1399.

59. Kasten, R. W., Hansen, L. M., Hjinojoza, J., Biebger, D., Ruehl, W. W. and Hirsch, D. C. (1995) *Pasteurella multocida* produces a protein with homology to the P6 outer membrane protein of *Haemophilus influenzae. Infect. Immunol.* **63,** 989–993.

60. Christensen, H. Jørgensen, K. and Olsen, J. E. (1999) Differentiation of *Campylobacter coli* and *C. jejuni* by length and DNA sequence of the 16S–23S rRNA internal spacer region. *Microbiology* **145,** 99–105.

19

Detection of *Salmonella* spp.

Burkhard Malorny and Reiner Helmuth

1. Introduction

Salmonella is still recognized as a major zoonotic pathogen for animals and humans *(1)*. In many countries, it is the leading cause of all food-borne outbreaks and infections *(2,3)*. Conventional cultural methods for the detection of *Salmonella* spp. in clinical human or animal material, food, feeding stuff or the environment require nonselective and selective incubation steps resulting in a 4 to 6 d isolation procedure *(4,5)*. A more rapid detection of *Salmonella* spp. can be expected by the application of the polymerase chain reaction (PCR). Detection of *Salmonella* from food by PCR can be achieved within 24 h. Generally, the procedure includes *(i)* a pre-enrichment step in a nonselective medium to allow recovery of damaged cells and multiplication of the target organism; *(ii)* nucleic acid extraction from the culture; *(iii)* in vitro amplification of *Salmonella* spp. specific nucleic acid target DNA; and *(iv)* detection of the PCR product by gel electrophoresis and/or hybridization techniques. Several nonselective media for pre-enrichment have been developed for various food matrices, but buffered peptone water is most commonly used and recommended in the ISO 6579 standard on *Microbiology of Food and Animal Feeding Stuffs—Horizontal Method for the Detection of Salmonella (6)* as the medium of choice for pre-enrichment. Since many substances occurring in food can inhibit the thermostable DNA polymerase *(7,8)*, the selection of a DNA extraction method depends on the food matrix. Several reviews were published summarizing the different DNA extraction and purification methods for various bacteria *(9–10)*. Commercially available extraction and purification systems can also be used for the efficient removal of PCR inhibitors.

Here we describe a simple method for eliminating a broad range of PCR inhibitors using Chelex® 100, a chelating resin that has a high affinity for poly-

From: *Methods in Molecular Biology, vol. 216: PCR Detection of Microbial Pathogens: Methods and Protocols*
Edited by: K. Sachse and J. Frey © Humana Press Inc., Totowa, NJ

valent metal ions. Chelex 100 protects the DNA from degradation by chelating metal ions.

Many primer sets for in vitro amplification of *Salmonella* DNA by PCR have been published *(11–21)*. However, they differ in sensitivity, accuracy (proportion of false negative and false positive results), and applicability. We have developed a PCR system for the detection of all *Salmonella* spp., including *Salmonella bongori*, which has a low detection limit (*see* **Fig. 1**). The method is highly specific, rapid and simple to use. The primer set used for amplification was previously published by Rahn et al. *(12)* and gives rise to amplification of a 284-bp sequence of the *invA* gene. The *invA* target gene is located on pathogenicity island 1 of *Salmonella* spp., which encodes proteins of a type III secretion system *(22)*.

2. Materials

All buffers and double-distilled water must be sterilized either by autoclaving or filtration.

2.1. Processing of Food Samples

1. Stomacher Lab Blender 400 (Seward Medical, London, UK).
2. Stomacher model 400 bags (Seward Medical).
3. Cell incubator.
4. 1-L Erlenmeyer culture flask.
5. Buffered peptone water: dissolve 10 g of enzymatic digest of animal tissues (1%), 5 g NaCl (0.5%), 9 g Na$_2$HPO$_4$·12 H$_2$O (0.9%), and 1.5 g KH$_2$PO$_4$ (0.15%) in 1000 mL of double-distilled water by heating, if necessary. Adjust the pH, so that after sterilization it is 7.0 ± 0.2. Autoclave.

2.2. DNA Extraction

1. Microcentrifuge (e.g., 5402, Eppendorf, Köln, Germany).
2. Thermal block (e.g., thermomixer 5436; Eppendorf) or water bath, capable of being heated to 100°C.
3. TE buffer: 1 mM EDTA, 10 mM Tris-HCl, pH 8.0.
4. Chelex 100 suspension, 5% (w/v): resuspend 5 g Chelex 100 resin (Bio-Rad, Hercules, CA, USA; Cat. no. 142-2832) in 100 mL of double-distilled water. Prepare a fresh suspension.

2.3. Amplification (PCR)

1. Thermal cycler (e.g., GenAmp® PCR system 9700; Applied Biosystems, Weiterstadt, Germany).
2. PCR buffer (10X): 500 mM KCl, 200 mM Tris-HCl, pH 8.4.
3. Deoxynucleotide triphosphate (dNTP) mixture (Roche Diagnostics, Mannheim, Germany) containing dATP, dCTP, dTTP, and dGTP (2 mM each).

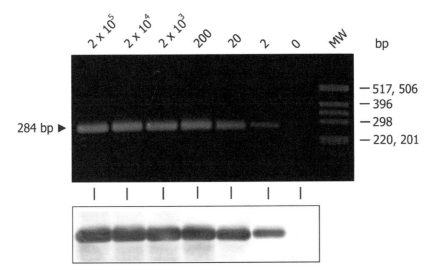

Fig. 1. Detection limit of the PCR after 38 cycles using the *Salmonella*-specific primer set 139/141 *(12)*. Reference strain *S.* Typhimurium 51K61 phagetype DT104 was grown in buffered peptone water overnight and serially 10-fold diluted. DNA extraction was performed as described in **Subheading 3.2.** PCRs were performed as described in **Subheading 3.3.** Ten microliters of PCR product was loaded on a gel and electrophoresed. Colony forming units (cfu)/reaction are given on top of the lanes. MW, molecular weight standard X. The detection limit was determined as 2–20 cfu/reaction. **(A)** agarose gel after staining with ethidium bromide. **(B)** Southern blot of the same gel, hybridized with the *invA* PCR probe as described in **Subheadings 3.4.–3.7.**

4. *Taq* DNA hot-start polymerase (e.g., 5 U/μL, Platinum *Taq* polymerase; Life Technologies, Karlsruhe, Germany; Cat. no. 10966-034).
5. MgCl$_2$: 50 m*M* solution.
6. Primers for the amplification of the target gene fragment *(12)*: 139: 5'-GTG AAATTATCGCCACGTTCGGGCAA-3', and 141: 5'-TCATCGCACCGTCA AAGGAACC-3.
7. Primers for the amplification of the amplification control: HB10: 5'-ATT CCACACAACATACGAGCCG-3', and HB11: 5'-GTTTCGCCACCTCTGAC TTGAG-3'.
8. pUC19 plasmid DNA (1 fg/μL).
9. *Salmonella* reference DNA (e.g., *S.* Typhimurium DT104 strain 51K61, BgVV, National *Salmonella* Reference Laboratory, or ATCC 15277, *S.* Typhiumurium LT2).
10. Fraction V bovine serum albumin (BSA) (e.g., Roth, Karlsruhe, Germany; Cat. no. 8076.8).
11. UV Transilluminator or UV light box (λ = 302 nm) and video or photo camera for documentation (e.g., Eagle Eye® II; Stratagene, La Jolla, CA, USA).

12. Apparatus for horizontal gel electrophoresis.
13. Low melting agarose (e.g., Life Technologies; Cat. no. 15510-019).
14. Ethidium bromide solution: 0.5 mg/L in water.
15. TBE buffer (10X): 450 mM boric acid, 450 mM Tris, 1 mM EDTA. Adjust to pH 8.0.
16. Loading buffer (6X): 40% (v/v) glycerol, 0.25% (w/v) bromophenol blue.
17. DNA molecular weight standard X (Roche Diagnostics; Cat. no. 1498037).

2.4. Southern Blot

1. Vacuum blotter (e.g., model 785; Bio-rad).
2. UV Transilluminator for nylon membrane (e.g., Stratalinker® 2400; Stratagene).
3. Nylon membrane, positively charged (Roche Diagnostics; Cat. no. 1417240).
4. Whatman 3MM filter paper (Whatman, Clifton, NJ, USA).
5. HCl: 0.25 M.
6. NaOH: 0.5 N.
7. SSC stock buffer (20X): 3 M NaCl, 0.3 M sodium citrate. Adjust to pH 7.0.

2.5. Labeling of the invA PCR Probe

1. Digoxygenin (DIG)-11-dUTP: 1 mM (Roche Diagnostics).
2. Deoxynucleotide triphosphate (dNTP) mixture containing dATP, dGTP, and dCTP (2 mM each).
3. dTTP: 10 mM solution.
4. Primers for amplification of the *invA* PCR probe: invA-int1: 5'-TTCGTT ATTGGCGATAGCCTGGC-3', and invA-int2: 5'-AGCTGGCTTTCCCTTT CCAGTAC-3'.

2.6. Hybridization

1. Hybridization oven for roller bottles (e.g., OV10; Biometra, Göttingen, Germany).
2. Hybridization roller bottles 150 x 35 mm (e.g., Biometra; Cat. no. 052-002).
3. Plastic boxes.
4. DIG Easy Hyb Granules (Roche Diagnostics; Cat. no. 1796895). Preparation of working solution: add 64 mL of double-distilled water in two portions to the plastic bottle, dissolve by stirring immediately for 5 min at 37°C. Store solution at –20°C.
5. Washing solution I: 2X SSC, 0.1% (w/v) sodium dodecyl sulfate (SDS).
6. Washing solution II: 0.5X SSC, 0.1% (w/v) SDS.

2.7. Detection

1. Maleic acid buffer: 0.1 M maleic acid, 0.15 M NaCl. Adjust to pH 7.5 with 2 N NaOH.
2. Washing buffer: maleic acid buffer with 0.3% (v/v) Tween® 20.
3. Detection buffer: 0.1 M Tris-HCl, 0.1 M NaCl. Adjust to pH 9.5.
4. Blocking solution (10X): dissolve blocking reagent (Roche Diagnostics; Cat. no. 1096176) in maleic acid buffer by heating to a final concentration of 10% (w/v).

Autoclave the stock solution. Store stock solution at 4°C.

5. Blocking solution (1X): dilute 10X blocking solution 1:10 in maleic acid buffer. Prepare always fresh 1X blocking solution.

6. TE buffer: 1 mM EDTA, 10 mM Tris-HCl. Adjust to pH 8.0.

7. Anti-DIG-AP-conjugate-solution: centrifuge the Anti-DIG-AP-conjugate-stock solution (Roche Diagnostics; Cat. no. 1093274) for 5 min at 10,000g in the original vial prior to each use. Dilute the centrifuged Anti-DIG-AP-conjugate-stock solution 1:5000 in 1X blocking buffer. Prepare always fresh conjugate solution.

8. Color substrate solution: dilute 45 μL of nitroblue tetrazolium stock solution (75 mg/mL in 70% dimethylformamide) and 35 μL of bromo-chloro-indolyl phosphate (50 mg/mL in dimethylformamide) in 10 mL of detection buffer. Always prepare fresh color substrate solution.

3. Methods

The PCR system described below comprises the following steps: *(i)* processing of the food matrix and pre-enrichment; *(ii)* simple DNA extraction method by thermal cell digestion after pre-enrichment; *(iii)* amplification of the *Salmonella* spp.-specific target gene; *(iv)* detection of PCR products by gel electrophoresis; and *(v)* confirmation of positive signals by Southern blot hybridization.

3.1. Processing of the Food Matrix and Pre-enrichment

1. Add 25 g of the food matrix to be investigated to 225 mL of buffered peptone water using a 1-L culture flask and homogenize the sample by vigorous stirring. If necessary, transfer the sample and buffered peptone water into a stomacher bag 400 and homogenize in a stomacher 400 for 30 s to 2 min using 800–1000 impulses/min (*see* **Note 1**). Transfer the suspension into a culture flask. If the sample weight differs from 25 g, adjust the vol of the buffered peptone water to an approximate ratio of 1:10 (w/v).

2. Incubate the pre-enrichment sample at 37°C for 16–20 h without shaking.

3.2. DNA Extraction (Thermal Cell Digestion)

1. Transfer 1 mL microbial pre-enrichment into a clean 1.5-mL microcentrifuge tube. Spin tube in a microcentrifuge for 5 min at 10,000g and 4°C (*see* **Note 2**).

2. Discard the supernatant carefully.

3. Resuspend the pellet in 300 μL of TE buffer or, for more efficient removal of PCR inhibitors, in 300 μL of 5% Chelex 100 suspension. Stir the suspension during pipeting.

4. Incubate the microcentrifuge tube for 15 min at 100°C. Briefly vortex mix the sample 2× during incubation. After incubation, chill it immediately on ice.

5. Centrifuge tube for 5 min at 13,000g and 4°C. Transfer the supernatant carefully into a fresh microcentrifuge tube.

6. Use a 5-μL aliquot of the supernatant as template DNA in the PCR.

7. Store the supernatant at –20°C.

3.3. Amplification (PCR)

General requirements for PCR according to standard laboratory practice have to be considered. Use pipet tips with filters only. The master mixture should be prepared in a room free of *Salmonella* DNA. Carefully avoid cross-contamination of DNA samples. During pipeting the reagents should be kept on ice. In order to avoid drops of fluid remaining on the wall of the reaction tube, spin the tubes shortly in a microcentrifuge before use.

3.3.1. Amplification of the Target Gene Fragment

1. Prepare the amount of master mixture according to the number of amplification reactions. Add 10% to the final vol to allow for small pipeting mistakes. **Table 1** gives the vol to be pipeted for 20 amplification reactions (*see* **Note 3**):
2. Pipet 20 µL of master mixture in a 0.2-mL PCR tube and complete each reaction mixture by adding 5 µL of DNA extraction template, respectively. If required, spin the tubes in a microcentrifuge in order to avoid liquid remaining on the wall.
3. Place the PCR tubes in a thermal cycler and perform the following thermal cycling program (maximal ramping rate): initial denaturation at 95°C for 1 min, then 38 cycles of: denaturation at 95°C for 30 s, annealing at 64°C for 30 s, extension at 72°C for 30 s, and then a final extension at 72°C for 4 min, and a final temperature at 4°C forever.
4. After finishing the PCR, spin the tubes again and cool them on ice until electrophoresis.

3.3.2. Control Reactions

For control of the performance of the PCR assay, suitable reagent controls have to be included (negative, positive, and amplification control). A negative control contains no template DNA extract and detects potential impurities of the reagents. A positive control contains DNA of a *Salmonella* reference strain extracted by the method used for sample preparation. The amplification control is used to verify the absence or presence of inhibitory substances which can be carried over from the food item or reagents during DNA extraction. The amplification control may be performed in the same reaction mixture or in a separate but parallel reaction mixture.

3.3.3. Amplification Control

PCR methods designed for detection of food-borne pathogens should not be performed without an external amplification control suitable for detection of inhibitory effects, in order to avoid false negative results. In the present assay, we propose an external control consisting of plasmid vector pUC19 in combination with primers specific for pUC19. The specific amplicon has a size of 429 bp. The thermal cycling program is identical to the program for amplifica-

Table 1
Volumes for Amplification Reactions

Components (stock concentration)	Vol/reaction (final concentration)	Vol for 22X master mixture (for 20 reactions)
PCR-grade water	9.6 µL	211.2 µL
10X PCR buffer	2.5 µL	55 µL
dNTP Mixture (2 mM)	2.5 µL (200 µM)	55 µL
Primer 139 (10 pmol/µL)	1.0 µL (0.4 µM)	22 µL
Primer 141 (10 pmol/µL)	1.0 µL (0.4 µM)	22 µL
Magnesium chloride (50 mM)	0.75 µL (1.5 mM)	16.5 µL
Taq DNA polymerase (5 U/µL)	0.15 µL (0.75 U)	3.3 µL
Fraction V BSA (10 mg/mL)	2.5 µL (1 µg/µL)	55 µL
DNA extraction template	5.0 µL	
Total vol	25.0 µL	

tion of the *invA* target gene fragment (*see* **Subheading 3.3.1.**) and has to be performed in the same run.

1. Prepare a master mixture according to the number of reactions. Add 10% to the final vol to allow for small pipeting mistakes. **Table 2** gives the vol to be pipeted for 20 amplification reactions:
2. Pipet 20 µL of master mixture into 0.2-mL PCR tubes and complete each reaction mixture by adding 5 µL of DNA extraction template, respectively. If required, spin the PCR tubes in a microcentrifuge in order to avoid liquid remaining on the wall.
3. Place the PCR tubes in a thermal cycler and perform the following thermal cycling program (maximal ramping rate): Initial denaturation at 95°C for 1 min, then 38 cycles of denaturation at 95°C for 30 s, annealing at 64°C for 30 s, extension at 72°C for 30 s, and then a final extension at 72°C for 4 min, and a final temperature at 4°C forever.
4. After the run, spin tubes again and cool on ice until electrophoresis.

3.3.4. Gel Electrophoretic Separation and Visualization of PCR Products

1. Prepare a 1.6% agarose gel (1.6 g/100 mL of 1X TBE buffer).
2. Mix 10 µL of each PCR product with 2 µL of 6X DNA loading buffer and load it into the wells of the gel (*see* **Note 4**).
3. Run the gel in 1X TBE buffer for 1.5 h at 90 V (*see* **Note 5**).
4. Stain the gel in ethidium bromide solution (0.5 mg/mL) for 10–15 min at room temperature.
5. Detect the PCR products under UV light. Determine the size of the PCR products and document the results (e.g., by photography). An example illustrating the sensitivity of detection is given in **Fig. 1A**.

Table 2
Volumes for Amplification Controls

Components (stock concentration)	Vol/reaction (final concentration)	Vol for 22X master mixture (for 20 reactions)
PCR-grade water	8.6 µL	189.2 µL
10X PCR buffer	2.5 µL	55 µL
dNTP Mixture (2 m*M*)	2.5 µL (200 µ*M*)	55 µL
Primer HB10 (10 pmol/µL)	1.0 µL (0.4 µ*M*)	22 µL
Primer HB11 (10 pmol/µL)	1.0 µL (0.4 µ*M*)	22 µL
Magnesium chloride (50 m*M*)	0.75 µL (1.5 m*M*)	16.5 µL
Fraction V BSA (10 mg/mL)	2.5 µL (1 µg/µL)	55 µL
pUC19 DNA (1 fg/µL)	1.0 µL (1 fg)	22 µL
Taq DNA polymerase (5 U/µL)	0.15 µL (0.75 U)	3.3 µL
DNA extraction template	5.0 µL	
Total vol	25.0 µL	

3.4. Southern Blot

In order to check if the *Salmonella*-specific PCR product has been formed, Southern blotting, with subsequent hybridization, should be conducted as described in the following protocols.

PCR products separated by gel electrophoresis are transferred onto a positively charged nylon membrane and hybridized against a DIG-labeled PCR probe. The transfer of PCR products onto the membrane can be performed by capillary, vacuum, or electroblotting techniques, all of which are standard laboratory protocols *(23–24)*. Here, we describe a Southern transfer using a vacuum blotter.

1. After documentation, rinse the agarose gel with distilled water and incubate it in 500 mL of 0.25 *M* HCl for 15 min at room temperature with gentle agitation.
2. Pour off the HCl and rinse the gel with distilled water. Add 500 mL of 0.5 *N* NaOH and shake as in **step 1** for 30 min.
3. Pour off the NaOH solution and place the gel in 2X SSC buffer.
4. Cut a piece of the nylon membrane large enough to cover the gel surface.
5. Set up the vacuum blotter according to the manufacturer's instructions. Generally, a set-up includes a Whatman 3MM paper of appropriate size, the nylon membrane, a window plastic gasket, the agarose gel, and the sealing frame. Overlay the arrangement with 10X SSC buffer. Apply vacuum no more than 5 mm Hg for 45 min.
6. After the transfer, rinse the nylon membrane in 2X SSC buffer (*see* **Note 6**). Dry the membrane between Whatman 3MM paper and place it DNA-side-down on a

Table 3
Incorporation of DIG-labeled dUTP

Components (stock concentration)	Vol/reaction (final concentration)		
PCR-grade water	11.88	μL	
10X PCR buffer	2.5	μL	
dATP, dCTP, and dGTP-solution (2 m*M* each)	2.0	μL	(200 μ*M* each)
dTTP solution (10 m*M*)	0.475	μL	(190 μ*M*)
DIG-11-dUTP (1 m*M*)	0.25	μL	(10 μ*M*)
Primer invA-int1 (10 pmol/μL)	1.0	μL	(0.4 μ*M*)
Primer invA-int2 (10 pmol/μL)	1.0	μL	(0.4 μ*M*)
Magnesium chloride (50 m*M*)	0.75	μL	(1.5 m*M*)
Taq DNA Polymerase (5 U/μL)	0.15	μL	(0.75 U)
Salmonella reference DNA (0.2 ng/μL)	5.0	μL	(1 ng)
Total vol	25.0	μL	

UV transilluminator to cross-link the transferred DNA. Expose the membrane to UV light no longer than 2 min.

7. Store the membrane until hybridization between Whatman 3MM paper at room temperature.

3.5. Preparation of the Labeled invA PCR Probe

The PCR probe (231 bp) consists of an internal *invA* sequence flanked by primers 139/141 and is produced by incorporation of labeled deoxynucleotide triphosphate during the amplification reaction. Several labeling kits are available on the market. Here, we describe the incorporation of DIG-labeled dUTP.

1. Prepare two identical amplification reactions according to **Table 3.**
2. Perform the PCR as described in **Subheading 3.3.1.**
3. Separate the PCR product by gel electrophoresis as described in **Subheading 3.3.4.** (*see* **Note 7**).
4. Excise the PCR product with a lancet from the agarose gel. Store the agarose block until use at –20°C.

3.6. Hybridization

1. Place the membrane into a roller bottle (150 × 35 mm) and add 15 mL of preheated DIG Easy-Hyb solution. Incubate the bottle for 1 h at 42°C with gentle rotation (*see* **Note 8**).
2. Denature the PCR product enclosed by the agarose block in the presence of 1 mL pre-hybridization by boiling for 10 min.

Table 4
PCR Results

Specific Amplicon (284 bp)	Amplification Control (429 bp)	Result
positive	positive	positive
negative	positive	negative
negative	negative	not interpretable (inhibition)

3. Replace the DIG Easy Hyb solution with 10 mL of fresh DIG Easy Hyb solution containing 500 µL of denatured PCR probe solution (20 ng PCR probe/mL hybridization solution) and incubate for at least 6 h or, even better, overnight at 42°C with slow rotation.
4. Pour off the hybridization solution and wash the membrane 2× with 100 mL of washing solution I for 5 min at room temperature (*see* **Note 9**).
5. Wash the membrane 2× with 100 mL of washing solution II for 15 min at 68°C.

3.7. Detection

Colorimetric detection is performed with the DIG Nucleic Acid Detection Kit (Roche Diagnostics; Cat. no, 1175041). All components for detection can be ordered separately.

1. Place the membrane into an appropriate plastic box and wash with 50 mL of washing buffer for 5 min at room temperature.
2. Incubate the membrane for 30 min in 40 mL of 1X blocking solution.
3. Incubate the membrane for 30 min in 20 mL of anti-DIG-AP conjugate, Fab fragments solution at room temperature.
4. Remove the supernatant and wash the membrane 2× at room temperature for 15 min with 50 mL of washing buffer.
5. Equilibrate the membrane for 5 min in 20 mL of detection buffer.
6. Remove the supernatant and incubate the membrane in 10 mL of freshly prepared color substrate solution at room temperature in the dark without agitation. When the positive bands are colored sufficiently (approx 2 h) stop the reaction with 50 mL of TE buffer.
7. Dry and photograph the membrane. Store it at room temperature between Whatman 3 MM filter paper. An example illustrating the sensitivity of detection is given in **Fig. 1B**.

3.8. Evaluation and Documentation

All possible combinations of PCR results are shown in **Table 4**. A specific amplicon is defined as a PCR product of the expected size, which is confirmed by hybridization as described in **Subheadings 3.6.** and **3.7.** (*see* **Note 10**).

4. Notes

1. The duration and frequency for stomacher treatment depends on the kind of food. Samples for bacteriological analysis typically require 30 s, although products with a high fat content (e.g., bacon) require a longer treatment period up to 2 min.
2. For the separation of the food matrix from the microbial pre-enrichment, a centrifugation step of 50g for 10 min is recommended. The supernatant is transferred into a new microcentrifuge tube and further processed as described in **Subheading 3.2.**
3. The addition of BSA to the PCR mixture is recommended if inhibitory substances are expected. BSA can reduce the inhibitory effects of a broad range of substances.
4. An appropriate molecular weight standard, e.g., Marker X (Roche Diagnostics), should be run with the samples in order to determine the correct size of the specific amplicon. The wells of the comb should not be larger than 1 × 7 mm.
5. The electrophoresis is performed at a voltage of 5 V/cm of electrode distance. The duration of the run depends on the required distance of 7 to 8 cm of the bromphenol blue front.
6. In order to check the efficient transfer of DNA from the gel to the membrane, put the blotted agarose gel under UV light. Bands of PCR products should not be visible anymore.
7. Due to the incorporation of the DIG-11-dUTP, the labeled PCR product runs slower than an unlabeled product. The labeled 231-bp PCR probe appears at a position corresponding to 290 bp of the molecular weight standard.
8. Instead of roller bottles, hybridization bags (Roche Diagnostics; Cat. no. 1666649) can be used. A vol of 20 mL of DIG Easy Hyb /100 cm^2 membrane is recommended for prehybridization, and 3.5 mL of DIG Easy Hyb/100 cm^2 membrane for hybridization.
9. The hybridization solution containing the PCR probe can be stored at –20°C and reused.
10. If the result is not interpretable, inhibitory substances extracted from the food matrix were probably present in the reaction mixture. In this case, an alternative DNA extraction method should be tried.

Acknowledgments

B.M. was supported by the European Commission, proposal no. PL QLRT-1999-00226. We thank Cornelia Bunge and Ernst Junker for technical assistance.

References

1. Humphrey, T. (2000) Public-health aspects of *Salmonella* infection, in *Salmonella in domestic animals* (Way, C. and Way, A., eds.), CABI Publishing, Oxon, UK, pp. 245–263.
2. Tirado, C. and Schmidt, K. (2002) WHO Surveillance Programme for Control of Foodborne Infections and Intoxications: Results and trends across greater europe. *J. Infect.* **43,** 80–84.

3. Wallace, D. J., Van Gilder, T., Shallow, S., et al. (2000) Incidence of food-borne illnesses reported by the food-borne diseases active surveillance network (FoodNet)-1997. FoodNet Working Group. *J. Food Prot.* **63**, 807–809.

4. Waltman, W. D. (2000) Methods for the cultural isolation of *Salmonella*, in *Salmonella in Domestic Animals* (Wray, C. and Wray, A., eds.), CABI Publishing, Oxon, UK, pp. 355–372.

5. van der Zee, H. and Huis in't Veld, J. H. J. (2000) Methods for rapid detection of Salmonella, in *Salmonella in Domestic Animals* (Way, C. and Way, A., eds.), CABI Publishing, Oxon, UK, pp. 373–391.

6. Anonymous (1993) Microbiology of food and animal feeding stuffs—Horizontal method for the detection of Salmonella (ISO 6579:1993). *International Organization for Standardization.*

7. Rossen, L., Nørskov, P., Holmstrøm, K., and Rasmussen, O. F. (1992) Inhibition of PCR by components of food samples, microbial diagnostic assays and DNA-extraction solutions. *Int. J. Food Microbiol.* **17**, 37–45.

8. Wilson, I. G. (1997) Inhibition and facilitation of nucleic acid amplification. *Appl. Environ. Microbiol.* **63**, 3741–3751.

9. Lantz, P.-G., Hahn-Hägerdal, B., and Rådström, P. (1994) Sample preparation methods in PCR-based detection of food pathogen. *Trends Food Sci. Technol.* **5**, 384–389.

10. Lantz, P. G., al Soud, W. A., Knutsson, R., Hahn-Hagerdal, B., and Rådström, P. (2000) Biotechnical use of polymerase chain reaction for microbiological analysis of biological samples. *Biotechnol. Annu. Rev.* **5**, 87–130.

11. Widjojoatmodjo, M. N., Fluit, A. C., Torensma, R., Keller, B. H., and Verhoef, J. (1991) Evaluation of the magnetic immuno PCR assay for rapid detection of *Salmonella. Eur. J. Clin. Microbiol. Infect. Dis.* **10**, 935–938.

12. Rahn, K., De Grandis, S. A., Clarke, R. C., et al. (1992) Amplification of an *invA* gene sequence of *Salmonella typhimurium* by polymerase chain reaction as a specific method of detection of *Salmonella. Mol. Cell. Probes* **6**, 271–279.

13. Aabo, S., Rasmussen, O. F., Rossen, L., Sorensen, P. D., and Olsen, J. E. (1993) *Salmonella* identification by the polymerase chain reaction. *Mol. Cell. Probes* **7**, 171–178.

14. Doran, J. L., Collinson, S. K., Burian, J., et al. (1993) DNA-based diagnostic tests for Salmonella species targeting *agfA*, the structural gene for thin, aggregative fimbriae. *J. Clin. Microbiol.* **31**, 2263–2273.

15. Jones, D. D., Law, R., and Bej, A. K. (1993) Detection of *Salmonella* spp. in oysters using polymerase chain reaction (PCR) and gene probes. *J. Food. Sci.* **58**, 1191–1197.

16. Cohen, N. D., Neibergs, H. L., Wallis, D. E., Simpson, R. B., McGruder, E. D., and Hargis, B. M. (1994) Genus-specific detection of salmonellae in equine feces by use of the polymerase chain reaction. *Am. J. Vet. Res.* **55**, 1049–1054.

17. Chen, S., Yee, A., Griffiths, M., et al. (1997) A rapid, sensitive and automated method for detection of *Salmonella* species in foods using AG-9600 AmpliSensor Analyzer. *J. Appl. Microbiol.* **83**, 314–321.

18. Burkhalter, P. W., Müller, C., Luthy, J., and Candrian, U. (1995) Detection of *Salmonella* spp. in eggs: DNA analyses, culture techniques, and serology. *J. AOAC Int.* **78**, 1531–1537.

19. Kwang, J., Littledike, E. T., and Keen, J. E. (1996) Use of the polymerase chain reaction for *Salmonella* detection. *Lett. Appl. Microbiol.* **22**, 46–51.

20. Bäumler, A. J., Heffron, F., and Reissbrodt, R. (1997) Rapid detection of *Salmonella enterica* with primers specific for *iroB. J. Clin. Microbiol.* **35**, 1224–1230.

21. Makino, S., Kurazono, H., Chongsanguam, M., et al. (1999) Establishment of the PCR system specific to *Salmonella* spp. and its application for the inspection of food and fecal samples. *J. Vet. Med. Sci.* **61**, 1245–1247.

22. Collazo, C. M. and Galán, J. E. (1997) The invasion-associated type-III protein secretion system in *Salmonella*—A review. *Gene* **192**, 51–59.

23. Ausubel, F. M., Brent, R., Kingstone, R. E., et al. (1997) *Current Protocols in Molecular Biology,* John Wiley & Sons, Massachusetts, USA.

24. Sambrook, J., Fritsch, E. F., and Maniatis, T. (1989) *Molecular Cloning: A Laboratory Manual,* CSH Laboratory Press, Cold Spring Harbor, New York, USA.

20

Detection of *Toxoplasma gondii*

Jonathan M. Wastling and Jens G. Mattsson

1. Introduction

Toxoplasma gondii is an important intracellular protozoan that is the causative agent of toxoplasmosis in humans and animals. Toxoplasmosis is a zoonosis and is normally caught by eating undercooked infected meat or by ingestion of oocysts excreted by its definitive host, the cat. It is responsible for abortion and congenital defects in humans and is an important cause of abortion in domestic livestock, especially sheep, goats, and pigs. Infection in farm animals poses a risk to public health, as well as causing economic losses to the farming industry. *T. gondii* infection is established by rapid multiplication of the tachyzoite stage of the parasite. Although generally effective, the immune response does not completely eliminate the parasite; instead the tachyzoites differentiate into bradyzoites that form quiescent cysts in the brain and other tissues causing chronic infection that persists for the lifetime of the host *(1)*. Infection with *T. gondii* is extremely common (20–80% prevalence in Europe and North America), but the majority of people show no overt clinical symptoms. Immunosuppression, however, rapidly leads to the breakdown of the tissue cysts, recrudescence of infection, and development of toxoplasmic encephalitis (TE). Since no drugs are available that are effective against chronic infection, the tissue cysts remain a life-long risk for reactivation of acute toxoplasmosis. TE is one of the major opportunistic infections of the central nervous system, it is rapidly progressive and fatal if not treated. Rapid and accurate diagnosis of toxoplasmosis is essential for the management of human infections and is also an important research tool for understanding the role of animal reservoirs in the spread of infection. Indirect diagnosis by detection of antibodies to *T. gondii* is feasible, but is less useful in determining the timing of infection. The ability to detect low numbers of parasites rapidly in a range of tissues

From: *Methods in Molecular Biology, vol. 216: PCR Detection of Microbial Pathogens: Methods and Protocols*
Edited by: K. Sachse and J. Frey © Humana Press Inc., Totowa, NJ

is an important consideration for the management of congenital and postnatally acquired toxoplasmosis. Direct detection of *T. gondii* parasites by amplification of DNA using polymerase chain reaction (PCR) has, therefore, become a valuable asset *(2)*. The most widely used gene target was first described by Burg and colleagues and consists of the 35-fold repetitive B1 locus *(3)*. Since then, a large number of reports have described the use of the B1 target sequence, both for the clinical management of the disease *(4–8)* and also an important research tool for the study of the parasite in humans and animals *(6,9–11)*. Independent studies show that the B1 locus is the best target identified so far for the routine and sensitive detection of *T. gondii* in human and animal tissue *(9,12,13)*.

2. Materials

2.1. Sample Preparation

1. Phosphate-buffered saline (PBS): 58 mM Na$_2$HPO$_4$, 17 mM NaH$_2$PO$_4$, 68 mM NaCl, pH 7.3–7.4. Alternative PBS recipes may also be used.
2. Red blood cell (RBC)-lysis solution: 10 mM Tris-NH$_4$Cl.
3. Proteinase K stock solution 20 mg/mL. Molecular biology-grade, product no. P2308 (Sigma, St. Louis, MO, USA).
4. Tween® 20 (Amersham Biosciences, Uppsala, Sweden).
5. Lymhoprep™ (Nycomed, Oslo, Norway).
6. Suspension buffer: 10 mM Tris-HCl, pH 8.3, 40 mM KCl, 2.5 mM MgCl$_2$ containing proteinase K 100 µg/mL and Tween 20 (0.5%).
7. Tissue solution: 50 mM Tris-HCl, pH 8.5, 1 mM EDTA. containing proteinase K 100 µg/mL and Tween 20 (0.5%).
8. Microcentrifuge.
9. Water bath.

2.2. DNA Amplification

1. Oligonucleotide primers P1(5'-GGA ACT GCA TCC GTT CAT GAG-3'), P2 (5'-TGC ATA GGT TGC AGT CAC TG-3'), P3 (5' GGC GAC CAA TCT GCG AAT ACA CC-3') and P4 (5'-TCT TTA AAG CGT TCG TGG TC-3').
2. AmpliTaq® DNA polymerase (Applied Biosystems, Foster City, CA, USA.)
3. 10X PCR buffer: 100 mM Tris, pH 8.3, 400 mM KCl, 25 mM MgCl$_2$.
4. Gelatin-solution: 0.1% (w/v).
5. Deoxynucleotide triphosphates (dNTPs) 10 mM of each (Amersham Biosciences).
6. Nuclease free water.
7. DNA thermal cycler.

2.3. Gel Electrophoresis

1. DNA grade Agarose (Amersham Biosciences).

2. 50X TAE buffer: 242g Tris base, 57.1 mL glacial acetic acid, and 100 mL 0.5 M EDTA, pH 8.0 per L, final concentration 2 M Tris-acetate, 50 mM EDTA.
3. 10X TBE buffer: 108 g Tris base, 15.5 mL 85% phosphoric acid, and 40 mL 0.5 M EDTA, pH 8.0, per L, final concentration 0.45 M Tris-borate, 10 mM EDTA.
4. 30% Acrylamide/bis solution, 19:1; TEMED, and 10% Ammonium persulphate solution (Bio-Rad, Hercules, CA, USA).
5. Ethidium bromide, 10 mg/mL (Bio-Rad). Note: *Wear gloves when handling.*
6. Gel loading solution (GLS) (6X): 0.25% bromophenol blue, 0.25% xylene cyanol FF, 40% (w/v) sucrose in water.
7. Molecular weight marker: 50-bp ladder (Amersham Biosciences).
8. Agarose and/or polyacrylamide gel electrophoresis equipment.
9. UV-transilluminator.

2.4. Southern Blotting

1. Transfer solution: 3 M NaCl and 8 mM NaOH.
2. Positively charged Nylon membranes (Roche Molecular Biochemicals, Mannheim, Germany).
3. 20X SSC: 3 M NaCl, 0.3 M sodium citrate, pH 7.0.
4. PCR DIG Probe synthesis kit (Roche Molecular Biochemicals).
5. GENECLEAN® (Qbiogene, San Diego, CA,USA).
6. DIG Easy Hyb solution (Roche Molecular Biochemicals).
7. Wash solution, containing (2X SSC, 0.1% sodium dodecyl sulfate (SDS).
8. Wash solution (0.5X SSC 0.1% SDS).
9. DIG Nucleic Acid Detection Kit (Roche Molecular Biochemicals).

3. Methods
3.1. Sample Preparation

The methods described below outline sample preparation for *(i)* whole blood, *(ii)* lymph, and *(iii)* tissue samples. Note that vol and quantities of starting material can be scaled-up or -down according to sample type and availability (*see* **Note 1**).

3.1.1. Whole Blood

T. gondii is an intracellular parasite that can invade virtually any nucleated cell, but is unable to survive in RBCs. To detect the parasite in blood samples, it is usually best to isolate the white cells, removing as many RBCs as possible since their contents can be inhibitory to the PCR.

1. Dilute the fresh blood sample (5–10 mL) 50:50 with PBS and spin at 1000g for 10 min. This will partition the cells, so that an interface of white cells forms between the RBCs at the bottom of the tube and a straw-colored plasma layer at the top of the tube.

2. Remove the tube from the centrifuge and using a plastic pipet, carefully remove the interface containing the white cells. Mix this with 2 mL PBS.

3. Layer the diluted white cells over Lymphoprep (10 mL) and spin at 1000g for 15 min to clean the cells further. Again, carefully remove the white cell interface using a plastic disposable pipet and decant it into a clean tube.

4. At this stage, some RBCs may remain, and these are best removed by lysis with ammonium chloride. To do this, treat the sample with an equal volume of 10 mM Tris-NH$_4$Cl for 2 min at room temperature and then wash the resulting cell suspension 3× by resuspension in PBS and centrifugation at 1000g for 10 min. The pellet may now be stored frozen at –20°C for later processing if desired.

5. Resuspend the resulting pellet in 50 μL of 10 mM Tris, pH 8.3, 40 mM KCl, 2.5 mM MgCl$_2$ containing proteinase K 100 μg/mL and Tween 20 (0.5%).

6. Incubate the mixture at 55°C for 1 h and then inactivate the proteinase K by boiling for 5 min.

7. Store the samples at –20°C for later PCR analysis (*see* **Note 2**).

3.1.2. Lymph

1. Dilute the lymph sample 1 in 10 with PBS to prevent clotting and centrifuge at 1000g for 10 min. Resuspend cells in 1 mL PBS.

2. Lyse any contaminating RBCs by the addition of 1 mL of 10 mM Tris-NH4Cl and wash the remaining cells 3× by resuspension in PBS and centrifugation at 1000g for 10 min. The pellet may now be stored frozen at -20°C for later processing.

3. Resuspend the resulting pellet in 50 μL of 10 mM Tris, pH 8.3, 40 mM KCl, 2.5 mM MgCl$_2$ containing proteinase K 100 μg/mL and Tween 20 (0.5%).

4. Incubate the mixture at 55°C for 1 h and then inactivate the proteinase K by boiling for 5 min.

5. Store the samples at -20°C for later PCR analysis (*see* **Note 3**).

3.1.3. Tissue Samples

PCR can be performed either on fresh or frozen tissue. However, if tissue is to be frozen before analysis care should be taken to freeze samples at –20°C as soon as possible after collection as autolysis of the sample and degradation of DNA may occur quite rapidly.

1. For processing for PCR, finely chop the samples and wash with ice-cold PBS.

2. Digest the tissues overnight at 37°C in 100 μL of 50 mM Tris, pH 8.5, 1 mM EDTA, Tween 20 (0.5%) containing proteinase K 200 μg/mL.

3. After digestion, inactivate the proteinase K by boiling for 5 min.

4. Store the samples at –20°C for later PCR analysis.

3.2. DNA Amplification

Detection of the parasite is by amplification of the *T. gondii* repetitive B1 locus.

1. For the first round of PCR set up a 50-mL volume reaction for each sample containing 5 µL of 10X PCR buffer, 0.01% gelatin, 100 mM of each dNTP, 0.2 mM of primer P1 and P4, respectively, and 2.5 U of Ampli*Taq* DNA polymerase. Add 5 µL of the test sample. Overlay the reaction mixture with 50 µL mineral oil if the thermal cycler does not have a heated lid.

2. Transfer the reaction tubes to a thermal cycler and amplify the 193-bp product over 25 cycles using the following conditions: 93°C for 1 min, 50°C for 1.5 min and 72°C for 3 min (*see* **Note 4**). Complete with a final extension step of 10 min at 72°C.

3. After amplification, the samples can be analyzed at this stage, stored frozen, or used directly in the second round amplification (*see* **Note 5**).

4. For the second round of nested amplification, dilute the PCR products from the first amplification 1 in 20 in distilled water to reduce amplification of nonspecific products.

5. Set up the second PCR mixture as above, but use the nested primers P2 and P3. Add 1 µL of the diluted sample.

6. Transfer the second round of reaction tubes to a thermal cycler and amplify a 94-bp product over 15 cycles using the following conditions: 93°C for 1 min, 50°C for 1.5 min, and 72°C for 3 min. Complete with a final extension step of 10 min at 72°C.

PCR controls: any PCR detection method is susceptible to false negative and false positive results. Because of the sensitivity of PCR, false positives usually result from cross-contamination of samples with minute quantities of target DNA, and careful steps need to be taken to ensure this does not occur (*see* **Note 6**). It is essential to run appropriate controls in all experiments. Negative controls should consist of *(i)* distilled water (to ensure that none of the PCR reagents have become contaminated with parasite DNA); *(ii)* uninfected blood and/or tissue samples extracted alongside the test samples (also to monitor cross-contamination of materials). In addition to the negative control, a positive control, consisting of parasite DNA, should be added to ensure the integrity of the PCR (*see* **Note 7**).

3.3. Product Detection

Using control *T. gondii* DNA, sensitivity can be achieved to at least 0.1 pg of target DNA using ethidium bromide detection. Sensitivity can be increased by Southern blotting to detect 0.05 pg of DNA. Agarose gel electrophoresis can normally be used for direct detection, (*see* **Fig. 1**) but the resolution can be improved on a polyacrylamide gel.

3.3.1 Agarose Gel Electrophoresis

1. Withdraw 15 µL of the completed PCR and mix with 3 µL of GLS.

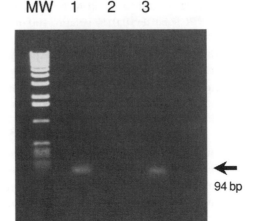

Fig. 1. B1-nested PCR products separated on an ethidium bromide-stained agarose gel and visualized under UV illumination. MW, molecular weight marker; lane 1, positive control; lane 2, negative control; lane 3, peripheral blood sample taken from a sheep 10 d after experimental infection with *T. gondii* tachyzoites.

2. Load all samples, including controls, on a 2.0% agarose gel containing 0.5 µg/mL ethidium bromide and 1X TAE buffer. Separate the samples at 5 V/cm for about 40 min.
3. Analyze and record the gel image under UV illumination.

3.3.2. Product Verification through Southern Blotting

1. Transfer DNA from the agarose gel in 3 M NaCl and 8 mM NaOH onto a nylon membrane by capillary blotting.
2. After the transfer mark the positions of each sample well and the positions of the molecular weight standards.
3. Wash the membrane with 2X SSC.
4. Cross-link DNA to the wet membrane according to the manufacturer's instructions. The membrane can be used immediately or can be stored dry at 4°C for future use.
5. To prepare the probe: assemble a PCR with primers P1 and P4 using the PCR DIG Probe synthesis kit according to the manufacturer's instructions. Use 100 ng *T. gondii* genomic DNA as template.
6. Separate the labeled PCR product on a 1% agarose gel. Use a clean scalpel to excise the gel segment containing the PCR product.
7. Purify the DNA from the agarose gel using GENECLEAN according to the manufacturer's instructions.
8. Prehybridize the membrane in DIG Easy Hyb solution for 2 h. Use 20 mL prehybridization solution per 100 cm^2 membrane.

9. Denature the probe by boiling for 10 min. Chill immediately on ice.
10. Dilute the probe in fresh DIG Easy Hyb solution. Replace the prehybridization solution and hybridize overnight at 42°C.
11. Wash the membrane 2×, 5 min/wash in 2X Wash solution at room temperature.
12. Wash the membrane 2×, 15 min/wash in 0.5X Wash solution at 65°C.
13. The probe that is hybridized to the amplified DNA can now be detected with the DIG Nucleic Acid Detection Kit. The correct PCR product will be indicated by a strong hybridization signal at 193 or 94 bp.

3.3.3. Polyacrylamide Gel Electrophoresis

1. Cast an 8% polyacrylamide gel in 1X TBE buffer.
2. Prepare the samples as above, in **Subheading 3.3.1**, and load them on the gel.
3. Run the gel in 1X TBE buffer at a voltage gradient between 1 and 8 V/cm.
4. Run the gel until the marker dyes have migrated the desired distance. Xylene cyanol FF will migrate as a 160 bp fragment and bromophenol blue as a 45 bp fragment in an 8% polyacrylamide gel.
5. Remove one glass plate and then gently submerge the gel and its attached glass plate in staining solution (0.5 µg/mL ethidium bromide in 1X TBE).
6. After staining for 30–45 min, analyze and record the gel image under UV illumination (*see* **Note 8**)

4. Notes

1. In low-level infections, the ability to detect *T. gondii* in any sample depends largely upon sampling strategy. If the parasite is very sparsely disseminated, then the more tissue that can be collected and processed, the greater chance there will be of detection by PCR. However, it becomes impractical to process very large quantities of tissue routinely, so detection thresholds will always be dependent on how much tissue is collected. When processing tissue, small quantities of tissue can be pooled from different part of an organ (like the brain) to maximize the chance of parasite detection.
2. The methods described here involve PCR performed directly on unpurified samples, without the specific extraction of DNA. These methods are especially suited for high-throughput analysis, while retaining excellent sensitivity. If desired, DNA can be extracted from the processed tissues using any number of standard methods *(14)* and the PCR performed on purified DNA. However, the potential increase in the PCR efficiency needs to be balanced against the extra processing steps required. Explicit DNA extraction does not always lead to increased sensitivity and, in some cases, can prove less sensitive because of the loss of starting material.
3. This method can be adapted for other fluid tissue samples, which contain small quantities of blood cells.
4. The number of cycles can be increased up to 35 in the first round amplification to try and achieve a greater sensitivity. However, increasing the number of cycles can lead to increased background and nonspecific amplification. Recently a real-

time PCR for quantitative detection of *T. gondii* targeting the BI locus has been described *(15)*. With this fully automatic system, a single 40-cycle PCR has a sensitivity consistent with the nested B1 PCR. Since the detection is done in real-time, the hands-on time is reduced significantly. However, for this application specialized real-time PCR instrumentation is necessary.

5. If desired, products can be analyzed at this stage. If the initial sample contains large amounts of parasite DNA, then a second of amplification will not be required. However, in practice, for routine analysis, it may be desirable to proceed immediately to the second round of amplification.

6. General steps to avoid DNA cross-contamination should include:

   a. The use of pipet tips with integral filters. Although more expensive, filter tips are highly effective at preventing contamination of the barrel of pipets with DNA.

   b. Preparation and freezing of batches of PCR reagents. Each PCR experiment is then run with the same batch of materials, and excess material is discarded. This way, any contamination of stock solutions (such as dNTPs) is contained within a single experiment and can be eliminated by repeating with a fresh batch of reagents.

   c. Amplified DNA products (i.e., any material downstream of the PCRs) should be processed away from PCR preparation areas. In practice, this usually means keeping the product analysis equipment separate, preferably in a different area of the laboratory, or in a different room. Laminar flow hoods are useful for setting up PCRs in a clean environment. Pipets used for loading gels with product DNA should never be used for sample preparation or for setting up PCRs. Ideally, dedicated sets of pipets should be used for the different processes. When performing a nested PCR, the second round of amplification should be prepared in area separated from where the first round was prepared.

7. Positive control *T. gondii* DNA can be obtained by extraction of DNA from cell culture derived *T. gondii* tachyzoites as follows. Lyse 10^9 *T. gondii* tachyzoites in 50 m*M* Tris, pH 8.0, 50 m*M* EDTA, SDS 1% containing proteinase K 100 µg/mL and incubate for 3 h. Extract the nucleic acids with phenol:chloroform, then precipitate with sodium acetate-ethanol at –20°C and pellet the DNA by centrifugation at 10,000*g* for 15 min. Wash the DNA pellet in ethanol 70% and resuspend in TE (10 m*M* Tris-HCl, 0.1 m*M* EDTA, pH 7.6). Remove RNA by incubation with RNAase 20 µg/mL for 30 min at 37°C and re-extract and precipitate the DNA as above. Finally, resuspend the DNA pellet in 100 µL of TE and quantify the DNA by measuring UV absorption at 260 nm. Use 10 ng DNA as a positive control.

8. Since polyacrylamide quenches the fluorescence of ethidium bromide, it is not possible to detect bands that contain less than about 10 ng of DNA by this method. It is possible to increase the sensitivity to 20–50 pg DNA/band by using any standard silver staining protocol *(14)* or silver staining kits, available from Amersham Biosciences.

References

1. Sibley, L. D., Mordue, D., and Howe, D.K. (1999) Experimental approaches to understanding virulence in toxoplasmosis. *Immunobiology,* **201,** 210–224.
2. Ellis, J. T. (1998) Polymerase chain reaction approaches for the detection of *Neospora caninum* and *Toxoplasma gondii. Int. J. Parasitol.,* **28,** 1053–1060.
3. Burg, J. L., Grover, C. M., Pouletty, P., and Boothroyd, J. C. (1989) Direct and sensitive detection of a pathogenic protozoan *Toxoplasma gondii,* by polymerase chain reaction. *J. Clin. Microbiol.* **27,** 1787–1792.
4. Hohlfeld, P., Daffos, F., Costa, J. M., Thulliez, P., Forestier, F., and Vidaud, M. (1994) Prenatal diagnosis of congenital toxoplasmosis with a polymerase-chain-reaction test on amniotic fluid. *N. Engl. J. Med.* **331,** 695–699.
5. Lamoril, J., Molina, J. M., de Gouvello, A., et al. (1996) Detection by PCR of *Toxoplasma gondii* in blood in the diagnosis of cerebral toxoplasmosis in patients with AIDS. *J. Clin. Pathol.* **49,** 89–92.
6. Parmley, S. F., Goebel, F. D., and Remington, J. S. (1992) Detection of *Toxoplasma gondii* in cerebrospinal fluid from AIDS patients by polymerase chain reaction. *J. Clin. Microbiol.* **30,** 3000–3002.
7. Sugane, K., Takamoto, M., Nakayama, K., Tada, T., Yuasa, T., and Kurokawa, K. (2001) Diagnosis of *Toxoplasma* meningoencephalitis in a non-AIDS patient using PCR. *J. Infect.* **42,** 159–160.
8. Vogel, N., Kirisits, M., Michael, E., et al. (1996) Congenital toxoplasmosis transmitted from an immunologically competent mother infected before conception. *Clin. Infect. Dis.* **23,** 1055–1060.
9. Wastling, J.M., Nicoll, S. and Buxton, D. (1993) Comparison of two gene amplification methods for the detection of *Toxoplasma gondii* in experimentally infected sheep. *J. Med. Microbiol.* **38,** 360–365.
10. Burney, D.P., Lappin, M.R., Spilker, M. and McReynolds, L. (1999) Detection of *Toxoplasma gondii* parasitemia in experimentally inoculated cats. *J. Parasitol.* **85,** 947–951.
11. Grigg, M. E., and Boothroyd, J. C. (2001) Rapid identification of virulent type I strains of the protozoan pathogen *Toxoplasma gondii* by PCR-restriction fragment length polymorphism analysis at the B1 gene. *J. Clin. Microbiol.* **39,** 398–400.
12. Pelloux, H., Guy, E., Angelici, M. C., et al. (1998) A second European collaborative study on polymerase chain reaction for *Toxoplasma gondii,* involving 15 teams. *FEMS Microbiol. Lett.,* **165,** 231–237.
13. Jones, C.D., Okhravi, N., Adamson, P., Tasker, S. and Lightman, S. (2000) Comparison of PCR detection methods for B1, P30, and 18S rDNA genes of *T. gondii* in aqueous humor. *Invest. Ophthalmol. Vis. Sci.* **41,** 634–644.
14. Sambrook, J., Fritsch, E. F., and Maniatis, T. (1989) *Molecular Cloning: A Laboratory Manual. 2nd ed.* CSH Laboratory, Cold Spring Harbor, New York, USA.
15. Lin, M. H., Chen, T. C., Kuo, T. T., Tseng, C. C., and Tseng, C.P. (2000) Real-time PCR for quantitative detection of *Toxoplasma gondii. J. Clin. Microbiol.* **38,** 4121–4125.

21

PCR-Derived Methods for the Identification of *Trichinella* Parasites from Animal and Human Samples

Edoardo Pozio and Giuseppe La Rosa

1. Introduction

Trichinella worms (family: Trichinellidae; phylum: Nematoda) are parasites that mainly infect mammals, including humans, although they have been found in birds and, recently, in African crocodiles *(1,2)*. The main reservoir is represented by carnivores with cannibalistic and scavenger behavior. These parasites are widespread on all continents but Antarctica, from frigid to torrid zones. The main distinguishing feature of their life cycle is that two generations occur in the same host. The first generation (from L_1 larva to adult) is present in the gut and the second generation (from a newborn larva of 80 μm in length to an infective larva of 0.6–1.0 mm in length) is present in the cell of striated muscles that is modified by the larva (referred to as the nurse cell) *(3,4)*.

Until 1972, *Trichinella spiralis* was the only known *Trichinella* species. At present, seven species and three additional genotypes are known to exist (**Table 1**) *(2)*. An additional *Trichinella* genotype has been recently identified in crocodiles from Zimbabwe, but its taxonomic status have yet to be established. The individual species or genotypes cannot be distinguished on the basis of morphological characters. Nonetheless, the species and genotypes can be categorized into two groups based on whether or not a collagen capsule surrounds the nurse cell in host muscles: when the collagen capsule is present, the species or genotype is referred to as encapsulated; if absent, the term nonencapsulated is used. The presence of the collagen capsule is important for the long-term survival of the larva in decaying muscle tissue (i.e., when the larva is not protected by the host homeothermy).

From: *Methods in Molecular Biology, vol. 216: PCR Detection of Microbial Pathogens: Methods and Protocols*
Edited by: K. Sachse and J. Frey © Humana Press Inc., Totowa, NJ

Table 1
Principal Features of *Trichinella* Species and Genotypes *(1,2)*

Trichinella species Genotype	Distribution	Cycle	Hosts	Collagen capsule
T. spiralis	Cosmopolitan[a]	Domestic and sylvatic	Swine, rats, carnivores	yes
T. nativa	arctic and subarctic areas of Holoarctic region[b]	Sylvatic	Terrestrial and marine carnivores	yes
Trichinella T6	Rocky mountains[c]	Sylvatic	Carnivores	yes
T. britovi	temperate areas of Palearctic region [d]	Sylvatic, seldom domestic	Carnivores, seldom swine	yes
Trichinella T8	South Africa[e]	Sylvatic	Carnivores	yes
Trichinella T9	Japan	Sylvatic	Carnivores	yes
T. pseudospiralis	Cosmopolitan[f]	Sylvatic, seldom domestic	Mammals and birds	no
T. murrelli	temperate areas of Nearctic region	Sylvatic	Carnivores	yes
T. nelsoni	Ethiopic region	Sylvatic	Carnivores, seldom swine	yes
T. papuae	Papua New Guinea	Sylvatic, seldom domestic	Swine	no
Trichinella T11	Zimbabwe	Domestic?	Crocodiles	no

[a] This species has not been detected in arctic regions.
[b] The isotherm $-5°C$ in January is the southern limit of distribution.
[c] From Alaska to Idaho, USA.
[d] The isotherm $-6°C$ in January is the northern limit of distribution.
[e] Probably imported from Europe in the 17–18th century.
[f] Three different populations have been identified in the Nearctic region (Alabama, USA), Palearctic region (many foci), and in the Australian region (Tasmania) (La Rosa et al. 2001).

The natural transmission cycle of all species and genotypes of *Trichinella* occurs among wildlife (i.e., a sylvatic cycle), although *T. spiralis* is primarily transmitted by a domestic cycle involving domestic pigs, synanthropic rats, and, on rare occasions, horses *(2)*. A domestic cycle involving pigs and rats has also been reported, though rarely, for *T. britovi* and *T. pseudospiralis (2)*. The prevalence of infection among carnivores (e.g., fox, wolf, racoon dog, lynx, bear, and mustelid) ranges from 0.01%–90%, according to the characteristics of the animal's habitat and the extent of human influence *(5,6)*. *Trichinella* infection in sylvatic and domestic animals is mainly asymptomatic, even when thousands of larvae per gram of muscle tissue are present.

Humans acquire trichinellosis by eating raw or undercooked meat from pigs, horses and game animals (e.g., wild boar, bear, and walrus). The clinical picture and the prognosis of human infection depend on the number of infective larvae ingested, the *Trichinella* species, and the allergic reaction of the host *(7)*. The number of human deaths ranges from 0.1–1.0 per 100 infections, and most deaths are due to complications of the respiratory system and the cardiovascular system and neurological disorders.

The lack of morphological markers and the fact that the biological and biochemical markers (allozymes) do not allow the species to be easily or rapidly identified stress the importance of methods based on polymerase chain reaction (PCR) in identifying *Trichinella* parasites at the species level. Species identification is of great importance in tracing back the source of infection, in determining and predicting the clinical course of infection, in estimating the potential risk for pigs, in establishing appropriate strategies for control and eradication, and in better understanding the epidemiology of the infection. The use of PCR-derived methods also allows the species to be identified based on a single larva, which is important because, frequently, only one larva is detected in human biopsies and in muscle samples of animal hosts. Furthermore, the identification of single larvae allows more than one species of *Trichinella* to be detected in the same host (mixed infections) *(8)*.

The molecular identification of single larvae of *Trichinella* is carried out with a multiplex-PCR analysis *(9, 10)*, which allows the identification of seven species (*T. spiralis, T. nativa, T. britovi, T. pseudospiralis, T. murrelli, T. nelsoni,* and *T. papuae*), two genotypes (*Trichinella* T6 and the crocodile strain *Trichinella* T11), and three populations of *T. pseudospiralis* (from the Australian, Nearctic, and Palearctic regions). In addition, two other genotypes (*Trichinella* T8 and *Trichinella* T9) can be distinguished from *T. britovi* by a PCR-restriction fragment length polymorphism (RFLP) analysis of the gene encoding for a 43-kDa protein, which is one of the main excretory/secretory (E/S) antigens produced by larvae in vitro *(11)*. Both multiplex-PCR (*see* **Note 1**) and PCR-RFLP analyses are described in detail here. These two PCR-derived analyses are routinely used at the International *Trichinella* Reference Center *(12)* (www.trichi.iss.it).

2. Materials

2.1. Hosts and Preferential Muscles

Large carnivores with scavenger and cannibalistic behavior that are at the top of the food chain are the best candidates for the detection of *Trichinella*. Specifically, the most common candidates are: polar bear, brown bear, black bear, wolf, red fox, lynx, bob cat, and racoon dog in Nearctic and Palearctic regions; spotted hyena and lion in the Ethiopian region; and carnivorous marsupials in Tasmania. At the farm level (domestic cycle), the best candidate is the domestic pig; other candidates include synanthropic animals such as rats, foxes, mustelids, armadillos, domestic cats, and dogs, and farm crocodiles in Zimbabwe. *Trichinella* infection has also been detected in horses and sheep that were the source of human infection *(2,8)*.

The identification of *Trichinella* worms is most commonly based on muscle larvae, which are very easy to collect from both animals and humans. The preferential muscles (i.e., those with the highest density of larvae) vary according to the specific host species, but as a general rule, the tongue can be considered as the preferential muscle. Other important muscles are: the pillar of the diaphragm for swine and rodents, the anterior tibial for foxes and wolves, and the masseter for carnivores and horses. Human biopsies are generally taken from the deltoid muscle.

2.2. Parasite Isolation and Preservation

1. Commercial blender with a vol of at least 500 mL.
2. Suction pump (e.g., a water pump).
3. Incubator (37–45°C) with a capacity of at least 100 L and with an inner electrical socket.
4. Magnetic stirrer and magnets.
5. Precision scale.
6. Dissection microscope (20–40x).
7. Thermometer.
8. Automatic pipets: 1–20 µL and 10–200 µL range.
9. Beakers: capacity of at least 1 L.
10. Scissors and forceps.
11. Conical vials (0.5 and 50 mL) and racks.
12. Cooler.
13. Disposable gloves.
14. Petri dishes (5 and 6 cm diameter).
15. Pepsin 1:10,000 (*see* **Note 2**).
16. Hydrochloric acid.
17. Ethyl alcohol, anhydrous.
18. Tap water (37–45°C).
19. Sterile H_2O at 4° C.
20. Phosphate-buffered saline (PBS): 137 mM NaCl (8 g/L), 7 mM K_2HPO_4 (1.21 g/L), KH_2PO_4 (0.34 g/L), prewarmed to 37–45°C.
21. Digestion fluid: 1% pepsin (w), 1% HCl (v), tap water (37–45°C)

Table 2
Multiplex-PCR Amplicon Sizes
of Primer Sets of 10 *Trichinella* Genotypes

Primer pair	Ts	Tna	Tb	Tps-Ne	Tps-Pa	Tps-Au	Tm	T6	Tne	Tpa	T11
I	173	127	127	310	340	360	127	127	155	240	264
II		253									
III									210		
IV						316					
V										404	

Abbreviations: *T. spiralis* (Ts); *T. nativa* (Tna); *T. britovi* (Tb); *T. pseudospiralis* (Tps) of Nearctic (Ne), Palearctic (Pa), and Australian (Au) regions; *T. murrelli* (Tm); *Trichinella* T6 (T6); *T. nelsoni* (Tne); *T. papuae* (Tpa); and *Trichinella* T11 (T11).

2.3. Primer Sets for Multiplex-PCR and for PCR-RFLP

2.3.1. Multiplex PCR (for amplicon sizes see **Table 2**):

Primer pair I: 5'-GTTCCATGTGAACAGCAGT-3';5'-CGAAAACATA CGACAACTGC-3'

Primer pair II: 5'-GCTACATCCTTTTGATCTGTT-3';5'-AGACACAAT ATCAACCACAGTACA-3'

Primer pair III: 5'-GCGGAAGGATCATTATCGTGT-3';5'-ATGGATTA CAAAGAAAACCATCACT-3'

Primer pair IV: 5'-GTGAGCGTAATAAAGGTGCAG-3';5'-TTCATCAC ACATCTTCCACTA-3'

Primer pair V: 5'-CAATTGAAAACCGCTTAGCGTGTTT-3';5'-TGATC TGAGGTCGACATTTCC-3'

Each primer is diluted at 100 pmol/µL in sterile H_2O.

Multiplex primer set concentration: combine the same vol of each primer; final concentration: 10 pmol/µL of each primer (*see* **Note 3**).

2.3.2. Primer Sets for PCR-RFLP, 43 kDa (E/S)

Ts43CAF: 5'- ATGCGAATATACATTTTTCTTA-3'
Ts43CAR: 5'- TTAGCTGTATGGGCAAGG-3'
Each primer is diluted at 100 pmol/µL in sterile H_2O.

2.4. Preparation and Amplification of Larva DNA

1. PBS washing buffer: 137 mM NaCl (8 g/L), 7 mM K_2HPO_4 (1.21 g/L), KH_2PO_4 (0.34 g/L).
2. Sterile Tris-HCl buffer: 1 mM Tris-HCl, pH 7.6.

3. Proteinase K: 20 mg/mL in sterile H_2O; store 0.5-mL aliquots at $-20°C$.
4. *Taq* DNA polymerase: Ex Taq™ from Takara (Otsu, Shiga, Japan) 5 U/µL.
5. ExTaq buffer (10X) containing 20 m*M* $MgCl_2$.
6. dNTPs solution: final concentration: 2.5 m*M* of each dNTP (from Takara).
7. Restriction endonucleases: *Ssp*I and *Dde*I with 10X restriction buffers (New England BioLabs, Beverly, MA, USA).
8. Primer mixture: 1 µL of multiplex primer set. Store in aliquots of 200 µL at $-20°C$.
9. Mineral oil: sterile, PCR-grade.
10. Dry bath.
11. PCR device: Perkin Elmer Model 2400 (Perkin Elmer, Norwalk, CT, USA), Perkin Elmer Model 9600, MJ Minicycler (MJ Research, Watertown, MA, USA) (*see* **Note 4**).
12. TBE buffer: Tris base (10.8 g/L), boric acid (5.5 g/L), 0.5 M EDTA (4 mL/L), pH 8.0.
13. TAE buffer: Tris base (4.85 g/L), glacial acetic acid (1.15 mL/L), 0.5 M EDTA (2 mL/L).
14. 0.5 M EDTA solution: 18.6 g/70 mL Na_2 EDTA $\times 2H_2O$, pH 8.0, with 10 N NaOH, H_2O to 100 mL.
15. Agarose: molecular biology standard grade (Fisher, Cincinnati, OH, USA)

3. Methods

3.1. Isolation of Larvae from Infected Muscles (see Notes 5–7)

1. Preparation of the digestion fluid: The ratio of muscle (w) and digestion fluid (v) should be 1:20–1:40. Using a blender, dissolve Pepsin (*see* **Note 2**) in a small amount of tap water (*see* **Subheading 2.2., step 18.**). Add additional tap water until reaching the final vol, then add 1% HCl (final concentration). The digestion fluid should be maintained at around 37–45°C for all steps.
2. Cut the muscle sample into small pieces (1–3 g), removing all nonmuscle tissue (tendons, fat, etc.). The most infected muscle tissue is that near the muscle insertion. Place the muscle sample and a small amount of digestion fluid in a blender and blend for 20–30 s. Add additional digestion fluid and blend again for about 10 s. Place the fluid in a beaker containing a magnet. To collect residual fluid from the blender, add additional digestion fluid and blend. Place residual fluid in the same beaker.
3. Place the beaker on the magnetic stirrer in the incubator at 37–45°C and stir for 20 min. Switch off the magnetic stirrer, collect 4 to 5 mL of digestion fluid from the bottom of the beaker, place the fluid in a Petri dish, and observe under a dissection microscope. If the larvae are free of muscle debris and are out of the capsule, stop the digestion. If larvae are still in the muscle and/or in the capsule, continue the digestion for another 10–20 min.
4. Allow the digestion fluid to sediment for 10–15 min, according to the height of the beaker (about 1 min for each centimeter of height). Remove the supernatant by placing the suction pump at about 2 cm from the bottom of the beaker, being

careful not to remove the sediment, which contains the larvae. Add PBS (37–45°C) (same quantity as supernatant removed) and then allow to sediment. Remove the supernatant and place the sediment in 50-mL conical vials (5 mL for each vial). Add PBS (37–45°C) and allow to sediment. Repeat this procedure until the supernatant is fairly transparent (i.e., you should be able to read newspaper text through the glass). Remove the supernatant and place the sediment in a Petri dish and place under dissection microscope.

5. If the larvae are dead (C-shaped or comma shaped) (*see* **Notes 5–7**), the following procedures should be carried out as rapidly as possible to avoid DNA destruction. Collect larvae with a 5-µL pipet and place them in a Petri dish containing cold sterile H_2O. Then collect the single larvae with 5 µL of the cold sterile H_2O and place each of them in a separate 0.5-mL conical vial. Freeze at –30°C or, if the larvae need to be shipped, store them in absolute ethyl alcohol at 4°C (the latter method allows for shipping without dry ice, although the larvae must be rehydrated through a graded alcohol series before molecular identification).

3.2. Preparation of Crude DNA from Single Larvae

1. Wash single larvae 10 × in PBS, place each larva, with 5 µL of PBS, in a 0.5-mL tube, and store at –20°C until use (*see* **Note 8**).
2. Add 2 µL Tris-HCl, pH 7.6.
3. Add 1 drop of sterile mineral oil.
4. Heat sample at 90°C for 10 min and then cool on ice.
5. Add 3 µL of proteinase K solution (final concentration 100 µg/mL) and spin sample.
6. Incubate sample at 48°C for 3 h.
7. Heat sample at 90°C for 10 min and cool on ice.
8. Store sample at –20°C until use.

3.3. Multiplex-PCR Protocol

1. Thaw a sample of crude DNA extraction on ice (at this point, each tube should contain 10 µL of the larva preparation).
2. To set up PCR (*see* **Note 9**), add sequentially 5 µL 10X PCR buffer, 4 µL dNTPs, 2 µL set of primers, 0.1 µL *Taq* DNA polymerase (*see* **Note 10**), 4 µL of crude DNA extraction (*see* **Note 11**), and H_2O up to 50 µL in a 0.2-mL thin-walled tube.
3. Place tubes on ice.
4. PCR cycle: pre-amplification cycle at 94°C for 5 min, followed by 35 cycles at 94°C for 20 s, 58°C for 30 s, and 72°C for 1 min, followed by an extension cycle at 72°C for 4 min, and then place on ice.
5. Hot start at 94°C: wait until the thermal cycler reaches 94°C and then place the tubes on the hot plate.
6. Electrophoresis: use 20 µL of each amplification reaction (*see* **Table 2** and **Fig. 1**).

Table 3
PCR-RFLP Amplicon Sizes
of Principal Bands of *Trichinella britovi* Genotypes

	Restriction enzymes	
Trichinella genotype	*Dde* I	*Ssp* I
T. britovi	700, 680, 560, 520	2100
T8	700, 650, 560, 330	2300, 2100
T9	700, 650, 560, 520	1500, 1200, 650

3.4. 43 kDa (E/S) PCR-RFLP Digestion Protocols

1. Thaw a sample of crude DNA extraction on ice (at this point, each tube should contain 10 µL of the larva preparation).
2. To set up PCR, add sequentially 5 µL 10X PCR buffer, 4 µL dNTPs, 2 µL set of primers, 0.2 µL *Taq* DNA polymerase, 10 µL of crude DNA extraction, and H$_2$O up to 50 µL in a 0.2-mL thin-walled tube.
3. Place tubes on ice.
4. PCR cycle: pre-amplification cycle at 94°C for 5 min, followed by 30 cycles at 98°C for 20 s, and 60°C for 15 min, followed by an extension cycle at 72°C for 4 min, and then place on ice.
5. Hot start at 94°C: wait until the thermal cycler reaches 94°C and then place the tubes on the hot plate.
6. Electrophoresis: use 10 µL of the amplification reaction. Select samples showing good single-band amplification for restriction analysis.
7. Restriction analysis: transfer 20 µL of the amplification reaction into a 1.5-mL conical tube, add 5 µL of the respective 10X restriction buffer, 10 U of the selected enzyme (*see* **Note 12**), and H$_2$O up to 50 µL.
8. Incubate at 37°C for 2 h.
9. Transfer on ice and stop the reaction with 5 µL of 0.5 M EDTA.
10. Electrophoresis: load all of the reaction onto the agarose gel (*see* **Table 3** and **Fig. 2**).

3.5. Electrophoresis Conditions

1. Standard agarose gel: follow standard procedures to prepare 1–1.5% agarose gel in TBE or TAE buffer and run at 10 V/cm.
2. High-resolution agarose gel: to have an adequate resolution of *T. pseudospiralis* isolates, run the amplification products on 3% metaphor agarose gel at 10 V/cm.

4. Notes

1. In the original description of multiplex-PCR *(7)*, a nested-PCR was used. However, this is not necessary with the present protocol.

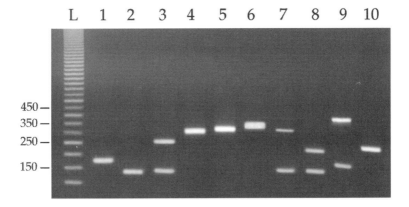

Fig. 1. Photograph of an ethidium bromide-stained 2.5% agarose gel under UV light illumination showing the multiplex PCR amplification (**Subheading 3.3.**) of single larvae of 10 genotypes of *Trichinella*. The samples are as follows: L (ladder 50), sizes are in base pairs; line 1, *T. spiralis*; line 2, *T. nativa*; line 3, *T. britovi*; line 4, *T. pseudospiralis* (Palearctic isolate); line 5, *T. pseudospiralis* (Nearctic isolate); line 6, *T. pseudospiralis* (Tasmanian isolate); line 7, *T. murrelli*; line 8, *Trichinella* T6; line 9, *T. nelsoni*; line 10, *T. papuae*.

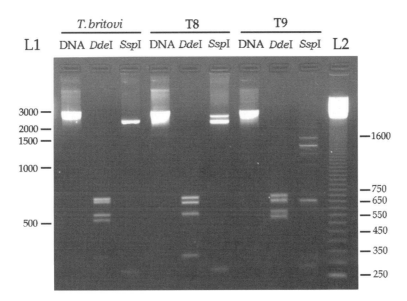

Fig. 2. Photograph of an ethidium bromide-stained 2.5% agarose gel under UV light illumination showing the PCR-RFLP identification (**Subheading 3.4.**) of single larvae of *T. britovi*, *Trichinella* T8, and *Trichinella* T9. The samples are as follows: DNA, 43 kDa (E/S)-PCR-amplified product; *Dde*I, 43 kDa (E/S)-PCR-amplified product *Dde*I-digested; *Ssp*I, 43 kDa (E/S)-PCR-amplified product *Ssp*I-digested; L1 (ladder 1000) sizes are in bp; L2 (ladder 50) sizes are in bp.

2. Pepsin should be stored in the dark at room temperature (20°C or less, but not below 4°C); avoid exposure to humidity. Pepsin should be no more than 6 mo old.

3. Balancing of primers. The primer-set mixture prepared with equimolar concentrations of all oligonucleotides generally provides good results; if results are not optimal, and the presence of *T. murrelli* is suspected, the concentration of the primer set IV can be doubled.

4. For the automatic amplification of DNA, other thermal cyclers could be used, but it could be necessary to first determine their efficiency in amplifying DNA.

5. Collection of worms from frozen samples: larvae from frozen muscle samples should be collected according to the protocol of **Subheading 3.1.,** yet to avoid DNA destruction, all procedures after digestion should be carried out very quickly, and sedimentation should be carried out on ice.

6. Collection of worms from formalin-fixed muscle samples: formalin-fixed muscle samples cannot be used to collect larvae, because formalin destroys the DNA.

7. Collection of worms from ethyl alcohol-fixed muscle samples: worms can be collected as follows: using a scalpel, cut the muscle sample into grain-size pieces. Crush the pieces between two trichinoscope slides (8-mm thick) and check for the presence of larvae among the muscle fibres under a dissection microscope at 20–40x. Mark the position of the larva on the bottom slide. Gently remove the upper slide and cut away the muscle surrounding the larva using a scalpel and one or two small needles (if the larva is encapsulated, remove it from the capsule with the scalpel and needles) under a dissection microscope at 20–40x. Place the larva in a 0.5-mL conical vial with 400 μL of cold H_2O. Wash the larva 3–4 times with cold H_2O, then store in 5 μL of H_2O at –20°C.

8. Pooled larvae: the protocol for the preparation of crude DNA of a single larva can also be used for pooled larvae by simply increasing the quantity of solution (for example, for 10 larvae, it is sufficient to double the quantity of solution used for a single larva). When using pooled larvae, it should be kept in mind that the presence of larvae belonging to two or more genotypes (mixed infections) could affect the interpretation of the results.

9. Precautions for PCR: use tip with barrier and gloves.

10. If using *Taq* DNA polymerases other than Takara, it is important to perform specific tests to evaluate their effectiveness.

11. Pipeting the sample for the PCR amplification: a sufficient quantity of DNA is critical for a successful amplification, thus, be sure to pipet the sample at the bottom of the tube and avoid collecting the mineral oil. To remove the oil from the tube, it is best to use pipeting, because chloroform or other organic solutions could remove part of the DNA sample.

12. If using restriction enzymes other than those of New England Biolabs for RFLP analysis, check the optimal working temperature of the enzyme.

Acknowledgments

We are very grateful to Marco Amati and Gianluca Marucci for their help in the preparation of figures.

References

1. Murrell, K.D., Lichtenfels, R.J., Zarlenga, D.S., and Pozio, E. (2000) The systematics of *Trichinella* with a key to species. *Vet. Parasitol.* **93**, 293–307.
2. Pozio, E. (2001) New patterns of *Trichinella* infections. *Vet. Parasitol.* **98**, 133–148.
3. Despommier, D.D. (1998) How does *Trichinella spiralis* make itself at home? *Parasitol. Today* **14**, 318–323.
4. Ljungström, I., Murrell, D., Pozio, E., and Wakelin, D. (1998) Trichinellosis, in *Zoonoses. Biology, Clinical Practice, and Public Health Control* (Palmer, S.R., Lord Soulsby, and Simpson, D.I.H., eds.), Oxford University Press, Oxford, UK, pp. 789–802.
5. Pozio, E. (1998) Trichinellosis in the European Union: Epidemiology, ecology and economic impact. *Parasitol. Today* **14**, 35–38.
6. Pozio, E., Casulli, A., Bologov, V.V., Marucci, G., and La Rosa, G. (2002). Hunting practices increase the prevalence of *Trichinella* infection in wolves from European Russia. *J. Parasitol.* **87**, 1498–1501.
7. Pozio, E., Sacchini, D., Sacchi, L., Tamburrini, A., and Alberici, F. (2001). Failure of mebendazole in the treatment of humans with *Trichinella spiralis* infection at the stage of encapsulating larvae. *Clin. Infect. Dis.* **32**, 638–642.
8. Pozio, E. (2000). Factors affecting the flow among domestic, synanthropic and sylvatic cycles of *Trichinella*. *Vet. Parasitol.* **93**, 241–262.
9. Zarlenga, D.S., Chute, M.B., Martin, A., and Kapel, C. M. O. (1999) A single multiplex PCR for unequivocal differentiation of six distinct genotypes of *Trichinella* and three geographical genotypes of *Trichinella pseudospiralis*. *Int. J. Parasitol.* **29**, 1859–1867.
10. Pozio, E., Owen, I.L., La Rosa, G., Sacchi, L., Rossi, P., and Corona, S. (1999) *Trichinella papuae* n. sp. (Nematoda), a new non-encapsulated species from domestic and sylvatic swine of Papua New Guinea. *Int. J. Parasitol.* **29**, 1825–1839.
11. Wu, Z., Nagano, I., Pozio, E., and Takahashi, Y. (1999) Polymerase chain reaction-restriction fragment length polymorphism (PCR-RFLP) for the identification of *Trichinella* isolates. *Parasitology* **118**, 211–218.
12. Pozio, E., La Rosa, G., and Rossi, P. (1989) *Trichinella* reference centre. *Parasitol. Today* **5**, 169–170.

22

Detection of Pathogenic *Yersinia enterocolitica* by a Swab Enrichment PCR Procedure

Rickard Knutsson and Peter Rådström

1. Introduction

The Gram-negative *Yersinia* genus belongs to the family of *Enterobacteriaceae*, in which 3 of the 11 *Yersinia* species are recognized as human pathogens, namely *Y. pestis* (the etiological factor of plague), *Y. pesudotuberculosis*, and *Y. enterocolitica* (the causative agents for yersiniosis) *(1)*. The rest of the genus is composed of *Y. ruckeri* (a fish pathogen) and the *Y. enterocolitica*-like organisms *(2,3)*, which have been biochemically classified as 7 species: *Y. fredriksenii*, *Y. kristensenii*, *Y. intermedia*, *Y. aldovae*, *Y. rohdei*, *Y. mollaretii*, and *Y. bercoveri* *(4–7)*. The heterogeneous composition of *Y. enterocolitica*, with respect to phenotypic and genotypic properties, is problematic from an analytical point of view; the species being distinguished by 6 biogroups *(8)* and more than 60 serotypes *(9)*.

Y. enterocolitica comprises both nonpathogenic and pathogenic members *(10)*, and the most common pathogenic bioserogroups include 4/O:3, 3/O:5.27, 2/O:9, and 1B/O:8. For the pathogenic members, it has been hypothesized that a correlation exists between serovars and their geographic origin, and in Europe, the serotypes O:3 and O:9 are predominant, whereas O:8, and O:5.27 are most common in North America *(3)*. Regarding the pathogenicity factors of the bacterium, it possesses a chromosomal and plasmoidal distribution of virulence genes *(11)*. A fully virulent strain of *Y. enterocolitica* carries a 70-kb plasmid, termed pYV (plasmid for *Yersinia* virulence) *(12)*. Recent DNA–DNA hybridization and 16S rRNA gene sequencing studies have suggested dividing the species of *Y. enterocolitica* into two subspecies, namely *Y. enterocolitica* subsp. *paleartica* and *Y. enterocolitica* subsp. *enterocolitica* *(13)*.

From: *Methods in Molecular Biology, vol. 216: PCR Detection of Microbial Pathogens: Methods and Protocols*
Edited by: K. Sachse and J. Frey © Humana Press Inc., Totowa, NJ

The reference methods employed today are based on traditional enrichment culturing followed by selective agar plating methods *(14,15)*. In Sweden, as in the other Nordic countries, the standard NMKL-177 method is based on cold enrichment in phosphate-sorbitol-bile salt (PSB) broth and modified-rappaport-broth (MRB) at 4°C. This method takes 21 d to complete and are rather laborious *(15)*. The isolation of *Y. enterocolitica* has, however, been associated with difficulties, and it has been suggested that the lack of appropriate analysis methods may mask the occurrence of the bacterium leading to underestimation of the true incidence *(16)*. Previous studies have shown that the growth of *Y. enterocolitica* O:3 is especially restricted in the presence of high concentrations of background flora *(17,18)*. Nevertheless, enrichment has been claimed to provide the most reliable way for isolating low numbers of *Y. enterocolitica* from the total microbial population *(19)*. The oral-pharyngeal area has been reported to be an important swab sampling site for high recovery of *Y. enterocolitica* from pigs *(20)*. Since no single isolation procedure is available for the recovery of all pathogenic strains of *Y. enterocoltica* *(21)*, the inadequacy of the insufficient methods available up until now might explain the underestimation of their prevalence. Reasons for the difficulties encountered may include the sensitivity of the bacterium to high concentrations of background flora present in food samples, but also the heterogeneity of the species.

Recently, it has been claimed that the development of a simple, sensitive, specific, rapid identification system applicable to the diagnosis of *Y. enterocolitica* for veterinary use is a challenge for the future *(22)*. It has been indicated that food-borne pathogens can be detected more rapidly and with better specificity through the use of DNA-based methods, especially polymerase chain reaction (PCR) *(23)*. PCR-based methods indicate a higher prevalence of pathogenic pYV+ *Y. enterocolitica* than traditional methods *(24)*. To avoid false negative PCR results for *Y. enterocolitica*, it is important that the PCR protocol be properly designed to take into account, for example, the presence of PCR-inhibitory components. In the establishment of diagnostic PCR, the pre-PCR processing constitutes a key step (*see* **Chapter 2**).

Today's PCR protocols for the detection of *Y. enterocolitica* rely on pre-enrichment prior to PCR analysis, and the following media have been used: irgasan-ticarcillin-chlorate (ITC) *(25)*, modified-trypticase-soy-broth (MTSB) *(26)*, PSB *(27)*, tryptone-soya-broth (TSB) *(28)*, and *Yersinia*-selective-enrichment (YSE) broth *(29)*. However, all these media contain components that inhibit or interfere with the PCR analysis *(29)*, making extensive sample preparation necessary. Thus, a medium that lacks PCR-inhibitory substances and is selective for *Y. enterocolitica* would simplify the analysis procedure and would be favorable for the establishment of PCR-based methods. Therefore, a new broth, called *Yersinia*-PCR-compatible-enrichment (YPCE) medium, has recently been developed to improve high-throughput PCR and the diagnosis of *Y. enterocolitica* *(30)*. The medium was developed through a mathematical approach by the use of screening factorial design experiments and confirmatory tests. Factors studied included PCR inhibition, growth of *Y. enterocolitica*, and growth of background flora (*see* **Fig. 1**). The formulation of the PCR-compatible enrichment medium is based on a number of requisites: *(i)* the buffer system of the medium should mimic the buffer system in the PCR mixture; *(ii)* neither the medium components nor the metabolites produced by bacterial growth should be PCR-inhibitory; and *(iii)* selective components must be included to restrict the growth of competing background flora.

The YPCE medium has been combined with a multiplex PCR assay that targets two genes, namely, the chromosomal 16S rRNA gene (for species identification) and the plasmid-borne virulence gene *yadA* (for virulence confirmation) *(29)*. This chapter describes a simple and rapid PCR protocol for the diagnosis of pathogenic pYV$^+$ *Y. enterocolitica* strains for veterinary use. An overview is presented in **Fig. 2**.

2. Materials

2.1. Swab Sampling

1. A wooden-stemmed cotton-tipped swab (15 cm) (SelefaTrade, Spånga, Sweden).
2. Swab sample dilution fluid (NaCl-Peptone-Water).

2.1.1. Preparation of Swab Sample Dilution Fluid (NaCl-Peptone-Water)

Add all the ingredients to a suitable flask: 1.0 g Peptone (Merck, Darmstadt, Germany), 8.5 g NaCl, and 1 L distilled water.

The pH should be around 7.0 ± 0.1, adjust if necessary. Sterilize the dilution fluid at 121°C for 15 min. Store the dilution fluid at 0–5°C until required.

2.2. Enrichment in YPCE Medium

1. 15-mL plastic test tubes (Sarstedt, Nümbrecht, Germany).
2. 10-mL YPCE medium.

2.2.1. Preparation of YPCE Medium

1. Add the following ingredients to a suitable flask: 10g Tryptone (peptone from casein, pancreatically digested) (Merck, Cat. no. 1.07213), 2.75 g MOPS (3-[*N*-morpholino] propanesulfonic acid) (Sigma, St Louis, MO, USA), 2.75 g Tris buffer (ICN Biochemicals, Costa Mesa, CA, USA), 20.0 g D-sorbitol (Sigma), and 1 L distilled water.

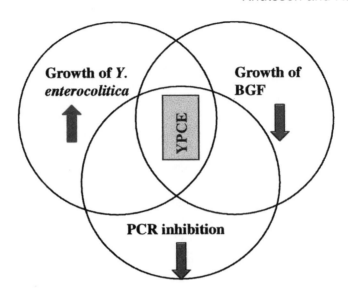

Fig. 1. Illustration of the development of the medium, which was based on screening and factorial design experiments, to find a medium composition that fulfills the requirements for a PCR-compatible enrichment medium. The medium was designed to optimize the growth of *Y. enterocolitica* and to minimize the presence of PCR-inhibitory components and growth of background flora.

Autoclave at 121°C for 15 min and allow to cool to room temperature (25°C). Transfer 983 mL aseptically to a flask of suitable size and add the following ingredients:
2. 2 vials *Yersinia*-selective-supplement (cefsulodin-irgasan-novobiocin, [CIN]) (Merck, Cat. no. 1.16466.001). The lyophilizate should be suspended in the vial by the addition of 1 mL sterile distilled water.
3. 5 mL sterilized potassium chlorate solution: 10 g $KClO_3$ (Merck) dissolved in 100 mL H_2O.
4. Finally, adjust the pH to 8.2 ± 0.1 by the addition of 1 M potassium hydroxide (KOH) solution (approx 10 mL).

2.3. PCR

1. PCR test tubes, 0.2-mL MicroAmp reaction tubes with cap (Applied Biosystems, Foster City, CA, USA).
2. *Taq* DNA polymerase, 5 U/μL (Roche Diagnostics, Mannheim, Germany).
3. 10x PCR buffer (Roche Diagnostics).
4. Primers; Y1, Y2, P1, and P2 (Scandinavian Gene Synthesis, Köping, Sweden). Store stock solution 10 μ*M* of each primer in aliquots at −20°C.
5. Nucleotides; dATP, dCTP, dGTP and dTTP (Roche Diagnostics) and the 10 m*M* dNTP mixture is stored in aliquots at −20°C.

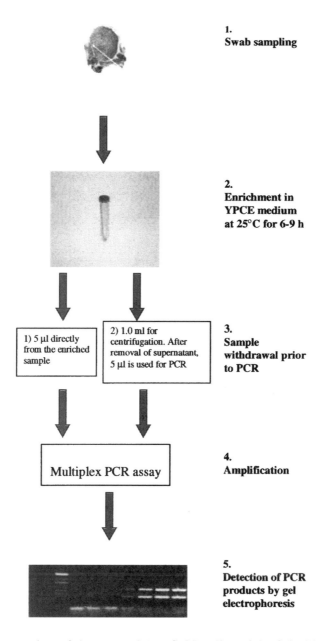

1.
Swab sampling

2.
Enrichment in
YPCE medium
at 25°C for 6-9 h

1) 5 µl directly from the enriched sample	2) 1.0 ml for centrifugation. After removal of supernatant, 5 µl is used for PCR

3.
Sample
withdrawal prior
to PCR

Multiplex PCR assay

4.
Amplification

5.
Detection of PCR
products by gel
electrophoresis

Fig. 2. An overview of the protocol (*see* **Subheadings 3.1.–3.4.**) illustrating the different steps. The protocol starts with swab sampling, followed by enrichment in YPCE medium at 25°C. After 6–9 h enrichment PCR samples are withdrawn; one sample is withdrawn directly from the enrichment broth, and another sample of 1.0 mL is withdrawn for centrifugation and removal of supernatant prior to PCR. DNA is amplified using a multiplex PCR assay for the detection of two PCR products originating from the 16S rRNA gene and the plasmid-borne virulence gene *yadA*. The PCR products are visualized by gel electrophoresis.

2.4. Electrophoresis and Visualization

1. 1.0 % Pronarose agarose DNA-grade (Labora, Sollentuna, Sweden).
2. Tris-borate EDTA electrophoresis buffer (TBE): 0.09 M Tris-borate, 0.002 M EDTA, pH 8.0. For 1 L of 10x TBE, mix 108 g Tris (ICN Biochemicals), 55 g boric acid (Prolabo, Fontenay-sous-Bois, France), and 40 mL 0.5 M EDTA (Merck). Dilute 1:10 before use.
3. Gel loading buffer: 0.25% (w/w) bromophenol blue (Sigma) and 40% (w/w) sucrose (ICN Biochemicals).
4. Gel stain: 0.5 μg/mL ethidium bromide (ICN Biochemicals). Ethidium bromide is mutagenic. Wear gloves and avoid direct contact with skin.
5. Gel electrophoresis equipment: Electrophoresis Power Supply—EPS 301 (Amersham Biosciences, Uppsala, Sweden).
6. DNA size marker: DNA Molecular Weight Marker XIV (100-bp ladder) (Roche Diagnostics).

2.5. General Equipment and Consumables

1. Incubator (Termaks, Bergen, Norway).
2. PCR thermal cycler, GeneAmp® 9700 Thermal Cycler (Perkin Elmer Cetus, Norwalk, CT, USA).
3. Vortex shaker, Vortex-Gene 2 (Scientific Industries, Bohemia, NY, USA).
4. Benchtop centrifuge with Eppendorf® rotor, HERMLE Z160M (Hermle Labortechnik, Wehingen, Germany).
5. Apparatus for horizontal gel electrophoresis: Electrophoresis Power Supply— EPS 301.
6. Gel documentation system: Gel Doc 1000 (Bio-Rad, Hercules, CA, USA).
7. Software to analyze gel images: Molecular Analyst Software (Bio-Rad).
8. Set of pipets covering the whole vol range from 0.1–1000 μM (Gilson, Villers le Bel, France).
9. Aerosol-resistant pipet tips with filter (Gilson).
10. 1.5-mL plastic tube for pre-PCR processing.

3. Methods

The protocol employed is based on: *(i)* swab sampling; *(ii)* enrichment in YPCE medium; *(iii)* PCR sample withdrawal; *(iv)* PCR using a multiplex assay for *Y. enterocolitica*; and *(v)* detection of PCR products by gel electrophoresis. The protocol with its conditions is presented in **Table 1**. The YPCE medium provides a system that enables pathogenic pYV$^+$ *Y. enterocolitica*; to multiply to PCR-amplifiable concentrations in the presence of up to, at least, three magnitudes higher concentration of background flora. The swab samples are incubated as described in the prepared YPCE medium. The preparation of the medium is a 4-step procedure (*see* **Subheading 2.2.1.**) and it is important to use the stated brands (*see* **Note 1**).

Table 1
Summary of the Multiplex PCR Assay and the Protocol
for the Detection of Pathogenic *Yersinia enterocolitica*

Swab sampling	Two wooden-stemmed cotton-tipped swabs (15 cm).
Enrichment in YPCE medium	10 mL are enriched at 25°C for 6–9 h.
Sample withdrawal prior to PCR	Two samples are withdrawn. A. 70 µL directly for PCR. B. 1.0 mL is transferred to an Eppendorf tube and centrifuged at 7000*g* for 5 min. The supernatant (930 µL) is removed leaving 70 µL in the Eppendorf tube. The sample is vortex mixed.
PCR sample vol	5 µL.
PCR master mixture vol	20 µL.
Total PCR vol	25 µL.
PCR test-tubes	0.2-mL MicroAmp reaction tubes with caps.
PCR thermal cycler	GeneAmp 9700 thermal cycler.
PCR reagents and concentrations	1x PCR buffer, 0.2 m*M* of each nucleotide dATP, dCTP, dGTP, and dTTP, 0.4 µ*M* of each primer P1, P2, Y1, and Y2, 0.75 U *Taq* DNA polymerase.
PCR program	Denaturation at 94°C for 5 min. 45 cycles: Denaturation: 94°C for 30 s, Annealing: 58°C for 30 s, Extension: 72°C for 40 s, Final extension at 72°C for 7 min.
Gel electrophoresis	1.0% agarose DNA-grade gel in TBE buffer. The samples are added to the gel and exposed to 100 V for 3 h, and then the PCR products are visualized.

3.1. Swab Sampling and Enrichment in YPCE Medium

The swab method is based on the ISO method *(31)* and a German DIN standard *(32)* for the determination of pathogenic microorganisms on the surface of carcasses of slaughter animals.

1. The surface to be investigated is delineated by a sterile frame pressed onto the sample (at least 20 cm²).
2. This area is first sampled using a wooden-stemmed swab moistened with a sterile NaCl-Peptone-Water dilution fluid (*see* **Subheading 2.1.2.**). Moistening is performed immediately before use by dipping the top of the swab into the dilution fluid. The swab is held at a 45° angle to the sample surface, and during sampling

the swab is rotated between the thumb and forefinger in two perpendicular directions, and the last movement is circular around the inner edges of the frame.
3. The top of the swab is immediately cut off, using a pair of sterile scissors, and put into a 15-mL plastic test tube containing 10 mL YPCE medium, which is then vortex mixed.
4. Without moving the frame, the surface is sampled again, in the same way as described above, but this time using a dry swab.
5. This swab is cut off and placed in the same manner as described above in the same 15-mL plastic test tube as the first swab.
6. The sample is vortex mixed once more.
7. The sample is transferred to an incubator, where it is incubated at 25°C for 6–9 h (*see* **Note 2**).

3.2. PCR Sample Withdrawal

Due to the various ecophysiological conditions of the different samples, the concentration of *Y. enterocolitica* and the composition of background flora should be considered. The time of sample withdrawal is crucial for the PCR detection. Since a growth batch is a dynamic system and highly dependent on the initial conditions, replicate sampling should be employed to reduce the risk of false negative results. Thus, to improve the reliability of PCR detection *(33)*, two PCR samples are withdrawn after 6–9 h enrichment and processed as follows.

1. Withdraw 70 µL directly from the enrichment culture for PCR detection, use 5 µL as the PCR template.
2. Withdraw 1.0 mL of the cell suspension, transfer to an Eppendorf tube and centrifuge at 7000g for 5 min.
3. Remove 930 µL of supernatant leaving a 70-µL sample. Vortex mix this suspension and use 5 µL as the PCR sample.

3.3. Multiplex PCR Assay

3.3.1. General Remarks (see **Note 3** and **Note 4**)

The pathogenesis of *Y. enterocolitica* makes use of both chromosomal and plasmid-borne virulence genes that are temperature-dependent *(11)*, and a fully virulent strain of *Y. enterocolitica* carries the pYV plasmid. It has been found that the pYV$^+$ *Y. enterocolitica* strains have a degree of DNA homology *(34)*. An important virulence gene, located on the pYV plasmid, is *yadA*, which codes for an outer membrane protein YadA (*Yersinia* adhesion A). This protein has several important functions in pathogenesis, including adhesion to epithelial cells *(35)* and resistance to attack by polymorphonuclear leukocytes *(36)*.

In **Table 2**, the sequences of the primers employed are given for the multiplex PCR assay. The primer pair P1:P2 codes for a 0.6-kb PCR product originating from the *yadA* gene, and the second primer pair Y1:Y2 codes for a 0.3-kb

**Table 2
Sequences and Characteristics of Primers
for the *Yersinia enterocolitica* Multiplex PCR Detection Assay**

Primer	Sequence (5'→3')	Size (bp)	GC Content (%)	$T_M{}^a$	Target gene	Location of target gene
P1	TGT TCT CAT CTC CAT ATG CAT T	22	36	60	*yadA*	pYV plasmid
P2	TTC TTT CTT TAA TTG CGC GAC A	22	32	58	*yadA*	PYV plasmid
Y1	GGA ATT TAG CAG AGA TGC TTT A	22	36	60	16S rRNA	Chromosomal
Y2	GGA CTA CGA CAG ACT TTA TCT	21	43	60	16S rRNA	Chromosomal

^{a} The theoretical melting temperature is calculated from the formula: $T_M = (4 \times [G+C]) + (2 \times [A+T])°C$.

PCR product from the chromosomal 16S rRNA gene. The location of the Y1 primer corresponds to the variable V3 region of the 16S rRNA gene, and the Y2 primer corresponds to the variable V9 region of the same gene according to the nomenclature *(37)*. A mismatch (C instead of G) was introduced during the development of the assay, at the second 3´end of the Y2 primer, to avoid detection of *Hafnia alvei*. More than 30 *Enterobacteriaceae* strains have been tested for specificity, and all were found to be negative. For detection to be considered positive, the two PCR products originating from both genes must be visualized on the gel (*see* **Fig. 3**). The 0.6-kb amplicon defines the virulence, while the 0.3-kb amplicon confirms the *Yersinia* origin. The specificity of the assay has been evaluated *(29)*. It was found that the Y1 and Y2 primers amplified the region of the 16S rRNA gene for some *Y. enterocolitica*-like organisms, such as *Y. kristensenii* and *Y. intermedia*.

3.3.2. Preparation of PCR Master Mixture

Each 25-µl reaction will contain the following reagents: 2.5 µL PCR buffer (10X), 2.5 µL dNTP mixture (0.2 m*M* of each nucleotide), 1 µL primer Y1 (0.4 µ*M*), 1 µL primer Y (0.4 µ*M*), 1 µL primer P1 (0.4 µ*M*), 1 µL primer P2 (0.4 µ*M*), 0.15 µL *Taq* DNA polymerase (0.75 U), and 10.85 µL double-distilled H$_2$O.

Calculate the total amounts of reagents according to the number of amplification reactions of the series. Alternative DNA polymerases and PCR mixtures can also be used (*see* **Note 5**).

Use of 5 µL cell suspension as PCR template.

3.3.3. PCR Program

1. The PCR program starts with a denaturation step at 94°C for 5 min.
2. The denaturation is followed by 36 cycles consisting of heat denaturation at 94°C for 30 s, primer annealing at 58°C for 30 s, and extension at 72°C for 40 s.

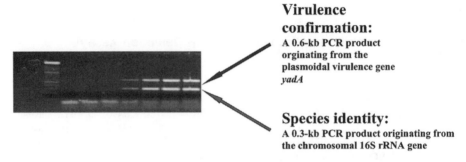

Virulence confirmation:
A 0.6-kb PCR product orginating from the plasmoidal virulence gene *yadA*

Species identity:
A 0.3-kb PCR product originating from the chromosomal 16S rRNA gene

Fig. 3. The figure shows the two PCR products for detection of the heterogeneous species of *Y. enterocolitica*. The 0.3-kb product originates from the 16S rRNA gene, whereas the 0.6-kb amplicon is a PCR product from the plasmid-borne virulence gene *yadA*. The latter product is only present in fully virulent strains of *Y. enterocolitica*. Multiplex PCR allows discrimination between pathogenic and nonpathogenic members of the *Y. enterocolitica* species.

3. A final extension step is performed at 72°C for 7 min to complete the synthesis of all strands.
4. Correct DNA amplification leads to the formation of two PCR products of 0.3 and 0.6 kb, respectively (*see* **Fig. 3**).

3.4. Detection of PCR Products by Gel Electrophoresis

After amplification, the PCR products are visualized by agarose gel electrophoresis *(38)*.

1. The 1.0% agarose gel is obtained by adding the agarose to 1x TBE buffer in an Erlenmeyer flask.
2. Liquify gel by microwave heating.
3. The mixture is allowed to cool, and before it forms a gel, it is stained with 0.5 µg/ mL ethidium bromide.
4. Pour the solution into an appropriate electrophoresis mold to form a gel for analysis in the electrophoresis equipment.
5. Add 2.5 µL of 10x loading buffer to each PCR sample.
6. Add 20 µL of the mixed sample solution to each well in the agarose gel.
7. Run gel electrophoresis at a constant voltage of 100 V for 3 h.
8. Include a DNA Molecular Weight Marker XIV (100-bp ladder) in the gel to determine the fragment size of the amplicons.
9. An UV table with a computer linkage (Gel Doc 1000 Documentation System) is used to visualize the PCR products.
10. Molecular Analyst Software is used to analyze the images.

4. Notes

1. The YPCE medium is compatible with other *Y. enterocolitica* assays in both conventional PCR and real-time PCR, but if the brands of reagents or swabs used are other than those stated in the protocol, it is advisable to screen for assay interference. To screen for possible PCR inhibition, 5 µL of the reagent of interest is added to 5 µL bacterial suspension with a high concentration. The total vol of the reagent mixture (without sample) is then reduced by omitting 5 µL ultrapure PCR water. Study the DNA amplification with the multiplex PCR assay as described.
2. If longer enrichment PCR procedures are used, it is recommended that the samples be diluted prior to PCR analysis. Too high a concentration of cells may interfere with the PCR amplification. It is, therefore, of importance to optimize the time of sample withdrawal to the conditions and sample the protocol is to be applied to.
3. Many other PCR assays for the detection of *Y. enterocolitica* have been developed. However, most of these assays detect either pathogenic or nonpathogenic members.
4. To improve the robustness of the method, it is recommended that DNA amplification of an external control be included each time the protocol is performed.
5. To improve the robustness, it is suggested that a DNA polymerase more robust to PCR-inhibitory components be used *(39)*, or that PCR amplification facilitators be included in the reaction *(40)*. For instance, *Taq* DNA polymerase, , which is an enzyme sensitive to PCR inhibitors, can be replaced by *Tth* DNA polymerase, which is more resistant. In addition, amplification facilitators such as 4% (w/w) glycerol (Merck) or 0.4% (w/w) bovine serum albumin (BSA) (Sigma) can be added to the reaction mixture.

Acknowledgments

We thank Prof. Michael Bülte and Dr. Amir Abdulmawjoodat at Justus-Liebig University, Giessen, Germany, for providing information about the swab sampling method. This work was supported by grants from VINNOVA, the Swedish Agency for Innovation Systems, and from the Commission of the European Community within the program "Quality of Life and Management of Living Resources," QLRT-1999-00226.

References

1. Schiemann, D.A. (1989) *Yersinia enterocolitica* and *Yersinia pseudotuberculosis*, in Foodborne Bacterial Pathogens 1st ed. (Doyle, M. P. ed.), Marcel Dekker, New York, USA, pp. 601–672.
2. Nesbakken, T. and Kapperud, G. (1985) *Yersinia enterocolitica* and *Yersinia enterocolitica*-like bacteria in Norwegian slaughter pigs. *Int. J. Food Microbiol.* **1**, 301–309.
3. Kapperud, G. (1991) *Yersinia enterocolitica* in food hygiene. *Int. J. Food Microbiol.* **12**, 53–65.

4. Aleksic, S., Steigerwalt, A.G., Bockemühl, J., G.P., Hartley-Carter, G. P., and Brenner, D.J. (1987) *Yersinia rohdei* sp. nov. isolated from human and dog feces and surface water. *Int. J. Syst. Bact.* **37,** 327–332.

5. Bercovier, H., Steigerwalt, A.G., Guiyoule, A., Huntley-Carter, G.P. and Brenner, D.J. (1984) *Yersinia aldovae* (formerly *Yersinia enterocolitica*-like group X2): a new species of *Enterobacteriaceae* isolated from aquatic ecosystems. *Int. J. Syst. Bact.* **34,** 166–172.

6. Brenner, H., Ursing, J., Bercovier, H., et al. (1980) Deoxyribonucleic acid relatedness in *Yersinia enterocolitica* and *Yersinia enterocolitica*-like organisms. *Curr. Microbiol.* **4,** 195–200.

7. Wauters, G., Janssens, M., Steigerwalt, A. G., and Brenner, D. J. (1988) *Yersinia mollaretii* sp. nov. and *Yersinia bercovieri* sp. nov., formerly called *Yersinia enterocolitica* biogroups 3A and 3B. *Int. J. Syst. Bact.* **38,** 424–429.

8. Wauters, G., Kandolo, K. and Janssens, M. (1987) Revised biogrouping scheme of *Yersinia enterocolitica*. *Contrib. Microbiol. Immunol.* **9,** 14–21.

9. Wauters, G. (1981) Antigens of *Yersinia enterocolitica*, in *Yersinia enterocolitica*. (Bottone, E .J., ed.), CRC Press, Boca Raton, FL, USA, pp. 41–53.

10. Mollaret, H. H., Bercovier, H., and Alonso, J. M. (1979) Summary of the data received at the WHO Reference Center for *Yersinia enterocolitica*. *Contrib. Microbiol. Immunol.* **5,** 174–184.

11. Bottone, E. J. (1997) *Yersinia enterocolitica*: the charisma continues. *Clin. Microbiol. Rev.* **10,** 257–276.

12. Cornelis, G. R., Boland, A., Boyd, A. P., et al. (1998) The virulence plasmid of *Yersinia*, an antihost genome. *Microbiol. Mol. Biol. Rev.* **62,** 1315–1352.

13. Neubauer, H., Aleksic, S., Hensel, A., Finke, E. J., and Meyer, H. (2000) *Yersinia enterocolitica* 16S rRNA gene types belong to the same genospecies but form three homology groups. *Int. J. Med. Microbiol.* **290,** 61–64.

14. ISO (1994) Microbiology: General Guidence for the Detection of Presumptive Pathogenic *Yersinia enterocolitica*. Draft International Organization for Standardization, ISO/DIS 10273.

15. NMKL-117 (1987) Nordic Committee on Food Analysis NMKL-117: *Yersinia enterocolitica*. Detection in food, 2nd ed. Nordic Committee on Food Analysis, Esbo, Finland.

16. Kotula, A. W., and Sharar, A. K. (1993) Presence of *Yersinia enterocolitica* serotype O:5,27 in slaughter pigs. *J. Food Prot.* **56,** 215–218.

17. Schiemann, D. A., and Olson, S. A. (1984) Antagonism by gram-negative bacteria to growth of *Yersinia enterocolitica* in mixed cultures. *Appl. Environ. Microbiol.* **48,** 539–544.

18. Kleinlein, N. and Untermann, F. (1990) Growth of pathogenic *Yersinia enterocolitica* strains in minced meat with and without protective gas with consideration of the competitive background flora. *Int. J. Food Microbiol.* **10,** 65–71.

19. Schiemann, D. A. (1982) Development of a two-step enrichment procedure for recovery of *Yersinia enterocolitica* from food. *Appl. Environ. Microbiol.* **43,** 14–27.

20. Funk, J. A., Troutt, H. F., Isaacson, R. E., and Fossler, C .P. (1998) Prevalence of pathogenic *Yersinia enterocolitica* in groups of swine at slaughter. *J. Food Prot.* **61,** 677–682.

21. de Boer, E. (1995) Isolation of *Yersinia enterocolitica* from foods, in *Culture Media for Food Microbiology,* vol. 34, Progress in Industrial Microbiology, (Corry, J. E. L., Curtis, G. D. W., and Baird, R. M., eds.), Elsevier Science B.V., Amsterdam, Netherlands, pp. 219–228.

22. Neubauer, H., Sprague, L. D., Scholz, H., and Hensel, A. (2001) The diagnostics of *Yersinia enterocolitica* infections: a review on classical identification techniques and new molecular biological methods. *Berl. Münch. Tierärztl. Wochenschr.* **114,** 1–7.

23. Hill, W. E. (1996) The polymerase chain reaction: applications for the detection of foodborne pathogens. *Crit Rev. Food Sci. Nutr.* **36,** 123–173.

24. Fredriksson-Ahomaa, M., Hielm, S. and Korkeala, H. (1999) High prevalence of *yadA*-positive *Yersinia enterocolitica* in pig tongues and minced meat at the retail level in Finland. *J. Food. Prot.* **62,** 123–127.

25. Jourdan, A. D., Johnson, S. C., and Wesley, I. V. (2000) Development of a fluorogenic 5' nuclease PCR assay for detection of the ail gene of pathogenic *Yersinia enterocolitica. Appl. Environ. Microbiol.* **66,** 3750–3755.

26. Bhaduri, S., and Cottrell, B. (1998) A simplified sample preparation method from various foods for PCR detection of pathogenic *Yersinia enterocolitica*: a possible model for other food pathogens. *Mol. Cell. Probes* **12,** 79–83.

27. Vishnubhatla, A., Fung, D. Y., Oberst, R. D., Hays, M. P., Nagaraja, T. G., and Flood, S. J. (2000) Rapid 5' nuclease (Taq Man) assay for detection of virulent strains of *Yersinia enterocolitica. Appl. Environ. Microbiol.* **66,** 4131–4135.

28. NMKL-163 (1998) Nordic committee on food analysis NMKL-163: Pathogenic *Yersinia enterocolitica*. PCR methods for detection in food. Esbo, Finland.

29. Lantz, P.G., Knutsson, R., Blixt, Y., Abu Al-Soud, W., Borch, E. and Rådström, P. (1998) Detection of pathogenic *Yersinia enterocolitica* in enrichment media and pork by a multiplex PCR: a study of sample preparation and PCR-inhibitory components. *Int. J. Food Microbiol.* **45,** 93–105.

30. Knutsson, R., Fontanesi, M., Grage, H. and Rådström, P. (2002) Development of a PCR-compatible enrichment medium for *Yersinia enterocolitica*: amplification precision and dynamic detection range during cultivation. *Int. J. Food Microbiol.* **72,** 185–201.

31. ISO 3100–2 (1988) Meat and meat products—Sampling and preparation of test— Part 2: preparation of test samples for microbiological examination. International Organization for Standardization, Switzerland.

32. DIN 10113-1. (1997) Enumeration of viable aerobic bacteria of surfaces of utensils and equipment in contact with food. Deutsches Institut für Normung e.V. Berlin, Germany.

33. Knutsson, R., Blixt, Y., Grage, H., Borch, E., and Rådström, P. (2002) Evaluation of selective enrichment PCR procedures for *Yersinia enterocolitica. Int. J. Food Microbiol.* **73,** 35–46.

34. Heesemann, J., Keller, C., Morawa, R., Schmidt, N., Siemens, H. J., and Laufs, R. (1983) Plasmids of human strains of *Yersinia enterocolitica*: molecular relatedness and possible importance for pathogenesis. *J. Infect. Dis.* **147,** 107–115.

35. Heesemann, J. and Grüter, L. (1987) Genetic evidence that outer membrane protein YOP1 of *Yersinia enterocolitica* mediates adherence and phagocytosis resistance to human epithelial cell. *FEMS Microbiol. Lett.* **40,** 37–41.

36. Ruckdeschel, K., Roggenkamp, A., Schubert, S., and Heesemann, J. (1996) Differential contribution of *Yersinia enterocolitica* virulence factors to evasion of microbicidal action of neutrophils. *Infect. Immun.* **64,** 724–733.

37. Gray, M. W., Sankoff, D., and Cedergren, R. J. (1984) On the evolutionary descent of organisms and organelles: a global phylogeny based on a highly conserved structural core in small subunit ribosomal RNA. *Nucleic Acids Res.* **12,** 5837–5852.

38. Sambrook, J., Fritsch, E. F., and Maniatis, T. (1989) Molecular cloning: a laboratory manual, 2nd ed., CSH Laboratory Press, Cold Spring Habor, New York, USA.

39. Abu Al-Soud, W. and Rådström, P. (1998) Capacity of nine thermostable DNA polymerases to mediate DNA amplification in the presence of PCR-inhibiting samples. *Appl. Environ. Microbiol.* **64,** 3748–3753.

40. Abu Al-Soud, W., and Rådström, P. (2000) Effects of amplification facilitators on diagnostic PCR in the presence of blood, feces, and meat. *J. Clin. Microbiol.* **38,** 4463–4470.

Index